ADVANCED ANALYSIS-II

DPH MATHEMATICS SERIES

ADVANCED ANALYSIS–II

(For M.A., M.Sc., I.A.S., P.C.S.)

By

A.K. Sharma

DISCOVERY PUBLISHING HOUSE
NEW DELHI-110002

First Published - 2004

Reprinted - 2017

ISBN: 978-81-7141-830-5

Advanced Analysis – II

Published by:

DISCOVERY PUBLISHING HOUSE PVT. LTD.

4383/4B, Ansari Road, Darya Ganj

New Delhi-110 002 (India)

Phone: +91-11-23279245, 43596064-65

Fax: +91-11-23253475

E-mail: discoverypublishinghouse@gmail.com

sales@discoverypublishinggroup.com

web: www.discoverypublishinggroup.com

Printed at:

Infinity Imaging Systems

Delhi

Preface

This book Analysis–II (for post graduate students) has been written to meet the requirement of post graduate students of Indian Universities. This book is also helpful for the candidates appearing at various competitive examination. The present publication has been written to present a logical and practical approach to the various complexities and intricacies of the subject at length. Inspite of comprehensiveness. The subject matter has been presented in such a way that it is easily expectable to students illustrated examples have been incorporated to help the students in getting the clear concept of subject and allied matters. The exercises given in the text form the integral part of the book and it is advised that the same should be attempted seriously.

Almost every aspect of the subject Power Series, Fourier Series, The Reimann-Stieltjes Integral, Integrations on R^3, Series of Arbitrary Terms and Double Series, The Lebesgue Integral, Function of Two and Three Variables, etc., is covered and illustrated by large number of examples. It is believed that the publication will serve as a useful guide not only to the students of post graduate of all Indian Universities but also for the candidate for appearing at various competitive examination.

Suggestion for the improvement of the book will always be most welcome.

A.K. Sharma

CONTENTS

Pages

Preface

1. **Power Series** 1
The Definition of Power Series, Radius of Convergence, Formula for the Radius of Convergence.

2. **Fourier Series** 26
Fourier Series, Periodic Function.

3. **The Riemann-Stieltjes Integral** 66
Introduction, Definition and Existence of the Integral, (A Relation Between Riemann Integral and RS-Integral), Properties of RS-Integrals.

4. **Integrations on R^3** 97
Line, Surface and Volume Integrals, Rectifiable Curves, Length of a Curve, Properties of Rectifiable Curves, Line Integrals, Line Integral with Respect to Arc Length, Surface, The Surface Area, Surface Integrals of the First Type, Reducing a Surface Integral of First to a Double Integral, Oriented Surface, Positive and Negative Sides, Surface Integral of the Second Type: Flux Across a Surface, Reducing a Surface Integral of the Second Type to a Double Integral, Vectorial Formation, Properties of Surface Integrals, Relation Between the Two Types of Surface Integrals, Stokes' Theorem, Gauss' Divergence Theorem, Integral of a Bounded Function Over a Rectangle, Calculation of Triple Integrals, Calculation of a Triple Integral on a Bounded Domain E, Change of Variables in a Triple Integral, Volume of a Solid by a Triple Integral.

5. Series of Arbitrary Terms and Double Series **212**

Introduction, Series of Arbitrary Terms, Rearrangement of Series, Double Sequence, Double Series, Some Modes of Convergence, A Fundamental Theorem for Positive Term Double Series, Absolute Convergence of a Double Series, Multiplication of Series.

6. The Lebesgue Integral **268**

Introduction, Measurable Sets, Outer and Inner Measure of a Bounded Set, Measure of an Open Set and a Closed Set, Measure of an Open Interval, Measure of a Closed Interval, Measure of a Rectangle, Measure of a Parallelopiped, Exterior and Interior Measure. (Lebesgue Measurable Set), Sets of Measure Zero, Borel Sets, Borel Measurability Function, Little Wood's Three Principles, Lebesgue Integral, Properties of Lebesgue Integral for Bounded Measurable Functions, Lebesgue Integral of a Bounded Function Over a Set of Finite Measure, Lebesgue Integral for Unbounded Functions, The General Integral, Some Fundamental Theorems, Lebesgue Theorem on Bounded Convergence, An Important Result for Integrable Functions.

7. Function of Two and Three Variable **354**

Explicit and Implicit Functions, An Explicit Function of Two Variables, The Neighbourhood of a Point, Limit Point, The Limit of a Function, Non-Existence of Limit, Continuity of a Function at a Point, Partial Derivatives, A Mean Value Theorem, A Sufficient Condition for Continuity, Differentiability at a Point, Schwartz's and Young's Theorems, Directional Derivatives, Implicit Function Theorem, Maxima and Minima of Functions of Two Variables, Maxima and Minima Subject to Constraints, Extreme Points of a Function of Three Variables, Legrange's Method of Undetermined Multipliers.

1

POWER SERIES

1.1 Power Series

A series of the form $\sum_{n=0}^{\infty} a_n z^n$ or $\sum_{n=0}^{\infty} a_n (z-a)^n$, whose terms are variable is called a power series where z is a complex variable and a_n, are complex constants.

The second form car be reduced to the first from merely by changing origin. So the consideration of the series of the second form is always reduced to that of the first form

The form $\sum_{n=0}^{\infty} a_n z^n$ being simpler, we have discussion of power series by considering the form $\sum_{n=0}^{\infty} a_n z^n$

1.2 Absolute Convergence of Power Series $\Sigma\ a_n\ z^n$

The power series $\Sigma\ a_n\ z^n$ is said to be absolutely convergent if series $\Sigma\ |a_n\ z^n|$ is convergent

If the series $\sum_{n=0}^{\infty} a_n z^n$ convergent but the series $\sum_{n=0}^{\infty} |a_n z|^n$ not converge, then the series $\sum_{n=0}^{\infty} a_n z^n$ is called conditionally convergent.

1.3 Radius of Convergent of Power Series

Applying Cauchy's not test to the series $\Sigma\ |a_n|\ |z|^n$ we find that $\Sigma\ a_n\ z^n$ is convergent or divergent according as limit of nth root of the nth term is less or greater than unity Thus.

$\Sigma\ |a_n|\ |z|^n$ convergent if $\lim_{n\to\infty} |a_n|^{1/n}\ |z| < |$ i.e. if $1/R\ .\ |z| < |$ where $\lim|a_n|^{1/n} = 1/R$ or if $|z| < R$ and divergent if $|z| > R$

Thus, for every power series $\Sigma\ a_x\ z^n$. There exist a non-negative real number R such that for every $|z| < R$ The series is absolutely convergent, and for every $|z| > R$ the series is not convergent.

Such a number R is called the Radius of convergence of the power series $\Sigma\ a_x\ z^n$. For if we draw a circle whose centre is origin and radius equal to R, then for all values of z with in this circle power series $\Sigma\ a_x\ z^n$ is absolutely convergent and for all values of z outside this circle the power series $\Sigma\ a_x\ z^n$ is divergent.

1.4 Some Useful Tests for Convergence

The following tests whose proofs are trivial should be recollected

Comparison Test : If $\sum|\sqrt{n}|$ converges and $|u_n|\ \Sigma\ |v_n|$ Then $\Sigma|u_n|$ converges absolutely.

Ratio Test : The series Σu_n converges (absolutely) or diverges according as $\lim_{n\to\infty}\left|\frac{u_{n-1}}{u_n}\right| < 1$ or > 1. If this limit equal to 1 the test fails.

Root Test : (n^{th} root test) the series Σu_n converges (absolutely) or diverge according as $\lim_{n\to\infty} \sqrt[n]{(|u_n|)} < 1$ or > 1

If this limit is equal to 1, the test fails.

Series absolutely convergent : The series Σu_n is convergent (absolutely) the series $|\Sigma u_n|$ is convergent.

Theorem 1:

If the power series $\Sigma a_n x^n$ is such that $a_n \neq 0$ for all n and $\lim_{n\to\infty}\left|\frac{a_{n+1}}{a_n}\right| = \frac{1}{R}$, *then $\Sigma a_n x^n$ is convergent (absolutely convergent) for $|x| < R$ and divergent for $|x| > R$..*

Proof:

Let $u_n = a_n x^n$ so that $u_{n+1} = a_{n+1} x^{n+1}$.

$$\therefore \lim_{n\to\infty}\left|\frac{u_{n+1}}{u_n}\right| = \lim_{n\to\infty}\left|\frac{a_{n+1}\,x}{a_n}\right| = \frac{|x|}{R}.$$

By D' Alembert's ratio test, $\Sigma a_n x^n$ converges absolutely, if

$$\frac{|x|}{R} < 1 \text{ i.e., } |x| < R.$$

Further $\Sigma a_n x^n$ diverges if $|x| > R$. Hence the theorem.

Remark 1:

By Cauchy's second theorem on limits, if $\{a_n\}$ is sequence of positive constants, then

$$\lim_{n\to\infty} (a_n)^{1/n} = \lim_{n\to\infty}\left(\frac{a_{n+1}}{a_n}\right),$$

provided the limit on the R.H.S. exists. It follows that :

If $\lim_{n\to\infty} |a_n|^{1/n} = \frac{1}{R}$, *Then the series $\Sigma a_n x^n$ is convergent (absolutely convergent) for $|x| < R$ and divergent for $|x| > R$.*

In view of the above discussion, the *radius of convergence R of a power series $\Sigma a_n x^n$ may be determined by the following :*

$$R = \lim_{n\to\infty}\left|\frac{a_n}{a_{n+1}}\right| = 1 \div \lim_{n\to\infty} |a_n|^{1/n}$$

$R = \infty$, when $\lim |a_n|^{1/n} = \infty$.

$R = 0$, when $\lim |a_n|^{1/n} = \infty$.

Remark 2:

Every power series converges absolutely within its interval of convergence.

Theorem 2:

If a power series $\Sigma a_n x^n$ converges for $|x| < R$, then it converges uniformly on $[-R + \varepsilon, R - \varepsilon]$, for each $\varepsilon > 0$.

Proof:

Let $x_0 = R - \varepsilon < R$ so that $\Sigma a_n x_0^n$ converges absolutely.

For $\quad |x| \le x_0$, $|a_n x^n| \le |a_n|\ x_0^n$.

Since $\sum a_n x_0^n$ converges absolutely, therefore, by Weierstrass's M-test, the series $\Sigma a_n x^n$ converges uniformly on

$[-x_0, x_0] = [-R+\varepsilon, R-\varepsilon]$. Hence the theorem.

Ex.:

Show that a power series converges uniformly on any closed and bounded subinterval of its interval of convergence. **(D.U., 1995)**

Theorem 3:

If a power series $\Sigma a_n x^n$ converges for $x = x_0$, then

(i) *it is absolutely convergent in the interval* $|x| < |x_0|$

(ii) *it is uniformly convergent in the interval* $|x| \leq |x_1|$ *where* $|x_1| \leq |x_0|$. **(D.U., 1994)**

Proof:

(i) Since $\sum a_n x_0^n$ converges, so $\lim_{n\to\infty} a_n x_0^n = 0$.

Consequently there exists a positive integer *m* such that

$$\left|a_n x_0^n = 0\right| < 1 \ \forall\ n \geq m \text{ or } |a_n| < \frac{1}{|x_0|^n}\ \forall\ n \geq m \qquad \ldots (1)$$

$$\Rightarrow |a_n x^n| = |a_n||x|^n < \frac{|x|^n}{|x_0|^n}\ \forall\ n \geq m, \text{ by } (1)$$

$$\Rightarrow |a_n x^n| < \left\{\frac{|x|}{|x_0|}\right\}^n \ \forall\ n \geq m.$$

The series on the R.H.S. converges for $|x| < |x_0|$ (being a geometric series with common ratio < (1), it follows by the comparison test that $\Sigma |a_n x^n|$ is convergent for $|x| < |x_0|$.

Hence $\Sigma a_n x^n$ is absolutely convergent for $|x| < |x_0|$.

(ii) Let $M_n = \dfrac{|x_1|^n}{|x_0|^n}$. Then ΣM_n converges, since $|x_1| < |x_0|$ (being a geometric series with common ratio <1).

Using (1), we obtain

$$|a_n x^n| = |a_n||x|^n < \frac{|x|^n}{|x_0|^n} \leq \frac{|x_1|^n}{|x_0|^n},$$

for $|x| \leq |x_1|$ and $n \geq m$

$\therefore |a_n x^n| < M_n \ \forall \ n \geq m$, where ΣM_n converges.

By the Weierstrass M-Test, $\Sigma a_n x^n$ is uniformly convergent in $|x| \leq |x_1|$.

Note : *If a power series whose radius of convergence is 1, then it converges absolutely in]– 1, 1[and uniformly in [–k, k], |k| < 1.*

Theorem 4:

(Cauchy's Product of Two Power Series)

Two power series $\sum_{n=0}^{\infty} a_n x^n$ *and* $\sum_{n=0}^{\infty} b_n x^n$ *can be multiplied to obtain the power series* $\sum_{n=0}^{\infty} c_n x^n$, *where*

$$c_n = a_0 b_n + a_1 b_{n-1} + a_2 b_{n-2} + \dots + a_n b_0,$$

the result being valid for each x within the common interval of convergence.

The proof of these theorems are omitted.

Some Important Power Series

Consider the following power series :

$$1 + \frac{x}{1!} + \frac{x^2}{2!} + \frac{x^3}{3!} + \dots + \frac{x^n}{n!} + \dots\dots$$

$$1 - \frac{x^2}{2!} + \frac{x^4}{4!} - \dots + (-1)^n \frac{x^{2n}}{(2n)!} + \dots\dots$$

$$x - \frac{x^3}{3!} + \frac{x^5}{5!} - \dots + (-1)^n \frac{x^{2n+1}}{(2n+1)!} + \dots\dots$$

For the first series, $\lim_{n\to\infty} \left|\frac{a_n}{a_{n+1}}\right| = \lim_{n\to\infty} \frac{(n+1)!}{n!} = \lim_{n\to\infty} (n+1) = \infty$.

Thus, the radius of convergence is R = ∞ so that the series is everywhere convergent (absolutely and uniformly). Similarly we can verify that the other two power series are also everywhere convergent (absolutely and uniformly). In this section we shall study their exclusive properties and show that the sums of these power series are actually the functions e^x, cos x and sin x respectively.

Theorem 5:

(Abel's Theorem)

If a power series converges at an end point of its interval of convergence, then the power series is uniformly convergent in the interval which includes this end point.

Remarks:

1. *If a power series with interval of convergence]–R, R[converges at both the end-points – R and R, then the power series is uniformly convergent in [–R, R].*
2. *If a power series converges at the end-point $x = R$ of the interval of convergence]–R, R[, then it is uniformly convergent in]–R, R] or [–R + ε, R], where $\varepsilon > 0$. In particular, the series is uniformly convergent in [0, R].*

Theorem 6:

(Abel's Limit Theorem)

But $\Delta\, a_n\, x^n$ be a power series with finite radius of convergence R that $r(x) = \Sigma a_n x^n$, $- R\Delta x<R$. If the series $\Sigma a_n R^n$ converges, be

$$\lim_{x \to R-} f(x) = \Sigma a_n\, R^n.$$

Remark 3 : *The case R = 1 is of much practical use :*

Let $\Sigma\, a_n x^n$ *be a power series with unit radius of convergence such that* $f(x) = \Sigma\, a_n\, x^n$, $-1\, \Delta x < 1$. *If the series* $\Sigma\, a_n\, x^n$ *converges for* $x = 1$ *i.e., if* Σa_n converges, then

$$\lim_{x \to 1-} f(x) = \Sigma a_n.$$

Theorem 7:

Suppose a power series $\Sigma a_n x^n$ converges for $|x| < R$ and that $f(x) = \Sigma a_n x^n$. Then

(i) *f(x) is continuous in]–R, R[.* **(D.U., 1995)**

(ii) *$\Sigma a_n x^n$ can be integrated term by term on]–R, R[and the integrated series has the same radius of convergence as $\Sigma a_n\, x^n$.*

(iii) *$\Sigma a_n x^n$ can be differentiated term and term on]–R, [R and $f'(x) = \Sigma n\, a_n x^{n-1}$, $|x| < R$. Further the differentiated series has the same radius of convergence as $\Sigma a_n\, x^n$.*

Proof:

(i) Since each term of the power series $\Sigma a_n x^n$ is continuous in]–R, R[and the series is uniformly convergent on]–R, R[, so the sum f(x) of the given series is continuous in]–R, R[. (See Theorem 6, p. 165).

(ii) Since each term of $\Sigma a_n x^n$ is continuous in]–R, R[and $\Sigma a_n x^n$ converges uniformly to f(x) in]–R, R[, so $\Sigma a_n\, x^n$ can be integrated term by termin] –R, R[and the integrated series is $\Sigma \dfrac{a_n}{n+1}\, x^{n+1}$. Further

$$\lim_{n\to\infty}\left|\frac{u_n}{u_{n+1}}\right| = \lim_{n\to\infty}\left(\frac{n+2}{n+1}\frac{a_n}{a_{n+1}}\right) = \lim_{n\to\infty}\frac{a_n}{a_{n+1}} = R,$$

which shows that $\Sigma a_n x^n$ and $\Sigma\frac{a_n x^{n+1}}{n+1}$ have the same radius of convergence.

(iii) We know each term of $\Sigma a_n x^n$ is continuous, has continuous derivative in $]-R, R[$ and $\Sigma a_n x^n$ is uniformly convergent in $]-R, R[$. Thus the series obtained by differentiating the power series $\Sigma a_n x^n$ is $\Sigma n a_n x^{n-1}$. Further

$$\lim_{n\to\infty}\left|\frac{u_n}{u_{n+1}}\right| = \lim_{n\to\infty}\left|\frac{n\, a_n}{(n+1)\, a_{n+1}}\right| = \lim_{n\to\infty}\left|\frac{a_n}{a_{n+1}}\right| = R.$$

Hence the differentiated series $\Sigma n\, a_n\, x^{n-1}$ is also a power series which has the same radius of convergence as $\Sigma\, a_n\, x^n$.

Note : If *$f(x) = \Sigma a_n x^n$, then $f(x)$ is called the* **sum function** *of the series.*

Theorem 8:

Define cosine and sine functions as sums of power series. Prove that

(i) $S(x + y) = S(x)C(y) + C(x)S(y)$,

(ii) $C(x + y) = C(x)C(y) - S(x)S(y)$,

where C, S denote cosine and since respectively. **(D.U., 1994)**

Proof:

The *cosine function* is defined as

$$C(x) = 1 - \frac{x^2}{2!} + \frac{x^4}{4!} - ... + (-1)^n \frac{x^{2n}}{(2n)!} + ... \quad \forall\ x \in \Re \qquad ...(1)$$

The *sine function* is defined as

$$S(x) = x - \frac{x3}{3!} + \frac{x^5}{5!} - ... + (-1)^n \frac{x^{2n+1}}{(2n+1)!} + ... \qquad ...(2)$$

$\forall\ x \in \Re$

These two power series are uniformly convergent $\forall\ x \in \Re$ and consequently, the functions C(x) and S(x) are continuous $\forall\ x \in \Re$.

Differentiating term by term the convergent series (2).

$$S'(x) = 1 - \frac{x^2}{2!} + \frac{x^4}{4!} + ...$$

which gives us the uniformaly convergent series (1),

Thus S(x) is variable, $\forall x \in \Re$ and $S'(x) = C(x)$ ... (3)

Similarly $C'(x) = -S(x)$. . . . (4)

From (1) and (2); $S(0) = 0$ and $C(0) = 1$. . . . (5)

Let y be any arbitrary but fixed real number. We write

$$f(x) = S(x + y) - S(x)C(y) - C(x)S(y), \quad \ldots (6)$$

$$g(x) = C(x + y) - C(x)C(y) + S(x)S(y). \quad \ldots (7)$$

Differentiating w.r.t. x, we obtain

$$f'(x) = C(x + y) - C(x)C(y) + S(x)S(y) = g(x), \quad \ldots (6)$$

$$g'(x) = -S(x + y) - S(x)C(y) + C(x)S(y) = -f(x). \quad \ldots (7)$$

[Using (3), (4)]

Now $\frac{d}{dx}[f^2(x) + g^2(x)] = 2f(x)f'(x) + 2g(x)g'(x)$

$$= 2f(x)g(x) - 2g(x)f(x) = 0 \quad \forall x$$

If follows that $f^2(x) + g^2(z)$ is a constant $\forall x$

$\Rightarrow f^2(x) + g^2(x) = f^2(0) + g^2(0) \quad \forall x$

$$= [S(y) - S(0)C(y) - C(0)S(y)]^2 + [C(y) - C(0)C(y) + S(0)S(y)]^2$$

$$= 0 \qquad \text{[Using (5)]},$$

$\therefore f^2(x) + g^2(x) = 0 \ \forall x \in \Re$ implies the $f(x) = 0$ and $g(x) = 0$.

Hence (6) and (7) yield

$$S(x + y) = S(x)C(y) + C(x)S(y),$$

$$C(x + y) = C(x)C(y) - C(x)S(y).$$

Cor. 1. $S(-x) = -S(x)$; $C(-x) = C(x) \quad \forall x \in \Re$.

We have $S(-x) = -x - \frac{(-x)^8}{3!} + \frac{(-x)^5}{5!} - \ldots\ldots$

$$= -\left[x - \frac{x^3}{3!} + \frac{x^5}{5!} - \ldots\ldots\right] = -S(x).$$

Similarly we can prove the second part.

Cor. 2. $S(x - y) = S(x)C(y) - C(x)S(y)$,

$C(x - y) = C(x)C(y) + S(x)S(y)$.

Replacing y by –y in (i) and (ii)

$$S(x-y) = S(x)C(-y) + C(x)S(-y),$$

$$C(x-y) = C(x)C(-y) - S(x)S(-y).$$

The desired results follow by Cor. 1.

Cor. 3. $C^2(x) + S^2(x) = 1$ or $\cos^2 x + \sin^2 x = 1 \quad \forall x \in \Re$.

We have $1 = C(0) = C(x - x)$

$= C(x)\, C(-x) - S(x)\, S(-x)$

$= C^2(x) + S^2(x)$, by Cor. 1.

Hence $\cos^2 x + \sin^2 x = 1 \; \forall x \in \Re$.

Cor. 4. $| C(x) \le 1, | S(x) | \le 1, \quad \forall x \in \Re$.

OR

$| \cos x | \le 1, | \sin x | \le 1 : \forall x \in \Re$.

This follows by Cor. 3.

Cor. 5. *$\sin 2x = 2 \sin x \cos x$, $\cos 2x = \cos^2 x - \sin^2 x$.*

Replacing y by x in (i) and (ii) of Theorem 8, we obtain

$S(2x) = 2S(x)\, C(x)$, $C(2x) = C^2(x) - S^2(x)$

Hence the results follow.

Remark:

It may be noted that the above properties of C(x) and S(x) are similar to those of the trigonometric functions cos x and sin x respectively which the students do at the elementary, stage. Thus we have

$\mathbf{C(x) \equiv \cos x}$ **and** $\mathbf{S(x) \equiv \sin x}$.

Theorem 9:

Let f be a non-zero function continuous on R^+ such that $f(xy) = f(x) + f(y)$ $\forall x, y > 0$. . . . (1)

Show that $f(x) \log_a x$, $x \in \Re^+$ $(a > 0, a \ne 1)$.

Proof:

For $x > 0$, let us take $t = \log_e x \Leftrightarrow x = e^t, t \in \Re$. . . . (2)

Let $g(t) = f(e^t) \; \forall t \in \Re$. Then g is continuous on $\Re$ and

$g(t + s) = f(e^{t+s}) = f(e^t e^3) = f(e^t) + f(e^s)$ by (1)

or $g(t + s) = g(t) + g(s)$.

Here $t, s \in \Re$. $g(t) = ct, \forall t \in \Re$, c being a constant.

Thus $f(x) = c \log_e x \; \forall x \in \Re^+$, where $c \ne 0$ as f is non-zero.

Let $a = e^{1/e}$. Then $f(x) = \log_a x \; \forall x \in \Re^+$ $(a > 0, a \ne 1)$.

Ex.:

See the exponential function as the sum of a power series and determine its range. Hence define the logarithmic function and deduce that it is continuous. **(D.U., 1996)**

[**Hint.** It can be verified that the exponential function $E : R \to R^+$ is strictly increasing on R i.e., E is one-one and onto. Thus E has inverse function, say $L : R^+ \to R$. L is called the *logarithmic function,* defined as

$$E\{L(y)\} = y \ (y > 0) \text{ or } L\{E(x)\} = x \ (x \in R).$$

In other words, E (x) = y ⇒ I. (y) = x or $e^x = y \Rightarrow \log_e y = x$.

Since the function E is differentiable, so its inverse function L is false differentiable and hence continuous.

Theorem 10:

Define exponential function E(x) as the sum of power series. Show that the domain is the set of all real number Prove that

$$E(x + y) = E(x)\, E(y) \text{ for all } x, y \in \Re.$$

If e denotes E(1), prove that E (x) = e^2 *for all real x.*

(D.U., 1992, 91, 88)

Proof:

The exponential function E(x) is defined as the sum of the power series

$$E(x) = 1 + \frac{x}{1!} + \frac{x^2}{2!} + \ldots + \frac{x^n}{n!} + \ldots.. \qquad \ldots (1)$$

$$\text{We have } R = \lim_{n\to\infty} \left|\frac{a_n}{a_{n+1}}\right| = \lim_{n\to\infty} \frac{(n+1)!}{n!} = \lim_{n\to\infty} (n+1) = \infty.$$

Thus, the radius of convergence of the above power series is R = ∞ that the power series is everywhere convergent i.e., E(x) is defined ∀ x ∈ ℜ. This shows that the domain of the exponential function is ℜ.

It is clear that the function E(x), defined by (1), is continuous and has derivatives of all orders, for each x ∈ ℜ.

Differentiating (1), we get

$$E'(x) = E(x) \text{ and so } E'(x) = E(x),\ E'(x) = E(x) \text{ etc.}$$

In fact $E^{(n)}(x) = E(x)\ \forall\ n \in N$.

By Taylor's Theorem, for all real values t and x

$$E(t) = E(x) + \frac{E'(x)}{1!}(t - x) + \frac{E''(x)}{2!}(t - x)^2 + \ldots\ldots$$

$$\text{or } E(t) = E(x) + \frac{E(x)}{1!}(t-x) + \frac{E(x)}{2!}(t-x)^2 + ...$$

Replacing t by x + y, we get

$$E(x+y) = E(x)\left\{1 + \frac{y}{1!} + \frac{y^2}{2!} + ...\right\} = E(x)E(y)$$

Thus, $E(x + y) = E(x)\, E(y)\ \forall\ x, y \in \Re$. . . . (2)

From (1), $E(1) = 1 + \frac{1}{1!} + \frac{1}{2!} + ... + \frac{1}{3!} + ...$

The above series on the right converges to a number which lies between 2 and 3.

We denote this number by e so that E(1) = e.

The property (2) can be extended as

$E(x_1 + x_2 + ... + x_n) = E(x_1)\, E(x_2)...E(x_n)$ . . . (3)

for any positive integer n and $\forall\ x \in \Re$.

Taking $x_1 = x_2 = ... = x_n = x$ in (3), we obtain

$E(nx) = \{E(x)\}^n \quad \forall\ x \in \Re, \forall\ n \in N.$ . . . (4)

For x = 1, we obtain

$E(n) = e^n \quad \forall\ n \in N$ $(\because E(1) = e)$ . . . (5)

Taking x = m/n (m and n being positive integers) in (4),

$E(m) = \{E(m/n)\}^n$ or $e^m = \{E(m/n)\}^n$, by (5)

$\therefore\ E(m/n) = (e^m)^{1/n} = e^{m/n}$.

Thus, $E(q) = e^n$ for all positive rationals q. . . .(6)

Let x be any *positive irrational* number. Then there always exists a sequence $\{x_n\}$ of positive rational numbers such that $x_n \to x$.

Using (6), $E(x_n) = e^{xn}\ \forall\ n \in N$.

Since $x_n \to x$ as $n \to \infty$ and E is a continuous function $\forall\ x \in \Re$, therefore

$E(x) = e^n$ for all positive irrationals. . . . (7)

From (6) and (7), we conclude that

$E(x) = e^x$ for all positive reals. . . . (8)

Taking y = – x in (2), we get

$E(0) = E(x)E(-x)$ or $1 = E(x)\, E(-x)\ (x \geq 0)$

$[\because E(0) = 1$, by (1)].

or $E(-x) = \frac{1}{E(x)} = \frac{1}{e^x} = e^{-x}$, using (8)

Hence, $E(x) = e^x \quad \forall\, x \in \Re$.

Theorem 11:

Let f be a continuous function on $\Re$ such that

$f(x + y) = f(x) + f(y)\ \forall\, x, y \in \Re$. . . . (1)

Show that $f(x) = e^x$ for some constant c and $\forall\, x \in \Re$.

Proof:

Putting y = x in (1), f(2x) = 2 f(x).

In fact $f(nx) = nf(x)\ \forall\, x \in \Re, \forall\, n \in N$.

Replacing x by x/n, we get

$f(x) = nf(x/n)$ or $f(x/n) = (1/n) = (1/n) f(x)\ \forall\, x \in \Re$.

For any integers p and p (q > 0), it follows that

$$f\left(\frac{p}{q}x\right) = \frac{p}{q} f(x).\ \forall\, x \in \Re.$$

Thus, f(rx) = r f(x), for any rational number r. . . . (2)

Let us write f(1) = c. Taking x = 1 in (2), we get

f(r) = cr, for every rational number r. . . . (3)

Let α be any irrational number. Then there exists a sequence $\{r_n\}$ of rational numbers such that $r_n \to a$

$\Rightarrow f(r_n) \to f(a)$, continuity of f

$\Rightarrow f(a) = \lim f(r_n) = \lim cr_n = c \lim r_n = ca$, using (3).

Hence $f(x) = cx\ \forall\, x \in \Re$.

Theorem 12:

Define cos x as a power series and determine its domain. Prove that there exists a least positive number π such that $\cos\left(\frac{1}{2}\pi\right) = 0$.

(D.U., 1994, 93, 90)

Proof:

cos x is defined as the sum of the power series

$$1 - \frac{x^2}{2!} + \frac{x^4}{4!} - ... + (-1)^n \frac{x^{2n}}{(2n)!} + ...$$

Its radius of convergence is

$$R = \lim_{n\to\infty}\left|\frac{a_n}{a_{n+1}}\right| = \lim_{n\to\infty}\left|\frac{(2n+2)!}{(2n+1)!}\right| = \lim_{n\to\infty}(2n+2) = \infty.$$

Thus, this power series is convergent for $\forall\ x \in \Re$ and so the domain of cos x is $\Re$.

$$\cos x = 1 - \frac{x^2}{2!} + \frac{x^4}{4!} - \frac{x^6}{6!} + \frac{x^8}{8!} + \dots\dots\dots \quad \forall\ x \in \Re \qquad \dots (1)$$

$\therefore$ cos 0 = 1 > 0 and

$$\cos 2 = 1 - \frac{2^2}{2!} + \frac{2^4}{4!} - \frac{2^6}{6!} + \frac{2^8}{8!} + \dots\dots\dots\dots$$

$$= 1 - \frac{2^2}{2!}\left(1 - \frac{2^2}{3.4}\right) - \frac{2^6}{6!}\left(1 - \frac{2^2}{7.8}\right) - \dots\dots\dots$$

Since the brackets are all positive, therefore

$$\cos 2 < 1 - \frac{2^2}{2!}\left(1 - \frac{2^2}{3.4}\right) = -\frac{1}{3} \text{ and so } \cos 2 < 0.$$

Thus, we see that cos x is a continuous function in the closed interval [0, 2], where cos 0 and cos 2 are of opposite signs. By Intermediate value theorem, there exists at least one number $\alpha \in\ ]0, 2[$ such that $\cos \alpha = 0$. We now proceed to show that α is unique.

Let, if possible, $\beta \in\ ]0, 2[$ satisfy $\cos \beta = 0$ so that

$$\cos \alpha = \cos \beta = 0,\ 0 < \alpha, \beta < 2.$$

By Rolle's theorem, there exists some number λ, $\alpha < \lambda < b$, such that $-\sin \lambda = 0$.

$$\therefore \sin \lambda = 0,\ 0 < \lambda < 2. \qquad \dots (2)$$

By definition,

$$\sin x = x - \frac{x^3}{3!} + \frac{x^5}{5!} - \frac{x^7}{7!} + \dots\dots \quad \forall\ x \in \Re$$

$$\therefore \quad \sin \lambda = \frac{\lambda}{1!}\left(1 - \frac{\lambda^2}{2.3}\right) + \frac{\lambda^5}{5!}\left(1 - \frac{\lambda^2}{6.7}\right) + \dots\dots > 0, \text{ for } 0 < \lambda < 2.$$

This contradicts (2).

Thus, there exists one and only one that of the equation cos x = 0 lying between 0 and 2. i.e., $\cos \alpha = 0,\ 0 < \alpha < 2$.

We denote α as $\frac{1}{2}\pi$. Hence $\frac{1}{2}\pi$ is the *least positive root* of the equation cos x = 0 i.e., $\cos \frac{1}{2}\pi = 0$.

Also, therefore, we have $\cos x > 0$ when $0 \leq x < \frac{1}{2}\pi$.

Theorem 13:

Let f be a non-zero continuous function on $\mathfrak{R}$ such that

$f(x + y) = f(x) f(y) \ \forall \ x, y \in \mathfrak{R}$. . . . (1)

Show that $f(x) = e^{cx}$, for some constant $c \neq 0$. **(D.U., 1995)**

Proof:

It is given that $f(x) \neq 0 \ \forall \ x \in \mathfrak{R}$. From (1)

$$f(x) = f\left(\frac{1}{2}x\right) f\left(\frac{1}{2}x\right) \text{ or } f(x) = \left\{f\left(\frac{1}{2}x\right)\right\}^2 > 0 \ \forall \ x \in \mathfrak{R}.$$

Let $g(x) = \log_e f(x) \ \forall \ x \in \mathfrak{R}$. Then by (1),

$g(x + y) = g(x) + g(y) \ \forall \ x, y \in \mathfrak{R}$. . . . (2)

Now g, being the composition of two continuous functions f(x) and $\log_e x$, is itself continuous on R and g satisfies (2).

It follows that

$g(x) = cx \ \forall \ x \in \mathfrak{R}$, c being a constant

$\Rightarrow \log_e f(x) = cx \qquad \Rightarrow f(x) = e^{cx} \ \forall \ x \in \mathfrak{R}$.

Example 1:

Show that

(i) $\log(1 - x) = -x - \frac{x^2}{2} - \frac{x^3}{3} - \ldots, \ -1 \leq x < 1.$

(ii) $\log 2 = 1 - \frac{1}{2} + \frac{1}{3} - \frac{1}{4} + \ldots\ldots$

(D.U., 1992, 90; Physics (H) 1995)

(iii) $\frac{1}{2}[\log(1-x)]^2 = \frac{x^2}{2} + \left(1 + \frac{1}{2}\right)\frac{x^3}{3} + \left(1 + \frac{1}{2} + \frac{1}{3}\right)\frac{x^4}{4} + \ldots, \ -1 < x < 1.$

Solution:

(i) We know that

$(1 - x)^{-1} = 1 + x + x^2 + x^3 + \ldots, \ |x| < 1.$ ...(1)

Its radius of convergence is 1. So the above power series converges absolutely for $|x| < 1$ and uniformly for [– k, k\, $|k| < 1$. The integrated series is also convergent absolutely for $| x | < 1$ and uniformly for $[-k, k]$, $| k | < 1$,

Integrating, we get

$$-\log (1 - x) = x + \frac{x^2}{2} - \frac{x^3}{3} + ... + c$$

Putting $x = 0$, we get $c = 0$.

$$\therefore \log (1 - x) = - x - \frac{x^2}{2} - \frac{x^3}{3} - ......, \ -1 < x < 1.$$

By Leibnitz Test, the power series on the R.H. S. converges at $x = -1$. So by Abel's Theorem.

$$\log (1 - x) = -x - \frac{x^2}{2} - \frac{x^3}{3} - ......, \ -1 < x < 1. \qquad ...(2)$$

(ii) At $x = -1$, by Abel's Limit Theorem

$$\log 2 = \lim_{x \to -1-} \log (1 - x) = 1 - \frac{1}{2} + \frac{1}{3} - \frac{1}{4} + ...$$

(iii) both the series given on the R.H.S. of (1) and (2) are absolutely convergent in $]-1, 1[$, so their Cauchy's product will converge to $(1 - x)^{-1} \log (1 - x)$. Thus

$$(1 - x)^{-1} \log (1 - x) = - [x + (1 + \frac{1}{2}) x^2 + (1 + \frac{1}{2} + \frac{1}{2}) x^2 + ...], \quad -1 < x < 1$$

Integrating, we get

$$\frac{1}{2} [\log (1 - x)]^2 = \frac{x^2}{2} + \left(1 + \frac{1}{2}\right) x^3 + \left(1 + \frac{1}{2} + \frac{1}{3}\right) \frac{x^4}{4} + ... \quad -1 < x < 1,$$

where the constant of integration vanishes.

Since the series on the right converge at $x = -1$, so by Abel's theorem, the above expansion is valid for $-1 < x < 1$.

Example 2:

Find the radius of convergence and the exact interval of convergence of the following power series :

(i) $\sum \frac{(n+1)}{(n+2)(n+3)} x^{n,}$

(ii) $\sum \frac{(n!)^2 x^{2n}}{(2n)!}$.

Solution:

(i) $a_n = \dfrac{(n+1)}{(n+2)(n+3)},\ a_{n+1} = \dfrac{(n+2)}{(n+3)(n+4)}$

$$\therefore\ R = \lim_{n\to\infty}\left|\frac{a_n}{a_{n+1}}\right| = \frac{(n+1)(n+4)}{!\,(n+2)^2} = 1.$$

Thus, the given series converges for $|x| < 1$ and diverges for $|x| > 1$. We now check at the end points $x = 1, -1$.

For $x = 1$, the nth term of the series is

$$u_n = \frac{(n+1)}{(n+2)(n+3)} \sim \frac{n}{n^2} = \frac{1}{n}, \text{ which diverges.}$$

However, for $x = -1$, the given series is an alternating series for which $u_{n+1} < u_n\ \forall\ n$ and $\lim_{n\to\infty} u_n = 0$. Thus, by Leibnitz's test, the series converges for $x = -1$. Hence the interval of convergence is $[-1, 1[$.

(ii) We have $a_n = \dfrac{(n!)^2}{(2n)!},\ a_{n+1} = \dfrac{[(n+1)!]^2}{(2n+2)!}$.

Then $\dfrac{a_n}{a_{n+1}} = \dfrac{(n!)^2}{\{(n+1)!\}^2}\,\dfrac{(2n+2)!}{(2n)!} = \dfrac{(2n+2)(2n+1)}{(n+1)^2}$

$$R = \lim_{n\to\infty}\left|\frac{a_n}{a_{n+1}}\right| = \lim_{n\to\infty}\frac{\left(2+\frac{2}{n}\right)\left(2+\frac{1}{n}\right)}{(1+1/n)^2} = 4.$$

It follows that the given series converges for $|x| < 4$ diverges for $|x| > 4$. Hence the interval of convergence is $]-4, 4[$.

Example 3:

Show that the series $\Sigma x^n/n^2$ is uniformly convergent in $[-1, 1]$.

[D.U. Physics (H) 1994]

Solution:

We have $a_n = \dfrac{1}{n^2},\ a_{n+1} = \dfrac{1}{(n+1)^2}$ so that

$$\lim_{n\to\infty}\left|\frac{a_n}{a_{n+1}}\right| = \lim_{n\to\infty}\frac{(n+1)^2}{n^2} = \lim_{n\to\infty}\left(1+\frac{1}{n}\right) = 1$$

The radius of convergence is $R = 1$ and so the given series is uniformly convergent in $]-1, 1[$.

For x = 1, given series is $\sum \frac{1}{n^2}$, which is convergent.

For x = –1, given series becomes

$$-1+\frac{1}{2^2}-\frac{1}{3^2}+\frac{1}{4^2}-\ldots\ldots$$

This is an alternating series for which $u_{n+1} \le u_n \ \forall$ n and

$$\lim_{n\to\infty} u_n = \lim_{n\to\infty}\frac{1}{n^2} = 0.$$

So by Leibnitz's Test, the series to convergent or x = –1. Hence by Abel's Theorem, given power series is uniformly convergent in [–1, 1].

Example 4:

Show that the following series

$$x+\frac{1}{2}\frac{x^2}{3}+\frac{1.3}{2.4}\frac{x^5}{5}+\frac{1.3.5}{2.4.6}\frac{x^7}{7}+\ldots\ldots$$

is uniformly convergent in [–1, 1].

Solution:

Ignoring the first term, we have

$$a_n = \frac{1.3.5\ldots\ldots(2n-1)}{2.4.6\ldots\ldots2n\ (2n+1)},\ a_{n+1} = \frac{1.3.5\ldots\ldots(2n-1)\ (2n+1)}{2.4.6\ldots\ldots2n(2n+2)\ (2n+3)}$$

$$\therefore \lim_{n\to\infty}\frac{a_n}{a_{n+1}} = \lim_{n\to\infty}\frac{(2n+2)\ (2n+3)}{(2n+1)^2} = 1.$$

Since the radius of convergence is R = 1, given power series is uniformly convergent in]–1, 1[. For x = 1, we see that the D'Alembert's Ratio Test fails. Now

$$\lim_{n\to\infty}\left(\frac{a_n}{a_{n+1}}-1\right) = \lim_{n\to\infty}\frac{n(6n+5)}{(2n+1)^2} = \frac{6}{4} > 1.$$

By Raabe's Test, the series is convergent for x = 1.

Again for x = – 1, the series is convergent.

($\therefore \Sigma a_n$ is convergent $\Rightarrow -\Sigma a_n$ is convergent).

Hence, the given power series is uniformly convergent in [–1, 1] (Abel's Theorem).

Ex.:

Prove that $\sum_{n=1}^{\infty} \frac{1.3.5.......(2n-1)}{2.4.6.........2n} x^n$ *is uniformly convergence in* $[-1, 1[$.

Example 5:

Show that

$$\frac{1}{2}[\log(1+x)]^2 = \frac{x^2}{2} - \frac{x^3}{3}\left(1+\frac{1}{2}\right) + \frac{x^4}{4}\left(1+\frac{1}{2}+\frac{1}{3}\right) + ..., -1 < x < 1.$$

Solution:

By the above example,

$$\log(1+x) = x - \frac{x^2}{2} + \frac{x^3}{3} + \frac{x^4}{4} + ..., -1 < x < 1.$$

and $$(1+x)^{-1} = 1 - x + x^2 - x^3 + x^4 - ......, -1 < x < 1.$$

Both the series are absolutely convergent in $]-1, 1[$, ($\because$ R = 1), so their Cauchy's product will converge to $(1 + x)^{-1} \log (1 + x)$. Thus

$$(1+x)^{-1} \log(1+x) = x - x^2(1 + ½) + x^3(1 + ½ + ½) - ....., \quad -1 < x < 1$$

Integrating, we get

$$\frac{1}{2}[\log(1+x)]^2 = \frac{x^2}{2} - \frac{x^3}{3}\left(1+\frac{1}{2}\right) + \frac{x^4}{4}\left(1+\frac{1}{2}+\frac{1}{3}\right) - ...., -1 < x < 1$$

where the constant of integration vanishes.

Since the series on the right converges at x = 1, so by Abel's theorem the above expansion is valid for $-1 < x \leq 1$.

Example 6:

Find the radius of convergence of $\sum_{n=1}^{\infty} nx^{n-1}$.

Solution:

Here $\lim_{n\to\infty}\left|\frac{a_n}{a_{n+1}}\right| = \lim_{n\to\infty} \frac{n}{n+1} = 1$.

Hence the radius of convergence = 1.

Ex:

Find the radius of convergence of the series

$$x+\frac{x^2}{2^2}+\frac{2\,!}{3^3}x^2+\frac{3\,!}{4^6}x^6+......$$ **(D.U., 1994)**

Solution :

$$a_n=\frac{(n-1)\,!}{n^n},\ a_{n+1}=\frac{n\,!}{(n+1)^{n+1}}$$

$$\therefore\ R=\lim_{n\to\infty}\left|\frac{a_n}{a_{n+1}}\right|=\lim_{n\to\infty}\frac{(n-1)\,!\,(n+1)^{n+1}}{n\,!\qquad n^n}$$

$$=\lim_{n\to\infty}\left(\frac{n+1}{n}\right)^{n+1}=\lim_{n\to\infty}\left(1+\frac{1}{n}\right)^n\left(1+\frac{1}{n}\right)=e.$$

Hence the radius of convergence = e.

Example 7:

Show that the series

$$1+\frac{x}{2}+\frac{x^2}{3}+\frac{x^3}{4}+......$$

is uniformly convergent in [–1, k], 0 < k < 1.

Solution:

We have

$$\lim_{n\to\infty}\left|\frac{a_n}{a_{n+1}}\right|=\lim_{n\to\infty}\left(\frac{n+1}{n}\right)=\lim_{n\to\infty}\left(1+\frac{1}{n}\right)=1.$$

The radius of convergence is R = 1 and so the given series is uniformly convergent in]–1, 1[.

For x = 1, given series $\sum\frac{1}{n}$, which is divergent.

However for x = 1, given series becomes $1-\frac{1}{2}+\frac{1}{3}+\frac{1}{4}+........$, which is convergent by Leibnitz's Test. Hence by Abel's Theorem, given power series is uniformly convergent in [–1, 1[or [–1, k], where 0 < k < 1.

Example 8:

Prove that

$$\int_0^x\frac{dt}{1+t^n}=x-\frac{x^{n+1}}{n+1}+\frac{x^{(2n+1)}}{2n+1}-.........,\ -<x<1,\ n>0.$$

Solution:

For n > 0, we have

$(1 + t^n)^{-1} = 1 - t^n + t^{2n} - t^{2n} + \ldots..$

It is clear that the series on the right has its radius of convergence R = 1.

This series can be integrated term by term on]–1, 1[and the integrated series will also have the radius of convergence as unity. Thus

$$\int_0^x \frac{dt}{1+t^n} = x - \frac{x^{n+1}}{n+1} + \frac{x^{(2n+1)}}{2n+1} - \ldots\ldots..., - < x < 1. \qquad \ldots.(1)$$

At x = 1, the series on the right is convergent, by Leibnitz's test.

Hence, (1) is true for $-1 < x \leq 1$.

Example 9:

Show that

$$\log(1+x) = x - \frac{x^2}{2} + \frac{x^3}{3} - \frac{x^4}{4} + \ldots..., -1<x\leq 1$$

and deduce that

$$\log 2 = 1 - \frac{1}{2} + \frac{1}{3} - \frac{1}{4} + \ldots...$$ **(D.U., Physics (H) 1996, 1994)**

Solution:

We know

$$(1 + x)^{-1} = 1 - x + x^2 - x^3 + \ldots\ldots.., \quad -1 < x < 1. \qquad \ldots(1)$$

Its radius of convergence is R = 1.

Thus the power series is absolutely convergent for $|x| < 1$ and uniformly convergent for [–k, k], $|k| < 1$.

The integrated series is also convergent absolutely in]–1, 1[and uniformly in [–k, k], $|k| < 1$.

Integrating (1), we get

$$\log(1+x) = c + x - \frac{x^2}{2} + \frac{x^3}{3} - \frac{x^4}{4} + \ldots..., 1 < x < 1.$$

Putting x = 0 gives c = 0. This

$$\log(1+x) = x - \frac{x^2}{2} + \frac{x^3}{3} - \frac{x^4}{4} + \ldots..., 1 < x < 1.$$

The series on the R.H.S. is convergent for x = 1 (by Leibnitz's test). Hence by Abel's theorem,

$$\log(1+x) = x - \frac{x^2}{2} + \frac{x^3}{3} - \frac{x^4}{4} + \ldots\ldots, 1 < x < 1.$$

At x = 1, by Abel's limit theorem, we get

$$\log 2 = \lim_{x\to 1-} \log(1+x) = 1 - \frac{1}{2} + \frac{1}{3} - \frac{1}{4} + \ldots\ldots$$

Example 10:

Show that

(i) $\tan^{-1} x = x - \frac{x^3}{3} + \frac{x^5}{5} \frac{x^7}{7} + \ldots\ldots, -1 < x \le 1$

(D. U., 1994; Physics (H) 1994)

(ii) $\frac{\pi}{4} = 1 - \frac{1}{3} + \frac{1}{5} - \frac{1}{7} + \ldots\ldots$ **(D.U., 1994)**

(iii) $\frac{1}{2}(\tan^{-1} x)^2 = \frac{x^2}{2} - \frac{x^4}{4}\left(1 + \frac{1}{3}\right)$

$$+ \frac{x^6}{6}\left(1 + \frac{1}{3} + \frac{1}{5}\right) + \ldots\ldots, -1 < x \le 1.$$

Solution:

We know

$$(1 + x^2)^{-1} = 1 - x^2 + x^4 - x^6 + \ldots\ldots, \ |x| \le 1. \qquad \ldots(1)$$

Radius of convergence is

$$R = \lim_{n\to\infty} \left|\frac{a_n}{a_{n+1}}\right| = 1.$$

Thus the series is convergent absolutely for all x such that $|x| < 1$, and is uniformly convergent in $[-k, k]$, $|k| < 1$.

The integrated series is also convergent absolutely in $]-1, 1[$ and uniformly in $[-k, k]$, $|k| < 1$.

Integrating (1), we get

$$\tan^{-1} x = \left(x - \frac{x^3}{3} + \frac{x^5}{5} - \frac{x^7}{7} + \ldots\ldots\right) + C, \ |x| < 1.$$

where C is constant of integration.

Putting x = 0, we get C = 0 and so

$$\tan^{-1} x = x - \frac{x^3}{3} + \frac{x^5}{5} - \frac{x^7}{7} + \ldots\ldots, \quad |x| < 1.$$

Now the series

$$x - \frac{x^3}{3} + \frac{x^5}{5} - \frac{x^7}{7} + \ldots\ldots \text{ is convergent at } x = \pm 1.$$

$\left[\because \text{ for } x = 1, 1 - \frac{1}{3} + \frac{1}{5} - \frac{1}{7} \ldots\ldots\right.$ is an alternating series which converges, by Leibnitz's test, Similarly it is convergent for $\left. x = -1\right]$.

By Abel's Theorem, it is uniformly convergent in [–1, 1].

Hence, $\tan^{-1} x = x - \frac{x^3}{3} + \frac{x^5}{5} - \frac{x^7}{7} + \ldots, -1 < x \le 1.$

(ii) Putting x = 1, by Abel's Limit Theorem,

$$\frac{\pi}{4} = \tan^{-1} 1 = \lim_{x \to 1-0} \tan^{-1} x = 1 - \frac{1}{3} + \frac{1}{5} - \frac{1}{7} + \ldots\ldots$$

(iii) We have

$$\tan^{-1} x = x - \frac{x^3}{3} + \frac{x^5}{5} - \frac{x^7}{7} + \ldots, -1 < x \le 1.$$

and $(1 + x^2)^{-1} \tan^{-1} x = x - x^2 + x^4 - x^6 + \ldots\ldots, 1 < x < 1.$

By part (i), both the series are absolutely convergent in]–1, 1[, so they can be multiplied and their Cauchy's product will be equal, to

$$(1 + x^2)^{-1} \tan^{-1} x = x - \left(1 + \frac{1}{3}\right)x^2 + \left(1 + \frac{1}{3} + \frac{1}{5}\right)x^2 \quad -1 < x < 1.$$

Integrating, we get

$$\frac{1}{2} (\tan^{-1} x)^2 = \frac{x^2}{2} - \frac{x^4}{4}\left(1 + \frac{1}{3}\right) + \frac{x^6}{6}\left(1 + \frac{1}{3} + \frac{1}{5}\right) - \ldots\ldots,$$

$$-1 < x < 1$$

Here the constant of integration vanishes on taking x = 0.

By Leibnitz's Test, the series on the R.H.S. converges for x = 1. Hence the above expansion is valid for $-1 < x < 1$.

Example 11:

Find the radius of convergence of the series

$$\frac{1}{2} x + \frac{1.3}{2.5} x^2 + \frac{1.3.5}{2.5.8} x^2 + \ldots\ldots$$

Solution:

$$R = \lim_{n\to\infty} \left|\frac{a_n}{a_{n+1}}\right|$$

$$= \lim_{n\to\infty} \frac{1.3.5\,.....(2n-1)}{2.5.8.....(3n-1)}, \frac{2.5.8......(3n-1)(3n+2)}{1.3.5.....(2n-1)(2n+1)}$$

$$= \lim_{n\to\infty} \frac{3n+2}{2n+1} = \frac{3}{2}.$$

Example 12:

Show that

$$\sin^{-1} x = x + \frac{1}{2}\frac{x^3}{3} + \frac{1.3}{2.4}\frac{x^5}{5} + \frac{1.3.5}{2.4.6}\frac{x^7}{7} + ..., 1<x<1.$$

Deduce that

$$\frac{\pi}{2} = 1 + \frac{1}{2}.\frac{1}{3} + \frac{1.3}{2.4}.\frac{1}{5} + \frac{1.3.5}{2.4.6}.\frac{1}{7} + ...$$

(D.U., Physics (H) 1995, Math (H) 1995, 1993, 89)

Solution:

We know that

$$\frac{1}{\sqrt{1-x^2}} = (1 - x^2)^{-1/2}$$

$$= 1 + \frac{1}{2}x^2 + \frac{1.3}{2.4}x^4 + \frac{1.3.5}{2.4.6}x^6 + ..., |x| < 1.$$

The series on the R.H.S. is a power series with radius of convergence 1 and so it is absolutely convergent in]–1, 1[and uniformly convergent for [–k, k], | k | < 1. The integrated series is also convergent in]–1, 1[and uniformly convergent for [–k, k], | k | < 1. Integrating, we get

$$\sin^{-1} x = c + x + \frac{1}{2}.\frac{x^3}{3} + \frac{1.3}{2.4}.\frac{x^5}{5} + \frac{1.3.5}{2.4.6}.\frac{x^7}{7} + ...,| x|< 1$$

Putting x = 0, we get c = 0.

$$\therefore \sin^{-1} x + \frac{1}{2}.\frac{x^3}{3} + \frac{1.3}{2.4}.\frac{x^5}{5} + \frac{1.3.5}{2.4.6}.\frac{x^7}{7} + ...,-1< x <1 \qquad ...(1)$$

For x = 1, the nth term of the series of the R.H.S. of (1) is

$$u_n = \frac{1.3.5..........(2n-1)}{2.4.6........2n(2n+1)}.$$

$$u_{n+1} = \frac{1.3.5..........(2n+1)}{2.4.6........(2n+2)(2n+3)}.$$

$$\therefore \frac{u_n}{u_{n+1}} = \frac{(2n+2)(2n+3)}{(2n+1)^2} . n\left(\frac{u_n}{u_{n+1}} - 1\right) = \frac{6n^2 + 5}{(2n+1)^2}$$

$$\lim_{n\to\infty} n\left(\frac{u_n}{u_{n+1}} - 1\right) = \frac{3}{2} > 1.$$

So by Raabe's Test, the series is convergent for x = 1. Hence

$$\sin^{-1} x = x + \frac{1}{2}.\frac{x^3}{3} + \frac{1.3}{2.4}\frac{x^5}{5} + \frac{1.3.5}{2.4.6}\frac{x^7}{7} + ..., - < x < 1.$$

(ii) At x = 1, by Abel's Limit Theorem

$$\frac{\pi}{2} = \lim_{x\to 1-} \sin^{-1} x = 1 + \frac{1}{2}.\frac{1}{3} + \frac{1.3}{2.4}.\frac{1}{5} + \frac{1.3.5}{2.4.6}\frac{1}{7} + ...$$

Example 13:

Prove that the power series

$$1 + \frac{a.b}{1.c} x + \frac{a(a+1)\, b(b+1)}{1.2c\,(c+1)} x^2 + ...$$ **(D.U., 1994)**

has unit radius of convergence. **(D.U., Physics (H) 1996)**

Solution:

$$a_n = \frac{a(a+1)\, ...(a+n-1)\; b(b+1)\, ...(b+n-1)}{1.2......(n+1)\; c(c+1)......(c+n-1)}$$

(omitting the first term)

$$a_{n+1} = \frac{a(a+1)\, ...(a+n)\; b(b+1)\, ...(b+n)}{1.2......(n+1)\; c(c+1).......(c+n)}$$

$$R = \lim_{n\to\infty} \left|\frac{a_n}{a_{n+1}}\right| = \lim_{n\to\infty} \frac{(n+1)\,(c+n)}{(a+n)\,(b+n)}$$

$$= \lim_{n\to\infty} \frac{(1+1/n)\,(1+c/n)}{(1+a/n)(1+b/n)} = 1.$$

Example 14:

Find the domain of convergence of the series

(i) $\sum \frac{x^n}{n^2}$, (ii) $\sum \frac{(-1)^n (x-1)^n}{2^n (3n-1)}$.

Solution:

(i) $a_n = \frac{1}{n^3}$, $a_{n+1} = \frac{1}{(n+1)}$ so that

$$\lim_{n\to\infty} \frac{a_n}{a_{n+1}} = \lim_{n\to\infty} \left(\frac{n+1}{n}\right)^3 = 1. \text{ Thus } R = 1.$$

For $x = 2$, the given series is $\sum \frac{(-1)^{n+1}}{n}$, which is convergent by Leibnitz's Test. Hence the exact interval of convergence is]0, 2].

(iii) The given power series is about the point $x = -2$.

$$\lim_{n\to\infty} \left|\frac{a_n}{a_{n+1}}\right| = \lim_{n\to\infty} \frac{n!}{n^n} \frac{(n+1)^{n+1}}{(n+1)!} = \lim_{n\to\infty} \left(\frac{n+1}{n}\right)^n$$

$$= \lim_{n\to\infty} \left(1+\frac{1}{n}\right)^n = e.$$

$\therefore R = e$ and the interval of convergence is $]-2-e, -2+e[$.

2

FOURIER SERIES

2.1 Fourier Series

A series of the form

$$\frac{1}{2}a_n + (a_1 \cos x + b_1 \sin x) + (a_2 \cos 2x + b_2 \sin 2x) + \ldots.$$

or more compactly

$$\frac{1}{2}a_0 + \sum_{n=1}^{\infty} (a_n \cos nx + b_n \sin nx)$$

is called a *trigonometric series,* and the constants a_0, a_n, b_n $(n = 1, 2, 3 \ldots)$ are called its *coefficients.* If the series converges for all values of x (in fact, in an interval $c \leq x \leq c + 2\pi$) then, in view of the periodicity of the trigonometric functions, its sum is a function f defined for all values of x and periodic with period 2π. To save tedious repetition, it may be mentioned once for all that the *functions considered in this chapter are all assumed to be periodic.*

We now proceed to obtain a relationship between the function f and the coefficients a_n, b_n, which was conjectured by *Euler.*

The following results of integral calculus form the basis of all subsequent work. Throughout we use m, n to *denote positive integers or zero*:

$$\int_{-\pi}^{\pi} \cos mx \cos nx \, dx = \begin{cases} 0, & \text{for } m \neq n \\ \pi, & \text{for } m \quad n > 0 \\ 2\pi, & \text{for } m = n = 0 \end{cases}$$

$$\int_{-\pi}^{\pi} \sin mx \sin nx \, dx = \begin{cases} 0, & \text{for } m \neq n \text{ and } m = n = 0 \\ \pi, & \text{for } m = n > 0 \end{cases}$$

$$\int_{-\pi}^{\pi} \cos mx \sin nx \, dx = 0 \, .$$

Fourier Series

If the numbers $a_0, a_1, \ldots, b_1, b_2, \ldots\ldots, b_n, \ldots$ are derived from a function f by means of Euler-Fourier formulas:

$$a_n = \frac{1}{n} \int_{-\pi}^{\pi} f(x) \cos nx \, dx \; (n = 0, 1, 2, \ldots\ldots),$$

$$b_n = \frac{1}{n} \int_{-\pi}^{\pi} f(x) \sin nx \, dx \; (n = 1, 2, \ldots\ldots), \qquad \ldots(1)$$

then the series

$$\tfrac{1}{2} a_0 + \sum_{n=1}^{\infty} (a_n \cos nx + b_n \sin nx) \qquad \ldots(2)$$

is called the *Fourier series of* f or the *Fourier series generated by f,* and the coefficients a_n, b_n defined by (1), the *Fourier coefficients** of f.

2.2 Periodic Function

We know that a period function of period λ is such that $f(x \pm \lambda) = f(x)$, for all x.

Theorem 1:

For a periodic function of period 2p, prove that

(i) $\int_{\alpha}^{\beta} f \, dx = \int_{\alpha+2\pi}^{\beta+2\pi} f \, dx$, (ii) $\int_{-\pi}^{\pi} f \, dx = \int_{\alpha}^{\alpha+2\pi} f \, dx$,

(iii) $\int_{-\pi}^{\pi} f(x) \, dx = \int_{-\pi}^{\pi} f(\gamma + x) \, dx$.

α, β γ being any numbers whatsoever.

Proof:

(i) For a periodic function of period 2π, we know

$f(t - 2\pi)$ $f(t)$

Hence, putting $t = x + 2\pi$, for all a and b, we get

$$\int_{\alpha}^{\beta} f(x) \, dx = \int_{\alpha+2\pi}^{\beta+2\pi} f(t - 2\pi) dt = \int_{\alpha+2\pi}^{\alpha+2\pi} f(t) dt = \int_{\alpha+2\pi}^{\beta+2\pi} f(x) \, dx$$

(ii) $\int_{\alpha}^{\alpha+2\pi} f dx = \int_{\alpha}^{-\pi} f dx + \int_{-\pi}^{\pi} f dx + \int_{\pi}^{\alpha+2\pi} f dx$

$= \int_{\alpha}^{-\pi} f dx + \int_{-\pi}^{\pi} f dx + \int_{-\pi}^{\alpha} f dx$ [using (i)]

$= \int_{-\pi}^{\pi} f dx$

(iii) Let $\gamma + x = t$

$\therefore \quad \int_{-\pi}^{\pi} f(\gamma + x)\, dx = \int_{\gamma-\pi}^{\gamma+\pi} f(t)\, dt$

$= \int_{\gamma-\pi}^{-\pi} f dt + \int_{-\pi}^{\pi} f\, dt + \int_{\pi}^{\gamma+\pi} f\, dt$

$= \int_{\gamma+\pi}^{\pi} f dt + \int_{-\pi} f dt + \int_{\pi}^{\gamma+\pi} f dt$ [using (i)]

$= \int_{-\pi}^{\pi} f\, dt = \int_{-\pi}^{\pi} f\, dx.$

These results, in fact mean that the integral of a periodic function over any interval whose length is equal to its period always has the same value.

Theorem 2:

(Dirichlet's Conditions) *: Let a function f be bounded, periodic with period 2π, and integrable in [−π, π] and let it be possible to divide [−π, π] into a finite number of open sub-intervals in each of which f is monotonic (i.e., f is piecewise monotonic in [−π, π]). Then*

$$\frac{1}{2}a_0 + \sum_{n=1}^{\infty}(a_n \cos nx + b_n \sin nx)$$

$$= \begin{cases} \frac{1}{2}[f(\xi - 0) + f(\xi + 0), & \text{for } -\pi < \xi < \pi \\ \frac{1}{2}[f(\pi - 0) + f(-\pi + 0) & \text{for } \xi - \pm\pi. \end{cases}$$

(Here $f(\xi + 0)$ and $f(\xi - 0)$ are the right and left hand limits of $f(x)$ at $(x = \xi)$.

The proof is omitted.

Corollary:

Suppose f is continuous at $x = \xi$ so that

$$f(\xi - 0) = f(\xi + 0) = f(\xi)$$

$$\Rightarrow \frac{1}{2}[f(\xi - 0) + f(\xi + 0)] = f(\xi).$$

If follows that if f satisfies the conditions of the above theorem, then

$$\frac{1}{2}a_0 + \sum_{n=1}^{\infty}(a_n \cos nx + b_n \sin nx)$$

$$= \begin{cases} f(x), \text{ if } x \text{ is a point of continuity,} \\ \frac{1}{2}[f(x-0)+f(x+0), \text{ if } x \text{ is a point of discontinuity.} \end{cases}$$

Ex.: *State the conditions under which $\phi(x)$ can be expressed as a Fourier series in $[-\pi, \pi]$.* ***(D.U., 1986)***

Remark:

While computing the Fourier coefficients, we can save a lot of time and labour by using certain properties of even and odd functions: A function f(x) is called even if f(–x) = f(x) for all x in the domain of f. A function g(x) is called odd if g(–x) = –g(x) for all x in the domain of g.

We know that if f is even and g is odd, then

$$\int_{-a}^{a} f(x)\,dx = 2\int_{0}^{a} f(x)\,dx, \quad \int_{-a}^{a} g(x)\,dx = 0.$$

Theorem 3:

If f is bounded and integrable on $[-\pi, \pi]$ and if a_n, b_n are its Fourier coefficients, then $\sum_{n=1}^{\infty}(a_n^2 + b_n^2)$ *converges.*

The integral

$$\int_{-\pi}^{\pi}\left(f - \sum_{n=1}^{m}(a_n \cos nx + b_n \sin nx)\right)^2 dx \geq 0$$

as its integrand is never negative, now

$$\int_{-\pi}^{\pi}\left(f - \sum_{n=1}^{m}(n_n \cos nx + b_n \sin nx)\right)^2 dx$$

$$= \int_{-\pi}^{\pi} f^2 dx + \int_{-\pi}^{\pi}[\sum_{n=1}^{m}(a_n \cos nx + b_n \sin nx)]^2\,dx$$

$$-2\Sigma\left[a_n \int_{-\pi}^{\pi} f \cos nx\,dx\right] - 2\Sigma\left[b_n \int_{-\pi}^{\pi} f \sin nx\,dx\right]$$

$$= \int_{-\pi}^{\pi} f^2 dx + \pi\Sigma a_n^2 + \pi\Sigma b_n^2 - 2\pi\Sigma a_n^2 - 2\pi\Sigma b_n^2$$

$$= \int_{-\pi}^{\pi} f^2 dx - \pi\sum_{n=1}^{m}\left(a_n^2 + b_n^2\right)$$

Since this is non-negative, we have

$$\sum_{n=1}^{m}\left(a_n^2 + b_n^2\right) \le \frac{1}{\pi}\int_{-\pi}^{\pi} f^2 dx,$$

taking limit as $m \to \infty$, we obtain

$$\sum_{n=1}^{m}\left(a_n^2 + b_n^2\right) \le \frac{1}{\pi}\int_{-\pi}^{\pi} f^2 dx \qquad \textbf{(Bessel's Inequality)}$$

Hence, the series $\sum_{n=1}^{m}\left(a_n^2 + b_n^2\right)$ is convergent, since f, and so f^2 is integrable on $[-\pi, \pi]$.

Corollary:

Since $\sum\left(a_n^2 + b_n^2\right)$ *converges, its general term* $\left(a_n^2 + b_n^2\right) \to 0$*, and so,* $a_n \to 0$*,* $b_n \to 0$*. Thus the Fourier coefficients* a_n *and* b_n *of an integrable (and bounded) function form a null sequence, i.e., when* $n \to \infty$*,*

$$a_n = \frac{1}{\pi}\int_{-\pi}^{\pi} f \cos nx\, dx \to 0, \text{ and } b_n = \frac{1}{\pi}\int_{-\pi}^{\pi} f \sin nx\, dx \to 0.$$

Theorem 4:

If a function f is bounded periodic with period 2π *and integrable on* $[-\pi, \pi]$*, and piecewise monotonic on* $[-\pi, \pi]$*, then*

$$\tfrac{1}{2}a_0 + \sum_{n=1}^{\infty}(a_n \cos n\xi + b_n \sin n\xi) = \begin{cases} \tfrac{1}{2}[f(\xi-) + f(\xi+)], & \text{for } -\pi < \xi < \pi, \\ \tfrac{1}{2}[f(\pi-) + f(-\pi+)], & \text{for } \xi = \pm\pi \end{cases}$$

where a_n*,* b_n *are Fourier coefficients of f.*

Solution:

Let $\frac{1}{2}a_0 + \sum_{n=1}^{\infty}(a_n \cos nx + b_n \sin nx)$ be the Fourier series of f, and ξ a point of $[-\pi, \pi]$.

The mth partial sum at the point ξ,

$$\tfrac{1}{2}a_0 + \sum_{n=1}^{\infty}(a_n \cos n\xi + b_n \sin n\xi)$$

$$= \frac{1}{2\pi}\int_{-\pi}^{\pi} f\, dx + \sum_{n=1}^{m}\frac{1}{\pi}\int_{-\pi}^{\pi} f[\cos nx \cos n\xi + \sin nx \sin n\xi]\, dx$$

$$= \frac{1}{2\pi}\int_{-\pi}^{\pi} f\,[1 + 2\sum_{n=1}^{m}\cos n(x - \xi)]\, dx$$

$$= \frac{1}{2\pi} \int_{-\pi}^{\pi} f(x + \xi) [1 + 2 \sum_{n=1}^{m} \cos nx)] dx,$$

$$= \frac{1}{2\pi} \int_{-\pi}^{\pi} f(x + \xi) \frac{\sin (m + \frac{1}{2}) x}{\sin \frac{1}{2} x} dx$$

$$= \frac{1}{2\pi} \int_{-\pi}^{0} f(x + \xi) \frac{\sin (m + \frac{1}{2}) x}{\sin \frac{1}{2} x} dx$$

$$= \frac{1}{2\pi} \int_{0}^{\pi} f(x + \xi) \frac{\sin (m + \frac{1}{2}) x}{\sin \frac{1}{2} x} dx$$

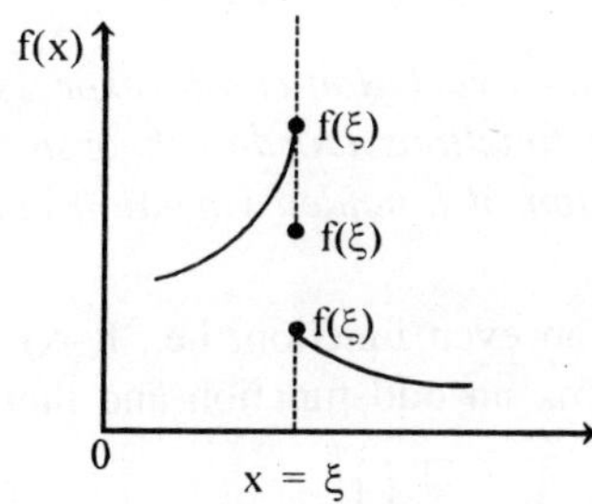

Fig. : Discontinuity at ξ.

$$= \frac{1}{\pi} \int_{0}^{\pi/2} f(-2t + \xi) \frac{\sin (2m + 1) t}{\sin t} dt$$

$$+ \frac{1}{\pi} \int_{0}^{\pi/2} f(2t + \xi) \frac{\sin (2m + 1) t}{\sin t} dt$$

Proceeding to limits when $m \to \infty$.

$$\frac{1}{2} a_0 + \sum_{n-1}^{\infty} (a_n \cos n\xi + b_n \sin n\xi) = \frac{1}{\pi} \left[\frac{1}{2} \pi f(\xi-) + \frac{1}{2} \pi f(\xi+) \right]$$

$$= \frac{f(\xi-) + f(\xi+)}{2}, \xi \in [-\pi, \pi]$$

Thus, the Fourier series of a (periodic) function f which is bounded, integrable and piecewise monotonic on $[-\pi, \pi]$, converges to ½ $[f(\xi-) + f(\xi+)]$ at a point ξ, $-\pi < \xi < \pi$, and (using periodicity of f) to ½$[f(\pi-) + f(-\pi+)]$ at the ends, $\pm \pi$.

Corollary 1:

At a point ζ of continuity of f, the sum of the Fourier series of f,

$$\tfrac{1}{2}a_0 + \sum_{n=1}^{\infty} (a_n \cos n\zeta + b_n \sin n\zeta) = \tfrac{1}{2}[f(\zeta-) + f(\zeta+1)]$$

$$= \tfrac{1}{2}[f(\zeta) + f(\zeta)] = f(\zeta)$$

Thus, for a function f satisfying the conditions of the theorem, *the sum of the Fourier series is actually f(x) at all points x where f is continuous;* while at points of (finite) discontinuity the sum of the series is 1/2 (f(x–) + f(x+)].

Corollary 2:

Since a function of bounded variation can be expressed as a difference of two monotone increasing functions (Jordan theorem), it may be easily shown that a periodic function of bounded variation can be expressed as a Fourier Series.

Even Function: If f is an even function, i.e., $f(-x) = f(x)$, $\forall$ x then f cos nx is an even and f sin nx an odd function and therefore

$$a_n = \frac{1}{\pi}\int_{-\pi}^{\pi} f \cos nx\, dx = \frac{1}{\pi}\left[\int_{-\pi}^{0} f \cos nx\, dx + \int_{0}^{\pi} f \cos nx\, dx\right]$$

$$= \frac{2}{\pi}\int_{0}^{\pi} f \cos nx\, dx \qquad ...(1)$$

$$b_n = \frac{1}{\pi}\int_{-\pi}^{\pi} f \sin nx\, dx = 0.$$

So, the Fourier Series of an even function consists of terms of cosines only and the coefficients a_n may be computed from (1).

Also, since for an even function.

$$f(0+) = f(0-) = f(0)$$

and

$$f(-\pi + 0) - f(\pi - 0)$$

the sum of the series of f(0) at 0 or ± (even multiple of π), and is f(π–) at ± π or ± (odd multiple of π).

Odd Function: If f is an odd function, i.e., $f(-x) = -f(x)$, $\forall$ x, then f cos nx is an odd and f sin nx and even function and therefore

$$a_n = \frac{1}{\pi}\int_{-\pi}^{\pi} f \cos nx\, dx = 0$$

$$b_n = \frac{1}{\pi}\int_{-\pi}^{\pi} f \sin nx \, dx = \frac{2}{\pi}\int_{0}^{\pi} f \sin nx \, dx \qquad ...(2)$$

So, the Fourier Series of an odd function consist of sine terms only and the coefficient b_n may be calculated from (2). Also for an odd function

$$f(0-) = 0 = f(0+)$$

and

$$f(-\pi + 0) = -f(\pi - 0)$$

the sum of the series is 0 at x = 0 (or any multiple of π).

Example 1:

Are the following series the Fourier Series?

(i) $\frac{\sin x}{\sqrt{1}} + \frac{\cos 2x}{\sqrt{2}} + \frac{\sin 3x}{\sqrt{3}} + \frac{\cos 4x}{\sqrt{4}} + \ldots\ldots$

(ii) $\frac{\cos x}{1^2} + \frac{\cos 2x}{2^2} + \frac{\cos 3x}{3^2} + \frac{\cos 4x}{4^2} + \ldots\ldots\ldots$ **(D.U., 1989)**

Solution:

(i) Suppose the given series is a Fourier Series

$$a_n = \frac{1}{\pi}\int_{-\pi}^{\pi} f(x) \cos nx = \frac{1}{\pi}\int_{-\pi}^{\pi} \sin nx \cos nx \, dx$$

$$= \begin{cases} 0, \text{ if n is odd.} \\ \frac{1}{\sqrt{n}}, \text{ if n is even} \end{cases}$$

$$\therefore \qquad a_n = \begin{cases} \frac{1}{\sqrt{n}}, \text{ if n is odd.} \\ 0, \text{ if n is even} \end{cases}$$

Similarly, $b_n = \begin{cases} \frac{1}{\sqrt{n}}, \text{ if n is odd.} \\ 0, \text{ if n is even} \end{cases}$

Then $a_n^2 + b_n^2 = \frac{1}{n}$ in both the cases and so

$$\sum_{n=1}^{\infty}(a_n^2 + b_n^2) = \sum_{n=1}^{\infty}\frac{1}{n}, \text{ which is divergent.}$$

This is a contradiction.

Hence the given series is not a Fourier Series.

(ii) The given series may be written as

$$\sum_{n=1}^{\infty}\left[\frac{\cos(2n-1)x}{(2n-1)^2}+\frac{\sin 2nx}{(2n)^2}\right]$$

We have

$$\left|\frac{\cos(2n-1)x}{(2n-1)^2}+\frac{\sin 2nx}{(2n)^2}\right| \le \frac{|\cos(2n-1)x|}{(2n-1)^2}+\frac{|\sin 2nx|}{(2n)^2}$$

$$\le \frac{1}{(2n-1)^2}+\frac{1}{(2n)^2} \le \frac{1}{n^2}+\frac{1}{n^2}=\frac{2}{n^2}.$$

Since $\sum M_n = \sum \frac{1}{n^2}$ is convergent, so by Weierstrass's M-test, the givens series is uniformly convergent on $[-\pi, \pi]$ and if f(x) is its sum, then the given series is the Fourier Series f f(x).

Example 2:

Find the Fourier Series generated by the periodic function | x | of period 2π.

Also compute the value of the series at 0, 2π, = 3π.

Solution:

The given function is monotonic and continuous on $[-\pi, \pi]$. Since f(x) = | x | is an even function, its Fourier series will consist of cosine terms only.

We have $f(x) = \begin{cases} -x, & \text{for } -\pi \le x \le 0, \\ x, & \text{for } 0 \le x \le \pi. \end{cases}$

We have

$$a_0 = \frac{1}{\pi}\int_{-\pi}^{\pi} f(x)\,dx = \frac{2}{\pi}\int_0^{\pi} f(x)\,dx = \frac{2}{\pi}\int_0^{\pi} x\,dx = \pi.$$

Now $a_n = \frac{2}{\pi}\int_0^{\pi} x\cos nx\,dx = \frac{2}{\pi n^2}(\cos n\pi - 1)$

$$\therefore \quad a_n = \begin{cases} 0, & \text{for } n \text{ even}, \\ \dfrac{-4}{\pi n^2}, & \text{for } n \text{ odd}. \end{cases}$$

Hence the Fourier series of | x | is

$$f(x) = \frac{1}{2}a_0 + \sum_{n=1}^{\infty} a_n \cos nx$$

$$= \frac{\pi}{2} + \frac{4}{\pi} \sum_{n=1}^{\infty} \frac{\cos nx}{n^2} \text{ (n odd)}$$

$$\therefore \quad f(x) = \frac{\pi}{2} - \frac{4}{\pi}\left[\frac{\cos x}{1^2} + \frac{\cos 3x}{3^2} + \frac{\cos 5x}{5^2} - \dots\dots\dots\right]$$

We shall now compute the value of the series at 0, 2π, -3π.

The above series converges at all points and so its sum is equal to the given function $f(x) = |x|$.

The sum of the series at 2π is the same as at 0.

At $x = 0$

$$|0| = 0 = \frac{\pi}{2} - \frac{4}{\pi}\left[\frac{1}{1^2} + \frac{1}{3^2} + \frac{1}{5^2} + \dots\dots\dots\right]$$

$$\therefore \quad 1 + \frac{1}{3^2} + \frac{1}{5^2} + \dots\dots = \frac{\pi^2}{8}.$$

The sum of the series at -3π is the same as at π and so

$$|\pi| = \pi = \frac{\pi}{2} + \frac{4}{\pi}\left[1 + \frac{1}{3^2} + \frac{1}{5^2} + \dots\dots\dots\right]$$

$$\therefore \quad 1 + \frac{1}{3^2} + \frac{1}{5^2} + \dots = \frac{\pi^2}{8}.$$

Example 3:

Find (a) the Fourier sine series and (b) the Fourier cosine series which represents $f(x) = (\pi - x)$ in $0 < x < \pi$. **(D.U., 1996)**

Solution:

Firstly we find sine series. We have

$$b_n = \frac{2}{\pi} \int_0^{\pi} (\pi - x) \sin nx \, dx$$

$$= \frac{2}{\pi}\left[\left|-(\pi - x)\frac{\cos nx}{n}\right|_0^{\pi} - \int_0^{\pi} \frac{\cos nx}{n} dx\right]$$

$$= \frac{2}{\pi}\left[\frac{\pi}{n} + 0\right] = \frac{2}{n}.$$

Since f is continuous in $0 < x < \pi$, we have

$$\sum b_n \sin nx = \tfrac{1}{2}[f(x+0) + f(x-0)] = [\frac{1}{2} f(x) + f(x)] = f(x)$$

or $$\pi - x = 2\Sigma \tfrac{1}{n} \sin nx = 2\left(\frac{\sin x}{1} + \frac{\sin 2x}{2} + \frac{\sin 3x}{3} + \ldots\right)$$

We know that the sum of the series must be 0 for $x = 0$ and $x = \pi$ and this fact can also be verified directly. The representation holds for $x = \pi$ but not for $x = 0$.

Now we find cosine series. We have

$$a_n = \frac{2}{\pi}\int_0^{\pi}(\pi - x)\cos nx\,dx = \frac{2[1-(-1)^n]}{\pi n^2},\ n \neq 0.$$

Also, $$a_0 = \frac{2}{\pi}\int_0^{\pi} f(x)\,dx = \frac{2}{\pi}\int_0^{\pi}(\pi - x)\,dx = \frac{2}{\pi}.\frac{\pi^2}{2} = \pi.$$

Since f is continuous in $0 < x < \pi$, we have

$$\pi - x = \tfrac{1}{2}\pi + \sum \frac{2[1-(-1)^n]}{\pi n^2}\cos nx$$

$$= \tfrac{1}{2}\pi + \frac{4}{\pi}\left(\frac{\cos x}{1^2} + \frac{\cos 3x}{3^2} + \frac{\cos 5x}{5^2} + \ldots\ldots\right)$$

We know that the sum of the series must be $f(0+) = \pi$ for $x = 0$ and $f(\pi - 0) = 0$ for $x = \pi$. At $x = 0$ or π, we obtain

$$\frac{\pi^2}{8} = 1 + \frac{1}{3^2} + \frac{1}{5^2} + \frac{1}{7^2} + \ldots\ldots$$

Example 4:

Expand in a series of sines and cosines of multiple angles of x, the periodic function f with period in defined as

$$f(x) = \begin{cases} -1, & \text{for } -\pi < x < 0, \\ 1, & \text{for } 0 \leq x \leq \pi. \end{cases}$$

Also calculate the sum f the series at $x = 0, \tfrac{1}{2}\pi, \pm\pi$.

Solution:

The given function is piecewise monotonic, bounded and integrable on $[-\pi, \pi]$.

We have $$a_0 = \frac{1}{\pi}\int_{-\pi}^{\pi} f(x)\,dx$$

$$= \frac{1}{\pi}\left[\int_{-\pi}^{0}(-1)\,dx + \int_{\pi}^{0} 1.dx\right] = 0. \qquad \therefore a_n = 0$$

Now $\quad a_n = \frac{1}{\pi}\int_{-\pi}^{\pi} f(x)\cos nx\,dx$

$$= \frac{1}{\pi}\left[\int_{-\pi}^{0} -\cos nx\,dx + \int_{0}^{\pi}\cos nx\,dx\right]$$

$$= \frac{1}{\pi}\left[-\left|\frac{\sin nx}{n}\right|_{-\pi}^{0} + \left|\frac{\sin nx}{n}\right|_{0}^{\pi}\right] = \quad \forall n.$$

Now $\quad b_n = \frac{1}{\pi}\int_{-\pi}^{\pi} f(x)\sin nx\,dx$

$$= \frac{1}{\pi}\left[\int_{\pi}^{0} -\sin nx\,dx + \int_{0}^{\pi}\sin nx\,dx\right]$$

$$= \frac{1}{\pi}\left[\left|\frac{\cos nx}{n}\right|_{-\pi}^{0} - \left|\frac{\cos nx}{n}\right|_{0}^{\pi}\right]$$

$$= \frac{1}{\pi}\left[\frac{1}{n}\{1-\cos n\pi - \cos n\pi + 1\}\right]$$

$$\therefore \quad b_n = \frac{2}{n\pi}(1-\cos n\pi) = \begin{cases} 0, & \text{for n even,} \\ \frac{4}{\pi n}, & \text{for n odd.}\end{cases}$$

Since the function f(x) is continuous at all points of $[-\pi, \pi]$ except $0, \pm \pi$, therefore

$$f(x) = \frac{1}{2}a_0 + \sum(a_n \cos nx + b_n \sin nx)$$

$$= (b_1 \sin x + b_2 \sin 3x + b_2 \sin 5x + ...)$$

$$(\therefore a_0 = 0, a_n = 0 \ \ \forall n \text{ and } b_n = 0 \text{ for n even})$$

Hence $\quad f(x) = \frac{4}{\pi}\left(\frac{\sin x}{1} + \frac{\sin 3x}{3} + \frac{\sin 5x}{5} + ...\right), \qquad ...(1)$

which is true for all x in $[-\pi, \pi]$ except t $0, \pm \pi$.

At $x = \pm \pi$ the sum of the series is equal to

$$\frac{1}{2}[f(\pi-) + f(-\pi+)] = \frac{1}{2}(1-1) = 0.$$

At x = 0, the sum of the series

$$= \frac{1}{2}\,[f\,(0-) + f(0+)] = \frac{1}{2}(-1+1) = 0.$$

At $x = \frac{1}{2}\pi$, the function is continuous and so by (1)

$$f(\pi/2) = \frac{4}{\pi}\left(\frac{\sin \pi/2}{1} + \frac{\sin 3\pi/2}{3} + \frac{\sin 5\pi/2}{5} + \ldots\ldots\ldots\right)$$

or
$$1 = \frac{4}{\pi}\left(1 - \frac{1}{3} + \frac{1}{5} - \ldots\ldots\ldots\ldots\right).$$

Hence $1 - \frac{1}{3} + \frac{1}{5} - \ldots. = \frac{\pi}{4}.$

Example 5:

Find the Fourier Series for the function | sin x | in the interval $[-\pi, \pi]$.

(D.U., 1994, 93, 87, 85)

Solution:

Since f(x) = | sin x | is an even function and so the Fourier Series of f(x) will have cosines terms only.

We know $|\sin x| = \begin{cases} -\sin x, & \text{for } -\pi \le x \le 0, \\ \sin x, & \text{for } \quad 0 \le x \le \pi. \end{cases}$

Now $a_0 = \frac{1}{\pi}\int_{-\pi}^{\pi} f(x)\,dx = \frac{2}{\pi}\int_0^{\pi} f(x)\,dx$

$$= \frac{2}{\pi}\int_0^{\pi} \sin x\,dx = -\frac{2}{\pi}\left|\cos x\right|_0^{\pi} = \frac{4}{\pi}.$$

Also $a_n = \frac{2}{\pi}\int_0^{\pi} f(x)\cos x\,dx$

$$= \frac{2}{\pi}\int_0^{\pi} \sin x \cos nx\,dx = \frac{2}{\pi} I_n,$$

where $I_n = \int_0^{\pi} \sin x \cos nx\,dx$

$$= \left|\sin x\,\frac{\sin nx}{n}\right|_0^{\pi} - \int_0^{\pi} \frac{\sin nx}{n}\cos x\,dx$$

$$= 0 - \frac{1}{n}\left[\left|\frac{-\cos nx}{n}\cos x\right|_0^{\pi} + \int_0^{\pi}\frac{\cos nx}{n}(-\sin x)\,dx\right]$$

$$\therefore \qquad I_n = \frac{1}{n^2}(-\cos nx - 1) + \frac{1}{n^2\, I_n}$$

$$\text{or} \quad \left(1 - \frac{1}{n^2}\right) i_n = -\frac{1}{n^2}(1 + \cos n\pi) = -\frac{1}{n^2}[1+(-1)^n]$$

$$\therefore \qquad I_n = -\frac{1}{(n^2-1)}[1+(-1)^n]$$

Thus,
$$a_n = \frac{2}{-\pi(n^2-1)}[1+(-1)^n]$$

$$\therefore \qquad a_n = \begin{cases} 0, & \text{if } n \text{ is odd,} \\ \dfrac{4}{-\pi(n^2-1)}, & \text{if } n \text{ is even} \end{cases}$$

Hence the Fourier Series of | sin x |

$$= \frac{1}{2}a_0 + \sum_{n=1}^{\infty} a_n \cos nx$$

$$= \frac{1}{2}\cdot\frac{4}{\pi} - \sum_{n=1}^{\infty}\frac{4\cos nx}{\pi(n^2-1)}, \text{ for } n \text{ even}$$

$$= \frac{4}{\pi}\left[\frac{1}{2} - \sum_{k=1}^{\infty}\frac{\cos 2kx}{(4k^2-1)}\right].$$

Example 6:

Expand the function f(x) = 2x in a series of sine and cosines in [0, π].

Solution:

(i) We have

$$b_n = \frac{2}{\pi}\int_0^{\pi} 2x \sin nx\, dx$$

$$= \frac{4}{\pi}\left[\left|-x\frac{\cos nx}{n}\right|_0^{\pi} + \int_0^{\pi}\frac{\cos nx}{n}\,dx\right]$$

$$= \frac{4}{\pi}\left(-\frac{\pi}{n}\cos n\pi + 0\right) = \frac{-4}{n}(-1)^n = \frac{4(-1)^{n+1}}{n}.$$

Since f(x) is continuous in [0, 1], therefore

$$\Sigma\, b_n \sin nx = \frac{1}{2}[f(x+0) + f(x-0)]$$

$$= \frac{1}{2}[f(x) + f(x)] = f(x).$$

Hence $2x = 4 \sum (-1)^{n+1} \frac{\sin nx}{n}$.

(ii) We have

$$a_0 = \frac{2}{\pi}\int_0^{\pi} f(x)\,dx = \frac{2}{\pi}\int_0^{\pi} 2x\,dx = 2\pi.$$

Now $a_n = \frac{2}{\pi}\int_0^{\pi} 2x \cos nx\, dx$

$$= \frac{4}{\pi}\left[\left|\frac{x \sin nx}{n}\right|_0^{\pi} - \int_0^{\pi} \frac{\sin nx}{n}\right]$$

$$= \frac{4}{\pi}\left[0 + \left|\frac{\cos nx}{n^2}\right|_0^{\pi}\right] = \frac{4}{\pi n^2}[\cos n\pi - 1]$$

$$= \frac{4}{\pi n^2}\left[(-1)^n - 1\right].$$

$$\therefore \quad a_n = \begin{cases} 0, & \text{for n even,} \\ -8/\pi n^2, & \text{for n odd.} \end{cases}$$

Since f(x) is continuous in [0, 1], therefore

$$2x = \frac{1}{2}a_0 + \Sigma\, a_n \cos nx$$

$$= \pi - \frac{8}{\pi}\left[\frac{\cos x}{1^2} + \frac{\cos 3x}{3^2} + \frac{\cos 5x}{5^2} + \ldots\right].$$

Ex.:

Find the Fourier series consisting of (i) sine terms only, (ii) cosine terms only, which represents the periodic function f(x) = x in [0, π]. Find the sum of the series at x = 0, π.

Example 7:

If f is bounded and integrable on [–p, p] and if a_n, b_n are its Fourier coefficients, then prove that

$$\sum_{n=1}^{\infty}(a_n^2 + b_n^2) \text{ converges.}$$ **(D.U., 1989)**

Solution:

We have

$$0 \le \int_{-\pi}^{\pi}\left[f - \sum_{n=1}^{m}(a_n \cos nx + b_n \sin nx)\right]^2 dx$$

$$= \int_{-\pi}^{\pi} f^2 dx + \int_{-\pi}^{\pi}\left[\sum_{n=1}^{m}(a_n \cos nx + b_n \sin nx)\right]^2 dx$$

$$- 2\sum_{n=1}^{m}\left[a_n \int_{-\pi}^{\pi} f \cos nx\, dx\right] - 2\sum_{n=1}^{m}\left[b_n \int_{-\pi}^{\pi} f \sin nx\, dx\right]$$

$$= \int_{-\pi}^{\pi} f^2 dx + \pi\sum_{n=1}^{m} a_n^2 + \pi\sum_{n=1}^{m} b_n^2 - 2\pi\sum_{n=1}^{m} a_n^2 - 2\pi\sum_{n=1}^{m} b_n^2$$

$$= \int_{-\pi}^{\pi} f^2 dx + \pi\sum_{n=1}^{m}(a_n^2 + b_n^2)$$

Thus, $\sum_{n=1}^{m}(a_n^2 + b_n^2) \le \frac{1}{\pi}\int_{-\pi}^{\pi} f^2\, dx.$

Taking limit as $m \to \infty$, we obtain

$$\sum_{n=1}^{m}(a_n^2 + b_n^2) \le \frac{1}{\pi}\int_{-\pi}^{\pi} f^2\, dx < \infty,$$

since f and so f^2 is integrable on $[-\pi, \pi]$

Hence, the series $\Sigma(a_n^2 + b_n^2)$ is convergent.

If, Define the Fourier coefficients a_n and b_n of the bounded and integrable function on $[-\pi, \pi]$. Show that $a_n, b_n \to 0$ as $n \to \infty$. **(D.U., 1995)**

Hint. Since $\Sigma(a_n^2 + b_n^2)$ converges, therefore

$(a_n^2 + b_n^2) \to 0$ as $n \to \infty$.

Hence $a_n \to 0$, $b_n \to 0$ as $n \to \infty$.

Example 8:

Find the Fourier sine series in $0 < x < \pi$, for

$$f(x) = \begin{cases} \pi/3, & \text{for } 0 < x < \pi/3 \\ 0, & \text{for } \pi/3 < x < 2\pi/3, \\ -\pi/3, & \text{for } 2\pi/3 < x < \pi. \end{cases}$$

Find the sum of the series when $x = \pi,\ 2\pi/3$. **(D.U., 1991)**

Solution:

We have

$$b_n = \frac{2}{\pi}\int_0^{\pi} f(x)\sin nx\,dx$$

$$= \frac{2}{\pi}\left[\int_0^{\pi/3} \frac{\pi}{3}\sin nx\,dx + \int_{\pi/3}^{2n/3} 0.\sin nx\,dx - \int_{2\pi/3}^{\pi} \frac{\pi}{3}\sin nx\,dx\right]$$

$$= \frac{2}{3}\left[\int_0^{\pi/3} \sin nx\,dx + (-1)^n \int_0^{\pi/3} \sin ny\,dy\,. \qquad y = \pi - x\right.$$

It is clear that $b_n = 0$, for n odd

and $b_n = \dfrac{4}{3}\displaystyle\int_0^{\pi/3} \sin nx\,dx$, for n even

$$= -\frac{4}{3}\left|\cos nx\right|_0^{\pi/3} = -\frac{4}{3\pi}\left[\cos\frac{n\pi}{3} - 1\right]$$

$\therefore \quad b_n = \dfrac{4}{3n}\left(2\sin^2\dfrac{n\pi}{6}\right)$, for n even.

Replacing n by 2n,

$$b_{2n} = \frac{4}{3n}\sin^2\left(\frac{n\pi}{3}\right)$$

$$= \begin{cases} 0, & \text{if } n = 3k \\ \dfrac{4}{3n}\sin^2\left(\dfrac{\pi}{3}\right), & \text{if } n \neq 3k \end{cases}$$

or $\quad b_{2n} = \begin{cases} 0, & \text{if } n = 3k, \\ 1/n, & \text{if } n \neq 3k. \end{cases}$

Hence the Fourier Series of f(x) is

$$\sigma(x) = \sum b_n \sin nx = \sum b_{2n} \sin 2nx \ (n \neq 3k)$$

$$= \sum_{1}^{\infty} \frac{\sin 2nx}{n} \quad (n \neq 3k)$$

$$= \sin 2x + \frac{\sin 4x}{2} + \frac{\sin 8x}{4} + \frac{\sin 10x}{5} + \frac{\sin 14}{7} + \ldots..$$

We know that the sum of the series must be 0 for $x = \pi$ and this fact can be verified directly ($\sigma(\pi) = 0$).

At $x = \frac{2\pi}{3}$, the sum of the series

$$= \frac{1}{2}\left[f\left(\frac{2}{3}\pi + 0\right) f\left(\frac{2}{3}\pi - 0\right)\right]$$

$$= \frac{1}{2}\left[-\frac{\pi}{3} + 0\right] = -\frac{\pi}{6}.$$

It can be shown that the sum of the series is $\pi/6$ at $x = \pi/3$.

Example 9:

Show that for all values of x in $[-\pi, \pi]$

$$|\cos x| = \frac{4}{\pi}\left[\frac{1}{2} + \sum_{n=1}^{\infty} \frac{(-1)^{n+1} \cos 2nx}{(4n^2 - 1)}\right]. \qquad \textbf{(D.U., 1989)}$$

Solution:

The function $f(x) = |\cos x|$ is piecewise monotonic, bounded and integrable in $[-\pi, \pi]$. Since $f(x)$ is an even function of x, so $b_n = 0\ \forall\ n$. Further cos x changes sign in $[0, \pi]$.

Thus,

$$a_0 = \frac{1}{\pi}\int_{-\pi}^{\pi} |\cos x|\, dx = \frac{2}{\pi}\int_{0}^{\pi} |\cos x\ dx$$

$$= \frac{2}{\pi}\int_{-\pi/2}^{\pi/2} |\sin y|\, dy,\ x = y + \tfrac{1}{2}\pi$$

$$= \frac{4}{\pi}\int_{0}^{\pi/2} \sin y\ dy = \frac{4}{\pi}. \qquad \therefore a_0 = \frac{4}{\pi}$$

Now, $$a_n = \frac{1}{\pi}\int_{-\pi}^{\pi} |\cos x| \cos nx\, dx$$

$$= \frac{2}{\pi}\int_{0}^{\pi} |\cos x| \cos nx\, dx$$

$$= \frac{2}{\pi} \int_{-\pi/2}^{\pi/2} |\sin y| \cos \left[n \left(\tfrac{1}{2}\pi + y\right)\right] dy, \quad \text{where } x = y + \tfrac{1}{2}\pi$$

Thus, $$a_{2n} = \frac{2}{\pi} (-1)^n \int_{-\pi/2}^{\pi/2} |\sin y| \cos 2ny \, dy \qquad \text{(Even integrand)}$$

$$= \frac{2(-1)^n}{\pi} \int_0^{\pi/2} \sin y \cos 2ny \, dy$$

$$= \frac{2(-1)^n}{\pi} \int_0^{\pi/2} [\sin (2n+1)y - \sin(2n-1) dy \; .$$

$$= \frac{2(-1)^n}{\pi} \left[\frac{1}{2n+1} - \frac{1}{2n-1} \right] = \frac{4(-1)^{n+1}}{\pi(4n^2 - 1)}.$$

Again $$a_{2n-1} = \frac{2(-1)^{n+1}}{\pi} \int_{-\pi/2}^{\pi/2} |\sin y| \sin (2n - 1) y \, dy = 0$$

(Odd integrand)

Since $|\cos x|$ is continuous for all x in $[-\pi, \pi]$

$$|\cos x| = \frac{1}{2} a_0 + \sum_{n=1}^{\infty} (a_n \cos nx + b_n \sin nx)$$

Hence, $$|\cos x| = \frac{4}{\pi} \left[\frac{1}{2} + \sum_{n=1}^{\infty} \frac{(-1)^{n+1} \cos 2nx}{(4n^2 - 1)} \right] \qquad (\because b_n = 0)$$

Example 10:

Find the Fourier Series of the periodic function f with period in defined as follows:

$$f(x) = \begin{cases} 0, & \text{for } -\pi \le x \le 0, \\ x, & \text{for } \quad 0 \le x \le \pi. \end{cases}$$

What is the sum of the series at $x = 0, \pm \pi$.

Solution:

The function is bounded, piecewise monotonic, and integrable on $[-\pi, \pi]$.

We have

$$a_0 = \frac{1}{\pi} \int_{-\pi}^{\pi} f(x) \, dx$$

$$= \frac{1}{\pi}\left[\int_{-\pi}^{0} 0.\,dx + \int_{\pi}^{0} x.dx\right] = \frac{\pi}{2}. \qquad \therefore a_0 = \frac{1}{2}\pi.$$

Now $a_n = \frac{1}{\pi}\int_{-\pi}^{\pi} f(x)\cos nx\,dx = \frac{1}{\pi}\left[0 + \int_0^{\pi} x\cos nx\,dx\right]$

$$= \frac{1}{\pi}\left[\left|\frac{x\sin nx}{n}\right|_0^{\pi} - \frac{1}{n}\int_0^{\pi}\sin nx\,dx\right]$$

$$= \frac{1}{n\pi}\left|\frac{\cos nx}{n}\right|_0^{\pi} = \begin{cases} 0, \text{ if } n \text{ even}, \\ \frac{-2}{\pi n^2}, \text{ for } n \text{ odd}. \end{cases} \qquad ...(1)$$

Again

$$b_n = \frac{1}{\pi}\int_{-\pi}^{\pi} f(x)\sin nx\,dx = \frac{1}{\pi}\int_{-\pi}^{\pi} x\sin nx\,dx$$

$$= \frac{1}{\pi}\left[\left|\frac{-x\cos nx}{n}\right|_0^{\pi} + \frac{1}{n}\int_0^{\pi}\cos nx\,dx\right]$$

$$= \frac{1}{\pi}\left[\left|\frac{-x\cos nx}{n}\right|_0^{\pi} + \frac{1}{n}\left|\frac{\sin nx}{n}\right|_0^{\pi}\right]$$

$$= \frac{1}{\pi}\left[\frac{-x\cos nx}{n}\right] + 0$$

$$= -\frac{\cos n\pi}{n}\begin{cases} -1/n, \text{ if } n \text{ is even}, \\ 1/n, \text{ if } n \text{ is odd}. \end{cases} \qquad ...(2)$$

In $[-\pi, \pi]$, the only points of discontinuity of f are $\pm\pi$.

For all $x \in\ ]-\pi, \pi[$.

$$f(x) = \frac{1}{2}a_0 + \Sigma(a_n\cos nx + b_n\sin nx)$$

$$f(x) = \frac{\pi}{4} - \frac{2}{\pi}\left\{\frac{\cos x}{1^2} + \frac{\cos 3x}{3^2} + \frac{\cos 5x}{5^2} + \ldots\ldots\right\}$$

$$+\left\{\frac{\sin x}{1} - \frac{\sin 2x}{2} + \frac{\sin 3x}{3} - \ldots\ldots\right\} \qquad ...(3)$$

Since f is continuous at $x = 0$,

$$f(0) = \frac{\pi}{4} - \frac{2}{\pi}\left[\frac{1}{1^2} + \frac{1}{3^2} + \frac{1}{5^2} + ...\right] + 0, \text{ by (3)}$$

$$\therefore \quad 1 + \frac{1}{3^2} + \frac{1}{5^2} + ... = \frac{\pi^2}{8} \qquad (\because f(0) = 0)$$

Since x = ± π are the points of discontinuity of f, so the sum of series is $= \frac{1}{2}[\,f(\pi-) + f(-\pi+)]$, where

$$f(n-) = \lim_{x \to \pi} f(x) = \lim_{x \to \pi} x = \pi,$$

$$f(-\pi+) = \lim_{x \to -\pi+} f(x) = 0.$$

Hence the sum of series at x = ± π is $\frac{1}{2}(\pi + 0) = \frac{1}{2}\pi$.

$$\therefore \quad \frac{\pi}{2} = \frac{\pi}{4} + \frac{2}{\pi}\left\{\frac{1}{1^2} + \frac{1}{3^2} + \frac{1}{5^2} + ...\right\}$$

$$\text{or} \quad 1 + \frac{1}{3^2} + \frac{1}{5^2} + ... = \frac{\pi^2}{8}.$$

Example 11:

Expand the periodic function x^2, $0 \le x \le 1$, of period 1 in a series of sines only. Also find the sum of the series at x = 0, 1. **(D.U., 1994)**

Solution:

The substitution y = πx *l* transforms the given function with period π on [0, π]. To get the series of sines only, we have

$$b_n = \frac{2}{l}\int_0^l x^2 \sin(n\pi x/l)\,dx$$

$$= \frac{2}{l}\left[-\frac{l}{n\pi}\left|x^2\cos\left(\frac{n\pi x}{l}\right)\right|_0^l + \frac{21}{n\pi}\int_0^l x\cos\left(\frac{n\pi x}{l}\right)dx\right]$$

$$= \frac{2}{l}\left[-\frac{l^3}{n\pi}\cos n\pi + \frac{2l^2}{n^2\pi^2}\left\{\left|x\sin\left(\frac{n\pi x}{l}\right)\right|_0^l - \int_0^l \sin\left(\frac{n\pi x}{l}\right)dx\right\}\right]$$

$$= \frac{2}{l}\left[-\frac{l^3}{n\pi}\cos n\pi + 0 + \frac{2l^3}{n^3\pi^3}\left|\cos\frac{n\pi x}{l}\right|_0^l\right]$$

$$= \frac{2}{l}\left[-\frac{l^3}{n\pi}\cos n\pi + \frac{2l^3}{n^3\pi^3}(\cos n\pi - 1)\right]$$

$$= -\frac{2l^2}{n\pi}\cos n\pi + \frac{4l^2}{n^3\pi^3}(\cos n\pi - 1)$$

$$\therefore \quad = -\frac{2(-l)^{n/2}}{n\pi} - \frac{4l^2}{n^3\pi^3}\{1-(-1)^n\}$$

Since the function x^2 is continuous in [0, 1], therefore

$$\therefore \quad x^2 = \sum_{n=1}^{\infty} b_n \sin\frac{n\pi x}{l} = \sum_{n=1}^{\infty}\left[-\frac{2(-1)^n)l^2}{n\pi} - \frac{4l^2(-1(-1)^n)}{n^2\pi^2}\right]$$

$$\times \sin\frac{n\pi x}{l} \quad \forall\ x \in [0, 1]$$

For x = 0 or l, the sum of the series is 0.

Example 12:

Show that the Fourier series which converges to f(x) in $-\pi \le x \le \pi$, where $f(x) = x + x^2$ when $-\pi < x < \pi$ and $f(x) = \pi^2$ when $x = \pm\pi$ is $\frac{\pi^2}{3}+4$ $\sum(-1)^n\left(\frac{\cos nx}{n^2} - \frac{\sin nx}{2n}\right)$.

Deduce that $\pi^2/6 = 1 + 1/2^2 + 1/3^2 + 1/4^2 + \ldots$.

Solution:

We have

$$a_0 = \frac{1}{\pi}\int_{-\pi}^{\pi}(x^2+x)\,dx$$

$$= \frac{1}{\pi}\left[\int_{-\pi}^{\pi}x^2\,dx + \int_{-\pi}^{\pi}x.dx\right] = \frac{2}{\pi}\cdot\frac{\pi^2}{3} = \frac{2\pi^2}{3}.$$

Now $$a_n = \frac{1}{\pi}\int_{-\pi}^{\pi}(x^2+x)\cos nx\,dx$$

$$= \frac{1}{\pi}\int_{-\pi}^{\pi}x^2\cos nx\,dx + 0 \quad (\because x\cos nx \text{ is an odd function})$$

$$= \frac{2}{\pi}\int_{0}^{\pi}x^2\cos nx\,dx \quad (\because x^2\cos nx \text{ is an even function})$$

$$= \frac{2}{\pi}\left[\left|\frac{x^2\sin nx}{n}\right|_0^{\pi} - \int_0^{\pi}\frac{2x\sin nx}{n}\,dx\right]$$

$$= \frac{2}{\pi}\left[0 - \frac{2}{\pi}\left\{\left|x\left(-\frac{\cos nx}{n}\right)\right|_0^{\pi} + \int_0^{\pi}\frac{\cos nx}{n}.1dx\right\}\right]$$

$$= \frac{2}{\pi}\left[\frac{2}{\pi^2}\left| x \cos nx \right|_0^\pi - \left|\frac{\sin nx}{n}\right|_0^\pi\right]$$

$$= \frac{4}{\pi^2 \pi}\left[\pi \cos n\pi\right] = \frac{4}{n^2}(-1)^n$$

$$\therefore \quad a_n = \frac{4}{\pi^2}(-1)^2.$$

Now $b_n = \frac{1}{\pi}\int_{-\pi}^{\pi}(x^2 + x) \sin nx \, dx$

$$= \frac{1}{\pi}\left[0 + 2\int_{-\pi}^{\pi} x \sin nx \, dx\right]$$

$$= \frac{2}{\pi}\left[\left| x\left(-\frac{\cos nx}{n}\right)\right|_0^\pi + \int_0^\pi 1.\frac{\cos nx}{x} dx\right]$$

$$= \frac{2}{\pi}\left[-\frac{\pi(-1)^n}{n}\right] = \frac{2}{n}(-1)^{n+1}$$

$$\therefore \quad b_n = \frac{2}{n}(-1)^{n+1}.$$

Since f is continuous at all points of $[-\pi, \pi]$ except at $\pm \pi$.

$$f(x) = \frac{1}{2} a_0 + \Sigma (a_n \cos nx + b_n \sin nx) \text{ in}] -\pi, \pi[.$$

or $\left| x^2 + x = \frac{1}{2} \cdot \frac{2}{3}\right|\pi^2 + \sum\left[\frac{4(-1)^n}{n^2}.\cos nx + \frac{2}{n}(-1)^{n+1} \sin nx\right]$

Hence, $x^2 + x = \frac{\pi^2}{3} + 4\,\Sigma(-1)^n\left[\frac{\cos nx}{n^2} - \frac{\sin nx}{2n}\right]$ in $]-\pi, \pi[$. ...(1)

Now, $f(n-) = \underset{h\to 0}{\text{Lt}} [(\pi - h)^2 + (\pi - h)] = \pi^2 + \pi.$

and $f(-\pi+) = \underset{h\to 0}{\text{Lt}} [(-\pi + h)^2 + (-\pi + h) = \pi^2 - \pi.$

At $x = \pm \pi$, the sum of the series

$$= \frac{1}{2}[f(\pi -) + f(-\pi +)] = ½ (\pi^2 + \pi + \pi^2 - \pi] = \pi^2.$$

From (1), we obtain

$$f(\pi-)=\frac{\pi^2}{3}+\lim_{h\to 0}\sum_{n=1}^{\infty}4(-1)^n\left[\frac{1}{n^2}.\cos n(\pi-h)-\frac{1}{2n}\sin n\,(\pi-h)\right]$$

and $$f(-\pi+)=\frac{\pi^2}{3}+\lim_{h\to 0}\sum_{n=1}^{\infty}4(-1)^n\left[\frac{1}{n^2}.\cos n(-\pi+h)\right.$$
$$\left.-\frac{1}{2n}\sin n(-\pi-h)\right]$$

$\therefore$ $$\frac{1}{2}\left[f(\pi-)+f(-\pi+)\right]$$
$$=\frac{1}{2}\left[\frac{\pi^2}{3}+\sum 4(-1)^n\left\{\frac{\cos n\pi}{n^2}-\frac{\sin n\pi}{2n}\right\}\right.$$
$$\left.+\frac{\pi^2}{3}+\sum 4(-1)^n\left\{\frac{\cos n\pi}{n^2}+\frac{\sin n\pi}{2n}\right\}\right]$$
$$=\frac{\pi^2}{3}+\sum\frac{4(-1)^n\,(-1)^n}{n^2}=\frac{\pi^2}{3}+\sum\frac{4(-1)^{2n}}{n^2}$$

$\therefore$ $$\pi^2=\frac{\pi^2}{3}+\sum_{n=1}^{\infty}\frac{4}{n^2}\Rightarrow\frac{2}{3}\pi^2=4\sum\frac{1}{n^2}$$

Hence, $$=\frac{\pi^2}{6}=\sum\frac{1}{n^2}.$$

Example 13:

Expand the periodic function, of period $2l > 0$.

$$f(x)=\left|\cos\left(\frac{\pi x}{l}\right)\right|,\text{ in a Fourier Series.}$$

Solution:

We have

$$f(x)=\begin{cases}-\cos(\pi x/l), & -l\le x\le -l/2,\\ \cos(\pi x/l), & -l/2\le x\le -l/2,\\ -\cos(\pi x/l), & -l/2\le x\le -l\end{cases}$$

Since f(x) is an even function, therefore $b_n = 0$, $\forall n$.

Now

$$a_0 = \frac{1}{l}\int_{-l}^{l} |\cos(\pi x/l)|dx$$

$$= -\frac{1}{l}\int_{-l}^{-l/2} \cos(\pi x/l)\, dx + \int_{-l/2}^{l/2} \cos(\pi x/l)\, dx$$

$$= \frac{1}{l}\int_{l/2}^{l} \cos(\pi x/l)\, dx$$

$$= -\frac{2}{l}\int_{l/2}^{l} \cos(\pi x/l)\, dx + \frac{2}{l}\int_{0}^{l/2} \cos(\pi x/l)\, dx = 4/\pi$$

Similarly,

$$a_n = -\frac{1}{l}\int_{-l}^{-l/2} \cos(\pi x/l)\ \cos(n\pi x/l)\, dx$$

$$+ \int_{-l/2}^{l/2} \cos(\pi x/l)\cos(n\pi x/l)\, dx - \int_{l/2}^{l} \cos(\pi x/l)\cos(n\pi x/l)\, dx$$

$$= -\frac{1}{l}\int_{l/2}^{l}\left[\cos\left\{\frac{(n+1)\, nx}{l}\right\} + \cos\left\{\frac{(n-1)\, nx}{l}\right\}\right]$$

$$+ \frac{1}{l}\int_{0}^{l/2}\left[\cos\left\{\frac{(n+1)\, nx}{l}\right\} - \cos\left\{\frac{(n-1)\, nx}{l}\right\}\right] dx$$

$$= -\frac{2\cos(n\pi/2)}{(n^2-1)\pi}$$

$$\therefore \quad a_{2n} = \frac{2(-1)^{n+1}}{(4n^2-1)\pi},\ \forall n, \text{ and } a_{2n-1} = 0,\ \forall n.$$ **Hence Proved.**

Example 14:

Expand the function f(x) = cos 2x in a series of sines in [0, π].

Solution:

We have

$$b_n = \frac{2}{\pi}\int_0^{\pi} \cos 2x \sin nx\, dx$$

$$= \frac{1}{\pi}\int_0^{\pi} [\sin(n+2)x + \sin(n-2)x]\, dx$$

$$= \frac{1}{\pi}\left[\frac{\cos(n+2)x}{n+2} + \frac{\cos(n-2)x}{n-2}\right]_0^{\pi}$$

$$= -\frac{1}{\pi}\left[\left\{\frac{(-1)^{n+2}}{n+2} + \frac{(-1)^{n-2}}{n-2}\right\} - \left\{\frac{1}{n+2} + \frac{1}{n-2}\right\}\right]$$

$$= -\frac{1}{\pi}\left[(-1)^n\left\{\frac{1}{n+2} + \frac{1}{n-2}\right\} - \left\{\frac{1}{n+2} + \frac{1}{n-2}\right\}\right]$$

$$= -\frac{2n}{\pi(n^2-4)}[(-1)^n - 1].$$

$$\therefore \qquad b_n = \begin{cases} 0, & \text{for n even,} \\ \dfrac{4n}{\pi(n^2-4)}, & \text{for n odd.} \end{cases}$$

Equivalently, $b_n = -\dfrac{4}{\pi}\dfrac{(2n-1)}{[2^2-(2n-1)^2]}.$

Since cos 2x is continuous in [0, π], we have

$$\cos 2x = \sum b_n \sin nx = -\frac{4}{\pi}\sum \frac{(2n-1)\sin(2n-1)x}{[2^2-(2n-1)^2]}.$$

Example 15:

Find the Fourier series of the function x^2, periodic with period 2l, on the interval [–1, 1].

Solution:

Let us substitute $y = \pi x/l$ so that the given function becomes a periodic function with period 2π on $[-\pi, \pi]$. Since x^3 is an even function, so $b_n = 0$ for all n. We have

$$a_0 = \frac{2}{l}\int_0^l x^2\,dx = \frac{2}{3}l^2.$$

Now $\quad a_n = \dfrac{2}{l}\displaystyle\int_0^l x^2 \cos\frac{n\pi x}{l}\,dx$

$$= \frac{2}{l}\left[\left|\frac{l}{n\pi}x^2 \sin\frac{n\pi x}{l}\right|_0^l - \frac{2l}{n\pi}\int_0^l x \sin\frac{n\pi x}{l}\,dx\right]$$

$$= \frac{2}{l}\left[0 + \frac{2l^2}{n^2\pi^2}\left\{\left|x\cos\frac{n\pi x}{l}\right|_0^l - \int_0^l \cos\frac{n\pi x}{l}\,dx\right\}\right]$$

$$= \frac{4l^2}{n^2\pi^2}\cos n\pi - 0 = \frac{(-1)^2 4l^2}{n^2\pi^2}$$

Since x^2 is continuous on [–1, 1], therefore

$$x^2 = \tfrac{1}{2}a_0 + \sum_{n=1}^{\infty} a_n \cos\frac{n\pi x}{l}.$$

$$\text{Hence } x^2 = \frac{l^2}{3} + \frac{4l^2}{\pi^2}\sum_{n=1}^{\infty}\frac{(-1)^n}{n^2}\cos\left(\frac{n\pi x}{l}\right).$$

Example 16:

Show that for all values f x in [–π, π], when k is not an integer,

$$\cos kx = \frac{\sin k\pi}{\pi}\left[\frac{1}{k} + \sum_{n=1}^{\infty}\frac{(-1)^n\, 2k\cos nx}{k^2 - n^2}\right]$$

Deduce that $\pi \cot k\pi = \dfrac{1}{k} + \sum_{n=1}^{\infty}\dfrac{2k}{k^2 - n^2}$. **(D.U., 1992)**

Solution:

The function f(x) = cos kx is piecewise monotonic, bounded and integrable for all x in [–π, π]. Now

$$a_0 = \frac{1}{\pi}\int_{-\pi}^{\pi} f(x)\,dx = \frac{2}{\pi}\int_0^{\pi}\cos kx\,dx = \frac{2}{k\pi}\sin k\pi.$$

$$a_n = \frac{1}{\pi}\int_{-\pi}^{\pi}\cos kx\cos nx\,dx = \frac{2}{\pi}\int_0^{\pi}\cos kx\cos nx\,dx$$

$$= \frac{1}{\pi}\int_0^{\pi}[\cos(k+n)\cos(k-n)x]\,dx$$

$$= \frac{1}{\pi}\left[\frac{\sin(k+n)x}{k+n} + \frac{\sin(k-n)x}{k-n}\right]_0^{\pi}$$

$$= \frac{1}{\pi}\left[\frac{\sin(k+n)\pi}{k+n} + \frac{\sin(k-n)\pi}{k-n}\right]$$

$$= \frac{\sin k\pi\cos n\pi}{\pi}\left[\frac{1}{k+n} + \frac{1}{k-n}\right] \qquad (\because \sin n\pi = 0)$$

$$= \frac{2(-1)^n k\sin k\pi}{n(k^3 - n^2)}$$

Now $b_n = \frac{1}{n}\int_{-\pi}^{\pi}\cos kx \sin nx\, dx = 0$ (integrand being odd)

Since cos kx is continuous in $[-\pi, \pi]$, therefore,

$$\cos kx = \frac{1}{2}a_0 + \sum_{n=1}^{\infty}(a_n \cos nx + b_n \sin nx)$$

$$\cos kx = \frac{\sin k\pi}{\pi}\left[\frac{1}{k} + \sum_{n=1}^{\infty}\frac{(-1)^n 2k \cos nx}{k^2 - n^2}\right] \qquad ...(1)$$

Putting $x = \pi$ in (1), we get

$$\pi \cot k\pi = \frac{1}{k} + \sum_{n=1}^{\infty}\frac{(-1)^n\ 2k \cos n\pi}{k^2 - n^2}$$

Hence $\pi \cot k\pi = \frac{1}{k} + \sum_{n=1}^{\infty}\frac{2k}{k^2 - n^2}.$

$$[\because\ (-1)^n \cos n\pi = (-1)^{n+n} = 1]$$

Example 17:

Expand | x | in a Fourier Series on [–1, 1].

Solution:

We have

$|x| = x$ for $0 \le x \le 1$, $|x| = -x$ for $-l \le x \le 0$.

On substituting $y = \pi x/l$, the given function becomes a periodic function with period 2π on $[-\pi, \pi]$. Since the given function is an even function, so $b_n = 0$ for all n. We have

$$a_0 = \frac{2}{l}\int_0^l x\, dx = l$$

Now $a_n = \frac{2}{l}\int_0^l x \cos\frac{n\pi x}{l}\, dx$

$$= \frac{2}{l}\frac{l}{n\pi}\left[\left|x \sin\frac{n\pi x}{l}\right|_0^l - \int_0^l \sin\frac{n\pi x}{l}\, dx\right]$$

$$= \frac{2}{n\pi}\left[0 + \frac{l}{n\pi}\left|\cos\frac{n\pi x}{l}\right|_0^l\right]$$

$$= \frac{2l}{n^2\pi^2}(\cos n\pi - 1) = \frac{2l}{n^2\pi^2}[(-1)^n - 1]$$

$$\therefore \quad a_n = \begin{cases} 0, & \text{for n even} \\ \dfrac{-4l}{n^2\pi^2}, & \text{for n odd.} \end{cases}$$

Hence the Fourier Series of | x | is

$$\frac{1}{2}a_0 + \sum_{n=1}^{\infty} a_n \cos\frac{n\pi x}{l}$$

$$= \frac{1}{2}l - \frac{4l}{\pi^2}\sum\frac{1}{n^2}\cos\left(\frac{n\pi x}{1}\right), \text{ for n odd}$$

$$= \frac{1}{2}l - \frac{4l}{\pi^2}\sum_{n=1}^{\infty}\frac{\cos\{(2n-1)\pi x/l\}}{(2n-1)^2}.$$

Example 18:

Find the Fourier Series f the function

$$f(x) = \begin{cases} x-\pi, & \text{when } -\pi < x < 0, \\ \pi - x, & \text{when } \ 0 < x < \pi. \end{cases}$$

that is the sum of the series for $x = 0, \pm\pi$?

Solution:

The given function is piecewise monotonic, bounded and integrable in $[-\pi, \pi]$. We have

$$a_0 = \frac{1}{\pi}\int_{-\pi}^{0}(x-\pi)\,dx + \frac{1}{\pi}\int_{0}^{\pi}(\pi - x)\,dx$$

$$= \frac{1}{\pi}\left[-\frac{\pi^2}{2} - \pi^2 + \pi^2 - \frac{\pi^2}{2}\right]$$

$$= -\pi. \qquad \therefore a_0 = -\pi.$$

$$a_n = \frac{1}{\pi}\int_{-\pi}^{0}(x-\pi)\cos nx\,dx + \frac{1}{\pi}\int_{0}^{\pi}(\pi-x)\cos nx\,dx$$

$$= \frac{1}{\pi}\left[\left|(x-\pi)\frac{\sin nx}{n}\right|_{-\pi}^{0} - \int_{-\pi}^{0}\frac{\sin nx}{n}dx\right.$$

$$\left. + \left|(\pi - x)\frac{\sin nx}{n}\right|_{0}^{\pi} + \int_{-\pi}^{0}\frac{\sin nx}{n}dx\right]$$

$$= \frac{1}{\pi}\left[0 + \frac{1}{n^2}\left|\cos nx\right|_{-\pi}^{0} + 0 - \frac{1}{n^2}\left|\cos nx\right|_{0}^{\pi}\right]$$

$$= \frac{1}{n^2\pi}\left[(1-\cos n\pi) - (\cos n\pi - 1)\right]$$

$$= \frac{2}{n^2\pi}\left[1-(-1)^n\right] \qquad [\because \cos n\pi = (-1)^n]$$

$$\therefore \quad a_n = \begin{cases} 0, & \text{if } n \text{ is even,} \\ \dfrac{4}{n^2\pi}, & \text{if } n \text{ is odd.} \end{cases}$$

Similarly,

$$b_n = \frac{1}{\pi}\int_{-\pi}^{\pi} f(x)\sin nx\,dx = \frac{2}{\pi}\left[1-(-1)^n\right]$$

$$\therefore \quad b_n = \begin{cases} 0, & \text{if } n \text{ is even,} \\ \dfrac{4}{n}, & \text{if } n \text{ is odd.} \end{cases}$$

In $[-\pi, \pi]$, the points $0, \pm\pi$ are the only points of discontinuity of f. For $x \neq 0, \pm\pi$, we have

$$f(x) = \tfrac{1}{2}a_0 + \sum_{n=1}^{\infty}(a_n \cos nx + b_n \sin nx)$$

$$= -\frac{1}{2}\pi + \frac{4}{\pi}\cdot\frac{\cos x}{1^2} + 4\,\frac{\sin x}{1} + \frac{4}{\pi}\,\frac{\cos 3x}{3^2} + 4\,\frac{\sin 3x}{3^2} + \ldots..$$

$$= -\frac{1}{2}\pi + \frac{4}{\pi}\cdot\left[\frac{\cos x}{1^2} + \frac{\cos 3x}{3^2} + \frac{\cos 5x}{5^2} + \ldots..\right]$$

$$+ 4\left[\frac{\sin x}{1} + \frac{\sin 3x}{3} + \ldots\ldots\right] \qquad \ldots(1)$$

For $x = 0$, the sum of the series $= \frac{1}{2}[f(0+) + f(0-)] = \frac{1}{2}[\pi - \pi] = 0.$

$$\therefore \quad 0 = -\frac{1}{2}\pi + \frac{4}{\pi}\left[1 + \frac{1}{3^2} + \frac{1}{5^2} + \ldots\ldots\ldots\right], \text{ by (1)}$$

or

$$\frac{\pi^2}{8} = 1 + \frac{1}{3^2} + \frac{1}{5^2} + \ldots\ldots$$

For $x = \pm\pi$, the sum of the series

$$= \frac{1}{2}[f(\pi - 0) + f(-\pi + 0] = \tfrac{1}{2}[0 + (-2\pi)] = -\pi.$$

Putting $x = \pm\pi$ in (1) and using $f(\pm\pi) = -\pi$, we again obtain

$$\frac{\pi^2}{8} = 1 + \frac{1}{3^2} + \frac{1}{5^2} + \ldots\ldots.$$

Example 19:

Expand the periodic function x^2, $0 \le x \le l$, of period 1 in a series of cosines only. Also find the sum of the series at $x = 0, 1$.

Solution:

The substitution $y = \pi x/1$ transforms the given function with period π on $[0, \pi]$. To get the series of cosines only, we have

$$a_n = \frac{2}{l}\int_0^l x^2 \cos\frac{n\pi x}{l}\, dx \quad \frac{(-1)^n 4}{n^2\pi^2}$$

Also $$a0 = \frac{2}{l}\int_0^l x^2\, dx = \frac{2}{3}l^2.$$

Since x^2 is continuous on [0, 1], therefore

$$x^2 = \tfrac{1}{2}a_0 + \sum_{n=1}^{\infty} a_n \cos\frac{n\pi x}{l}.$$

Hence $$x^2 = \frac{l^2}{3} + \frac{4l^2}{\pi^2}\sum_{n=1}^{\infty}\frac{(-1)^n}{\pi^2}\cos\frac{n\pi x}{l} \qquad \forall\, x \in [0, 1]$$

At $x = 0$, $$0 = \frac{l^2}{3} + \frac{4l^2}{\pi^2}\left(1 - \frac{1}{2^2} + \frac{1}{3^2} - \frac{1}{4^2} + \ldots\ldots\right)$$

$$\therefore \quad 1 - \frac{1}{2^2} + \frac{1}{3^2} - \frac{1}{4^2} + \ldots.. = \frac{\pi^2}{12}.$$

At $x = l$, l^2 $$= \frac{l^2}{3} + \frac{4l^2}{\pi^2}\sum_{n=1}^{\infty}\frac{(-1)^n}{n^2}\cos n\pi$$

$$= \frac{l^2}{3} + \frac{4l^2}{\pi^2}\sum_{n=1}^{\infty}\frac{1}{n^2} \qquad [\because \cos n\pi = (-1)^n]$$

$$1 + \frac{1}{2^2} + \frac{1}{3^2} + \ldots.. = \frac{\pi^2}{6}.$$

Example 20:

Expand f in a Fourier series in $[-\pi, \pi]$

$$f(x) = \begin{cases} 1, & \text{for } -\pi < x \le 0, \\ -2, & \text{for } 0 < x \le \pi. \end{cases}$$

Solution:

The given function is bounded, piecewise monotonic and integrable in $[-\pi, \pi]$.

We have

$$a_0 = \frac{1}{\pi}\int_{-\pi}^{\pi} f(x)\,dx = \frac{1}{\pi}\left[\int_{-\pi}^{0} 1.dx + \int_{0}^{\pi} -2\,dx\right]$$

$$= \frac{1}{\pi}[\pi - 2\pi] = -1, \quad \therefore a_0 = -1.$$

$$\text{Now} \quad a_n = \frac{1}{\pi}\int_{-\pi}^{\pi} f(x)\cos nx\,dx$$

$$= \frac{1}{\pi}\int_{-\pi}^{0} \cos nx\,dx - \frac{2}{\pi}\int_{0}^{\pi} \text{cs } nx\,dx$$

$$= \frac{1}{\pi}\left|\frac{\sin nx}{n}\right|_{-\pi}^{0} - \frac{2}{\pi}\left|\frac{\sin nx}{n}\right|_{0}^{\pi} = 0 \quad \therefore a_n = 0 \ \forall\ n$$

$$\text{Now} \quad b_n = \frac{1}{\pi}\int_{-\pi}^{\pi} f(x)\sin nx\,dx$$

$$= \frac{1}{\pi}\int_{-\pi}^{0} \sin nx\,dx - \frac{2}{\pi}\int_{0}^{\pi} \sin nx\,dx$$

$$= \frac{1}{\pi}\left|\frac{\cos nx}{n}\right|_{-\pi}^{0} + \frac{2}{\pi}\left|\frac{\cos nx}{n}\right|_{0}^{\pi}$$

$$= -\frac{1}{n\pi} + \frac{\cos n\pi}{n\pi} + \frac{2\cos n\pi}{n\pi} - \frac{2}{n\pi}$$

$$= \frac{3(-1)^n}{n\pi} - \frac{3}{n\pi}$$

$$\therefore \quad b_n = \begin{cases} 0, & \text{if n is even,} \\ -\dfrac{6}{n\pi}, & \text{if n is odd.} \end{cases}$$

Hence, the Fourier Series of the given function

$$= \frac{1}{2}a_0 + \Sigma(a_n \cos nx + b_n \sin nx)$$

$$= -\frac{1}{2} - \frac{6}{\pi}\sum_{n=1}^{\infty}\frac{\sin(2n-1)x}{2n-1}.$$

Example 21:

In the i interval [0, 2], expand the function

f(x) = x, when 0 < x ≤ 1; f(x) = 2 – x, when 1 < x ≤ 2

as a (i) series of cosines, (ii) series of sines. **(D.U., 1988)**

Solution:

(i) We have $l = 2$. To get the Fourier Series of cosine terms only in $[0, l]$, we have to extend the definition of f(x) to that it becomes even (i.e., f(–x) = f(x)) and then $b_n = 0 \ \forall$ n. Now

$$a_n = \frac{2}{l}\int_0^1 f(x)\,dx = \int_0^2 f(x)\,dx \qquad (\because l = 2)$$

$$= \int_0^1 x\,dx + \int_1^2 (2-x)\,dx = 1.$$

Again
$$a_n = \frac{2}{l}\int_0^1 f(x)\cos\frac{n\pi x}{l}\,dx = \int_0^2 f(x)\cos\left(\frac{nax}{2}\right)dx$$

$$= \int_0^1 x\cos\left(\frac{n\pi x}{2}\right)dx + \int_1^2 (2-x)\cos\left(\frac{n\pi x}{2}\right)dx$$

$$= \frac{2}{n\pi}\left[\left|x\sin\left(\frac{n\pi x}{2}\right)\right|_0^1 - \int_0^1 \sin\left(\frac{n\pi x}{2}\right)dx\right.$$

$$\left. + \left|(2-x)\sin\left(\frac{n\pi x}{2}\right)\right|_1^2 + \int_1^2 \sin\left(\frac{n\pi x}{2}\right)dx\right]$$

$$= \frac{2}{n\pi}\left[\sin\frac{n\pi}{2} + \frac{2}{n\pi}\left|\cos\left(\frac{n\pi x}{2}\right)\right|_0^1 - \sin\frac{n\pi}{2}\right.$$

$$\left. - \frac{2}{n\pi}\left|\cos\left(\frac{n\pi x}{2}\right)\right|_1^2\right]$$

$$= \frac{4}{n^2\pi^2}\left[\left\{\cos\frac{n\pi}{2} - 1\right\} - \left\{\cos n\pi - \cos\frac{\eta\pi}{2}\right\}\right]$$

$$= \frac{4}{n^2\pi^2}\left[2\cos\frac{n\pi}{2} - (-1)^n - 1\right]$$

It is clear that $a_n = 0$, for n odd

and $$a_{2n} = -\frac{2}{n^2\pi 2}\left[1+(-1)^{n+1}\right]$$

Thus, $$a_{4n} = 0,\ a_{4n-2} = -\frac{4}{(2n-1)^2\pi 2}.$$

Since the given function is continuous in [0, 2],

$$f(x) = \frac{1}{2}a_0 + \sum_{n-1}^{\infty} a_n \cos\left(\frac{n\pi x}{2}\right)$$

$$= \frac{1}{2} - \frac{4}{\pi^2}\sum \frac{\cos(2n-1)\pi x}{(2n-1)^2}.$$

(ii) To get the Fourier series of sine terms only, we have to extend the definition of f(x) so that it becomes odd and then

$$a_n = 0 \quad \forall n.$$

Now, $$b_n = \frac{2}{1}\int_0^1 f(x)\sin\left(\frac{n\pi x}{1}\right)dx$$

$$= \int_0^2 f(x)\sin\left(\frac{n\pi x}{1}\right)dx$$

$$= \int_0^1 x\sin\left(\frac{n\pi x}{2}\right)dx + \int_1^2 (2-x)\sin\left(\frac{n\pi x}{2}\right)dx$$

$$= \frac{2}{n\pi}\left[-x\cos\left(\frac{n\pi x}{2}\right)\Bigg|_0^1 + \int_0^1 \cos\left(\frac{n\pi x}{2}\right)dx\right.$$

$$\left. -\left|(2-x)\cos\left(\frac{n\pi x}{2}\right)\right|_1^2 - \int_0^1 \cos\left(\frac{n\pi x}{2}\right)dx\right]$$

$$= \frac{2}{n\pi}\left[-\cos\frac{n\pi}{2} + \frac{2}{n\pi}\left|\left(\frac{n\pi x}{2}\right)\right|_0^1 + \cos\frac{n\pi}{2}\right.$$

$$\left. - \frac{2}{n\pi}\left|\sin\left(\frac{n\pi x}{2}\right)\right|_1^2\right]$$

$$= \frac{4}{n^2\pi^2}\left[\sin\frac{n\pi}{2} - \left\{0 - \sin\frac{n\pi}{2}\right\}\right] = \frac{8}{n^2\pi^2}\sin\frac{n\pi}{2}$$

$\therefore$ $b_n = 0$, for n even.

and $$b_n = \frac{8}{n^2\pi^2}(-1)^{n-1}, \text{ for n odd.}$$

Hence the Fourier sine series is

$$f(x) = \sum_{n=1}^{\infty} b_n \sin\left(\frac{n\pi x}{2}\right) = \frac{8}{\pi^2}\sum_{n=1}^{\infty}(-1)^{n-1}\frac{\sin[(2n-1)\,\pi x/2]}{(2n-1)^2}.$$

Example 22:

Obtain the Fourier Series in $[-\pi, \pi]$ for the function:

$$f(x)=\begin{cases} x, \text{ if } -\pi < x \le 0, \\ 2x, \text{ if } 0 \le x < \pi. \end{cases}$$

Solution:

The given function is bounded, integrable and continuous in $[-\pi, \pi]$ except at the points $\pm\,\pi$. Now

$$a_0 = \frac{1}{\pi}\int_{-\pi}^{\pi} f(x)\,dx.$$

$$= \frac{1}{\pi}\int_{-\pi}^{0} x\,dx + \frac{1}{\pi}\int_{0}^{\pi} 2x\,dx$$

$$= \frac{1}{\pi}\left|-\frac{\pi^2}{2}\right| + \frac{1}{\pi}(\pi^2) = \frac{\pi}{2}.$$

We have $a_n = \dfrac{1}{\pi}\displaystyle\int_{-\pi}^{\pi} f(x)\cos nx\,dx$

$$= \frac{1}{\pi}\int_{-\pi}^{0} x\cos nx\,dx + \frac{1}{\pi}\int_{0}^{\pi} 2x\cos nx\,dx$$

$$= \frac{1}{\pi}\left|\frac{x\sin nx}{n}\right|_{-\pi}^{0} - \frac{1}{\pi}\int_{0}^{\pi} 2x\cos nx\,dx$$

$$+ \frac{2}{\pi}\left|\frac{x\sin nx}{n}\right|_{0}^{\pi} - \frac{2}{\pi}\int_{0}^{\pi}\frac{\sin nx}{n}\,dx$$

$$= 0 + \frac{1}{n^2\pi}\left|\cos nx\right|_{-\pi}^{0} + 0 + \frac{2}{n^2\pi}\left|\cos nx\right|_{0}^{\pi}$$

$$= \frac{1}{n^2\pi}(1-\cos n\pi) + \frac{2}{n^2\pi}(\cos n\pi - 1)$$

$$= \frac{\cos n\pi - 1}{n^2\pi}.$$

$$\therefore \quad a_n = \begin{cases} = 0, \text{ when n is even} \\ -\frac{1}{\pi n^2}, \text{ when n is odd} \end{cases}$$

Now, $$b_n = \frac{1}{\pi}\int_{-\pi}^{\pi} f(x) \sin nx\, dx$$

$$= \frac{1}{\pi}\int_{-\pi}^{0} x \sin nx\, dx + \frac{1}{\pi}\int_{0}^{\pi} 2x \sin nx\, dx$$

$$= \frac{1}{\pi}\left[\left| -x\frac{\cos nx}{n}\right|_{-\pi}^{0} + \int_{-\pi}^{0}\frac{\cos nx}{n} dx\right]$$

$$+ \frac{1}{\pi}\left[\left|\frac{-2x \cos nx}{n}\right|_{0}^{\pi} + 2\int_{0}^{\pi}\frac{\cos nx}{n} dx\right]$$

$$= \frac{1}{\pi}\left[\frac{-\pi \cos n\pi}{n} - \frac{2\pi \cos n\pi}{n}\right] = \frac{3(-1)^{n+1}}{n}.$$

$[\because \cos n\pi = (-1)^n]$

Hence the Fourier Series of f(x)

$$= \frac{1}{2}a_0 + \sum_{n=1}^{\infty}(a_n \cos nx + b_n \sin nx)$$

$$= \frac{\pi}{4} - \frac{2}{\pi}\left[\frac{\cos x}{1^2} + \frac{\cos 3x}{3^2} + \frac{\cos 5x}{5^2} = \ldots\ldots\right]$$

$$+ \left[\frac{\sin x}{1} + \frac{\sin 2x}{2} + \frac{\sin 3x}{3} = \ldots\ldots\right]$$

Example 23:

If $f(x) = cx$, *when* $-\pi < x < \pi$,

$f(-\pi) = f(\pi) = 0$ *and* $f(x + 2\pi) = f(x)\ \forall x$; *show that*

$f(x) = 2c\ [\sin x - \tfrac{1}{2}\ 2x + \tfrac{1}{2} \sin 3x - \ldots..]\ \ \forall x.$

Solution:

The given function is bounded, integrable and piecewise monotonic in $[-\pi, \pi]$.

Now $$a_0 = \frac{1}{\pi}\int_{-\pi}^{\pi} f(x)\, dx = \frac{1}{\pi}\int_{-\pi}^{\pi} cx\, dx = 0$$

$$a_n = \frac{1}{\pi}\int_{-\pi}^{\pi} f(x)\cos nx\, dx = \frac{c}{\pi}\int_{-\pi}^{\pi} x\cos nx\, dx$$

$$= \frac{c}{\pi}\left[\left|\frac{x\sin nx}{n}\right|_{-\pi}^{\pi} - \frac{1}{n}\int_{-\pi}^{\pi}\sin nx\, dx\right]$$

$$= \frac{c}{\pi}\left[0 + \frac{1}{n^2}\left|\cos nx\right|_{-\pi}^{\pi}\right] = 0. \qquad \therefore a_n = 0 \ \forall\ n.$$

$$b_n = \frac{1}{\pi}\int_{-\pi}^{\pi} f(x)\sin nx\, dx = \frac{c}{\pi}\int_{-\pi}^{\pi} x\sin nx\, dx$$

$$= \frac{c}{\pi}\left[\left|-x\frac{\cos nx}{n}\right|_{-\pi}^{\pi} + \frac{1}{\pi}\int_{-\pi}^{\pi}\cos nx\, dx\right]$$

$$= -\frac{c}{n\pi}\times 2\pi\cos n\pi = -\frac{2c}{\pi}(-1)^n = \frac{2c}{n}(-1)^{n+1}.$$

Since f is continuous for all x, therefore

$$f(x) = \tfrac{1}{2}a_0 + \sum_{n=1}^{\infty}(a_n\cos nx + b_n\sin nx)$$

or $$f(x) = \sum_{n=1}^{\infty} b_n \sin nx = b_1\sin n + b_2\sin 2x + \ldots..$$

Hence $f(x) = 2c\,[\sin x - \tfrac{1}{2}\sin 2x + \tfrac{1}{3}\sin 3x^{-} \ldots. 1.$

Example 24:

Let $\frac{1}{2}a_0 + \Sigma(a_n\cos nx + b_n\sin nx)$ *be a Fourier Series which converges uniformly to f(x) on* $[-\pi, \pi]$. *Show that*

$$\frac{1}{2}a_0^2 + \sum_{n=1}^{\infty}(a_n^2 + b_n^2) = \frac{1}{\pi}\int_{-\pi}^{\pi}\{f(x)\}^2\, dx. \qquad \textbf{(D.U., 1995)}$$

Solution:

We have $f(x) = \frac{1}{2}a_0 + \Sigma\,(a_n\cos nx + b_n\sin nx)$.

Multiplying both sides by f(x) an integrating term by term from $-\pi$ to π (which is justified since the series is uniformly convergent), we obtain

$$\int_{-\pi}^{\pi}\{f(x)\}^2\, dx = \tfrac{1}{2}a_0\int_{-\pi}^{\pi} f(x)\, dx +$$

$$\sum_{n=1}^{\infty}\left\{a_n \int_{-\pi}^{\pi} f(x)\cos nx\,dx + b_n \int_{-\pi}^{\pi} f(x)\sin nx\,dx\right\}$$

$$= \tfrac{1}{2} a_0^2 \pi + \pi \sum_{n=1}^{\infty}(a_n^2 + b_n^2), \quad \text{of a0, an, } b_n.$$

using the definitions

Hence the result is proved.

Ex. If $\phi(x) = \dfrac{1}{2}x$ for $0 < x < \alpha$, $\phi(x) = \dfrac{1}{2}\alpha$ for $0 < x < \pi - \alpha$ and $\phi(x) = \dfrac{1}{2}(\pi - \alpha)$ for $\pi - \alpha < x < \pi$; prove that

$$\phi(x) = \tfrac{2}{\pi}\left(\sin\alpha \sin x + \tfrac{1}{9}\sin 3\alpha \sin 3x + \ldots\right). \qquad \textbf{(D.U., 1995)}$$

Example 25:

Let $f(x) = \begin{cases} -\cos x, & \text{for } -\pi < x < 0 \\ \cos x, & \text{for } \ 0 < x < \pi. \end{cases}$

Show that the Fourier Series which converges to f(x) is

$$\frac{\pi}{4}\left(\frac{2}{1.3}\sin 2x + \frac{4}{3.5}\sin 4x + \frac{6}{5.7}\sin 6x + \ldots\ldots\ldots\right). \qquad \textbf{(D.U., 1990)}$$

Solution:

We have

$$c_0 = \frac{1}{\pi}\int_{-\pi}^{\pi} f(x)\,dx$$

$$= \frac{1}{\pi}\int_{-\pi}^{0} -\cos x\,dx + \frac{1}{\pi}\int_{0}^{\pi}\cos x\,dx = 0.$$

Now $a_n = \dfrac{1}{\pi}\displaystyle\int_{-\pi}^{\pi} f(x)\cos nx\,dx$

$$= \frac{1}{\pi}\int_{-\pi}^{0} -\cos x\cos nx\,dx + \frac{1}{\pi}\int_{0}^{\pi}\cos x\cos nx\,dx$$

$$= -\frac{1}{\pi}\int_{-\pi}^{0}\cos x\cos nx\,dx + \frac{1}{\pi}\int_{0}^{\pi}\cos x\cos nx\,dx$$

$\therefore \quad a_n = 0 \ \forall n.$

Now, $b_n = \dfrac{1}{\pi}\displaystyle\int_{-\pi}^{0}\cos x\sin nx\,dx + \dfrac{1}{\pi}\int_{0}^{\pi}\cos x\sin nx\,dx$

$$= \frac{2}{\pi} \int_0^\pi \cos x \sin nx \, dx$$

$$= \frac{i}{\pi} \int_0^\pi \{\sin (n+1)x + \sin (n-1)x\} \, dx$$

$$= \frac{-1}{\pi} \left[\frac{\cos (n+1)x}{n+1} + \frac{\cos (n-1)x}{n-1} \right]_0^\pi$$

$$= -\frac{1}{\pi} \left[\frac{1}{(n+1)} \{(-1)^{n+1} - 1) + \frac{1}{(n-1)} \{(-1)^{n-1} - 1\} \right]$$

$$= -\frac{1}{\pi} \left[(-1)^{n+1} \left\{ \frac{1}{n+1} + \frac{1}{n-1} \right\} - \left\{ \frac{1}{n+1} - \frac{1}{n-1} \right\} \right]$$

$$= -\frac{2n}{\pi(n^2 - 1)} [(-1)^{n+1} - 1]$$

$$\therefore \quad b_n = -\begin{cases} 0, & \text{when n is odd,} \\ \dfrac{4n}{\pi(n^2 - 1)}, & \text{when n is even.} \end{cases}$$

Hence the Fourier Series of | cos x | is

$$\frac{1}{2} a_0 + \sum_{n=1}^{\infty} (a_n \cos nx + b_n \sin nx)$$

$$= 0 + \sum \frac{4n}{n(n^2 - 1)} \sin nx, \text{ (n even)} \qquad [_n = 0 \ \forall \ n]$$

$$= \frac{4}{\pi} \left[\frac{2}{1.3} \sin 2x + \frac{4}{3.5} \sin 4x + \frac{6}{5.7} \sin 6x + \ldots\ldots\ldots \right].$$

EXERCISES

Find the sum of the series when $x = \pi$, $2\pi/3$.

1. Show that for all values of x in $[-\pi, \pi]$, when k is not an integer, cos

$$k x = \frac{\sin k\pi}{\pi} \left[\frac{1}{k} + \sum_{n=1}^{\infty} \frac{(-1)^n \, 2k \cos nx}{k^2 - n^2} \right]$$

Deduce that

$$\pi \cot k\pi = \frac{1}{k} + \sum_{n=1}^{\infty} \frac{2k}{k^2 - n^2},$$

$$\frac{\pi}{\sin k\pi} = \sum_{n=0}^{\infty}(-1)^n\left(\frac{1}{n+k}+\frac{1}{n+1-k}\right)$$

2. Show that the Fourier Series

$$\frac{\pi^2}{3} = 4\sum_{n=1}^{\infty}(-1)^{n-1}\left(\frac{\sin nx}{2n}-\frac{\cos nx}{n^2}\right)$$

converges to the periodic function f in $]-\pi, \pi[$, where $f(x) = x^2 + x$, for $-\pi < x < \pi$, and $f(x) = \pi^2$, for $x = \pm\pi$.

What is the value of the series at

$x = \pm\pi, 0, 9\pi, 19\pi/2, 10\pi$?

deduce that

$$\frac{\pi^2}{6} = 1 + \frac{1}{2^2} + \frac{1}{3^2} + \ldots\ldots$$

3. Find the Fourier Series in $[0, \pi]$ for the function

$$f(x) = \begin{cases} \pi/3, & \text{for } 0 < x < \pi/3 \\ 0, & \text{for } \pi/3 < x < 2\pi/3 \\ -\pi/3, & \text{for } 2\pi/3 < x < \pi \end{cases}$$

ANSWER

1. $\dfrac{\sin 2x}{1} + \dfrac{\sin 4x}{2} + \dfrac{\sin 8x}{4} + \dfrac{\sin 10x}{5} + \ldots..$

3

THE RIEMANN-STIELTJES INTEGRAL

3.1 Introduction

In this chapter, we shall give the definition of the Riemann-Stieltjes Integral (RS-Integral) and some example based on it may be mentioned that the Riemann-Stieltjes Integral is a generalisation of the Riemann Integral. Recall that a real valued functions is monotonically increasing in an interval *l* if

$f(x) \geq f(y)$, when $x, \geq y$ $(x, y \in l)$.

3.2 Definition and Existence of the Integral

Let *f* and α be bounded functions on [a, b] and α be *monotonic increasing* on [a, b], $b \geq a$.

Let $P = \{a = x_0, x_1, \ldots, x_n = b\}$, of [a, b]. Now we write as follow.

$\Delta\alpha_i = a(x_i) - \alpha(x_{i-1})$, $i = 1, 2, \ldots, n$.

It is clear that $\Delta\alpha_i \leq 0$, Here, we define two sums,

$$U(P, f, \alpha) = \sum_{i=1}^{n} M_i \, \Delta\alpha_i$$

$$L(P, f, \alpha) = \sum_{i=1}^{n} m_i \, \Delta\alpha_i$$

When m_i, M_i are the bounds (infimum and supremum respectively) of *f* in Δx_i. respectively called the *Upper* and the *Lower* sums of *f* corresponding to the partition *P*.

If *m*, *M* are respectively the lower and the upper bounds of *f* on [a, b], we have.

$$m \leq m_i \leq M_i \leq M$$

$\Rightarrow\ m\ \Delta\alpha_i \leq m_i\ \Delta\alpha_i \leq M_i\ \Delta\alpha_i \leq M\Delta\alpha_i, \Delta\alpha_i \geq 0$

Putting i = 1, 2,...., n and adding all inequalities, we get

$$m\{\alpha_i(b) - \alpha(a)\} \leq L(P, f, \alpha) \leq U(P, f, \alpha) \leq M\{\alpha(b) - \alpha(a) \qquad (1)$$

We define two integrals, which always exist by a similar reasoning,

$$\overline{\int_a^b} f\, d\alpha = \inf, U(P, f, \alpha) \qquad (2)$$

$$\underline{\int_a^b} f\, d\alpha = \sup, U(P, f, \alpha).$$

The infimum and supremum being taken over all partitions of [a, b], These are respectively called the *Upper* and the *Lower* integrals of *f* with respect to α.

There two integrals may or may not be equal. In cases these two integrals are equal, i.e.,

$$\overline{\int_a^b} f\, d\alpha = \underline{\int_a^b} f\, d\alpha$$

We say that *f* is integrable with respect to α in the Riemann sense and write $f \in \Re_\alpha$ [a, b] or simply $\Re(\alpha)$. Their common value is denoted by

$$\int_a^b f\, d\alpha$$

or sometimes by

$$\int_a^b f(x)\, d\alpha(x)$$

And is called the *Riemann-Stieltjes integral* (or simply the *Stieltjes integral* or *f* with respect to α, over [a, b].

From (1) and (2), it follows that

$$m\ \{\alpha(b) - \alpha(a)\} \leq L(P, f, \alpha) \leq \underline{\int_a^b} f\, d\alpha \leq \overline{\int_a^b} f\, d\alpha$$
$$\leq U(P, f, \alpha) \leq M\{\alpha(b) - \alpha(a)\}.$$

Remark:

The upper and lower integrals always exist for bounded functions but these may not be equal for all bounded functions. Such functions are not integrable. Thus the question of their equality and hence that of the integrability of the function is our main concern. It we take $\alpha(x) = x$ then the *Riemann-Stieltjes Integral* becomes the usual *Riemann-Integral.*

Some Deduction

(i) If $f \in \Re(\alpha)$, Then ∃ a number λ lying between the bound of *f* such that

$$\int_a^b f d\alpha = \lambda\{\alpha(b) - \alpha(a)\}$$

(ii) If f M continuous on [a, b] Then $\exists$ a number $\xi \in$ [a, b] such that

$$\int_a^b f d\alpha = f(\xi)[\alpha(b) - \alpha(a)]$$

(iii) If $f \in \mathfrak{R}(\alpha)$ and k is a number such that $|f(x)| \leq k$ for all $x \in [a, b]$ then

$$\left|\int_a^b f d\alpha\right| \leq k[\alpha(b) - \alpha(a)]$$

(iv) If $f \in \mathfrak{R}(\alpha)$, $g \in \mathfrak{R}(\alpha)$ over [a, b] such that $f(x) \geq g(x)$ Then

$$\int_a^b f d\alpha \geq \int_a^d g d\alpha, \; b \geq a$$

and $$\int_a^b f d\alpha \leq \int_a^d g d\alpha, \; b \leq a$$

(v) If $f \in R(\alpha)$ over [a, b] and $f(x) \geq 0$ for all $x \in [a, b]$ then

$$\int_a^b f d\alpha \geq \begin{cases} \geq 0 & b \geq a \\ \leq 0 & b \leq a \end{cases}$$

Since $f(a) \geq 0$ the lower bound $m \geq 0$.

Theorem 1.

If f *is monotonic on* [a, b], *and if a is continuous on [a, b], then* $f \in \mathfrak{R}(\alpha)$.

Proof:

Let $\varepsilon > 0$ be a given positive number.

For any positive integer n, choose a partition $P = \{x_0, x_1, \ldots, x_n\}$ of [a, b] such that

$$\Delta\alpha_i = \frac{\alpha(b) - \alpha(a)}{n}, \; i = 1, 2, \ldots, n$$

α is continuous and monotonic increasing on the closed interval [a, b] and thus assumes every value between its bounds, $\alpha(a)$ and $\alpha(b)$. Hence it is true.

Let f be monotonic increasing on [a, b], so that its lower and the upper bound, m_i, M_i in Δx_i are given by

$$m_i = f(x_{i-1}), \; M_i = f(x_i), \; i = 1, 2, \ldots, n$$

$$\therefore U(P, f\,\alpha) - L(P, f, \alpha) \;\; \sum_{i=1}^{n} (M_i - m_i)\Delta\alpha_i$$

$$\Rightarrow \frac{\alpha(b)-\alpha(a)}{n} \sum_{i=1}^{n}\{f(x_i)-f(x_{i-1})$$

$$\Rightarrow \frac{\alpha(b)-\alpha(a)}{n} \{f(b)-f(a)\}$$

$< \varepsilon$, for large n

$\Rightarrow f \in \Re$ (a) our [a, b], Hence the theorem.

Note: $\Rightarrow f \in \Re(a)$, i.e., $\int f\,d\alpha$ exists when either

(i) *f* is continuous and α is monotonic, or

(ii) *f* is monotonic and α is continuous; of course α is still monotonic.

Theorem 2:

If f is RS-integrable on [a, b], then so is |f| and

$$\left|\int_a^b f\,d\alpha\right| \le \int_a^b |f|\,d\alpha .$$

Proof:

Since $\in R(a)$, there exists a partition P of [a, b] such that

$U(P, f, \alpha) - L(P, f, \alpha) < \varepsilon$

whereε is an arbitrary positive number

i.e $$\left(\sum_{i=1}^{n} M_i \Delta_{\alpha i} - \sum_{i=1}^{n} m_i \Delta_{\alpha i}\right) < \varepsilon$$

or $$\sum_{i=1}^{n} (M_i - m_i)\ \Delta\alpha_i < \varepsilon$$

where M_i = l.u.b. f(x) $(x_{i-1} \le x \le x_i)$

and m_i = g.l.b. f(x) $(x_{i-1} \le x \le x_i)$

and $\Delta_{\alpha i} = \alpha(x_i) - \alpha(x_{i-1})$

Now, let for the same partition P we have

M_i' = l.u.b. | f(x) | $(x_{i-1} \le x \le x_i)$

m_i' = g.l.b. | f(x) | $(x_{i-1} \le x \le x_i)$

If ξ_1, ξ_2 are any two points in Δx_i, then

$$|\,|f(\xi_2)| - |f(\xi_1)|\,| \le |f(\xi_2) - f(\xi_1)|$$

$$\therefore\ M_i' - m_i' \le M_i' - m_i'$$

i.e. $$\sum_{i=1}^{n} (M_i - m_i)\ \Delta\alpha_i \le \sum_{i-1}^{n} (M_i - m_i)\Delta\alpha_i < \varepsilon$$

Hence, $\sum_{i=1}^{n}(M_i - m_i)\,\Delta\alpha_i < \varepsilon$

i.e. $U(P, |f|, \alpha) \mathbin{|} L(P, |f|, \alpha) < \varepsilon$

Thus, $|f| \in R(a)$

Again, since M_i can never exceed M_i'

$$\therefore \left|\sum_{i=1}^{n} M_i\,\Delta\alpha_i\right| \le \sum_{i-1}^{n} M'_i\,\Delta\alpha_i$$

$$\therefore \left|\int_a^b f\,d\alpha\right| \le \int_a^b |f|\,dx$$

Hence the theorem.

Theorem 3:

A function f is integrable with respect to α on [a, b] if and only f for every ε > 0 there exists a partition P of [a, b] such that

Proof:

$U(P, f, \alpha) - L(P, f\ \alpha) < \varepsilon$

Condition am Necessary: Let $f \in \mathfrak{R}(a)$ over $[a, b]$

$$\therefore \int_a^b f\,d\alpha = \overline{\int_a^b} f\,dx = \int_a^b f\,d\alpha$$

Let $\varepsilon > 0$ be any number.

Since the upper and the lower integrals are the infimum and the supremum, respectively, of the upper and the lower sums, therefore ∃ partitions P_1 and P_2 such that

$$\Rightarrow U(P_1, f, \alpha) > \overline{\int_a^b} f\,d\alpha + \frac{1}{2}\varepsilon = \int_a^b f\,d\alpha + \frac{1}{2}\varepsilon$$

$$\Rightarrow U(P_2, f, \alpha) > \overline{\int_a^b} f\,d\alpha - \frac{1}{2}\varepsilon = \int_a^b f\,d\alpha - \frac{1}{2}\varepsilon$$

Let $P = P_1 \cup P_2$ be the common refinement of P_1 and P_2.

$\therefore U(P\ f, \alpha) \le U(P_1\ f, \alpha)$

$$< \int_a^b f\,d\alpha + \frac{1}{2}\varepsilon < L(P_2, f, \alpha) + \varepsilon$$

$$\le L(P, f, \alpha) + \varepsilon$$

$\Rightarrow L(P, f, \alpha) - L(P, f, \alpha) < \varepsilon$

Condition M Sufficient

For $\varepsilon > 0$, let P be a partition for which

$$U(P, f, \alpha) - L(P, f, \alpha) < \varepsilon$$

For any partition P, we know that

$$L(P, f, \alpha) \leq \underline{\int_a^b} f\, d\alpha \leq \overline{\int_a^b} f\, d\alpha \leq L(P, f, \alpha)$$

$$\therefore \overline{\int} f\, d\alpha - \underline{\int} f\, d\alpha \leq U(P, f, \alpha) - L(P, f, \alpha) < \varepsilon$$

But a non-negative number can be less than every positive number, if it is zero.

$$\therefore \overline{\int} f\, d\alpha - \underline{\int_a^b} f\, d\alpha \quad \text{so that } f \in \Re(a) \text{ over } [a, b]$$

A Definition

(Integral as a Limit Sum)

As an analog to the Riemann sum, we introduce a sum which will lead to a sufficient condition for the existence of a Riemann-Stieltjes Integral.

Definition: Corresponding to a partition P of [a, b] and $t_i \in \Delta x_i$, consider the sum

$$S(P, f, \alpha) = \sum_{i=1}^{n} f(t_i)\, \Delta\alpha_i$$

we say that $S(P, f, \alpha)$ converges to A as $\mu(P) \to 0$, i.e.,

$$\lim_{\mu(P)\to 0} S(P, f, \alpha) = A$$

if, for every $\varepsilon > 0$ there exists $\delta > 0$ such that $| S(P, f, \alpha) - A | < e$, for every partition $P = \{a = x_0, x_1, x_2, \ldots, x_n = b\}$, of [a, b] with mesh $\mu(P) < \delta$ and every choice of t_i in Δx_i,

Theorem 4:

If lim $S(P, f, \alpha)$ exists as $\mu(P) \to 0$, then

$$f \in \Re(\alpha), \text{ and } \lim_{\mu(P)\to 0} S(P, f, \alpha) = \int_a^b f\, d\alpha$$

Let us suppose that lim $S(P, f, \alpha)$ exists as $\mu(P) \to 0$ and equal to A.

Therefore, for $\varepsilon > 0$, $\exists > 0$ such that for every partition P of [a, b] with mesh $\mu(P) < \delta$ and every choice of t_i in Δx_i we have

$$| S(P, f, \alpha) \mid A | < \frac{1}{2}\varepsilon$$

or

$$A - \frac{1}{2}\varepsilon < S(P, f, \alpha) < A + \frac{1}{2}\varepsilon. \qquad (1)$$

Let P be one such partition, If we let the points t_i range over the intervals Δx_i and take the infimum and the supremum of the sums $S(P, f, \alpha)$, Then from (1) we have

$$A - \frac{1}{2}\varepsilon < L(P, f, \alpha) \leq U(P, f, \alpha) < A + \frac{1}{2}\varepsilon \qquad (2)$$

$$\Rightarrow U(P, f, \alpha) - L(P, f, \alpha) < \varepsilon$$

$$\Rightarrow f \in \mathfrak{R}(\alpha) \text{ over } [a, b].$$

Again, since $S(P, f, \alpha)$ and $\int_a^b f\,d\alpha$ lie between $U(P, f, \alpha)$ and $L(P, f, \alpha)$

$$\therefore \left| S(P, f, \alpha - \int_a^b f\,d\alpha \right| \leq U(P, f, \alpha) - L(P, f, \alpha) < \varepsilon$$

$$\Rightarrow \lim_{\mu(P)\to 0} S(P, f, \alpha) = \int_a^b f\,d\alpha \text{ Hence the theorem}$$

Remark:

The theorem asserts that the existence of the limit of $S(P, f, \alpha)$ implies that $f \in \mathfrak{R}(\alpha)$. The existence of the limit is a sufficient condition for $f \in \mathfrak{R}(\alpha)$ but as shown in Example 3 it is not a necessary condition, i.e., functions exist which are integrable but for which limit of $S(P, f, \alpha)$

3.3 (A Relation Between Riemann Integral and RS-Integral)

Let f be Riemann integrable on [a, b,] and α monotonically increasing on [a, b] such that α' is Riemann integrable on [a, b], then $f \in \mathfrak{R}(\alpha)$ and

$$\int_a^b f\,d\alpha = \int_a^b f\alpha'dx.$$

A particular case of the above property is the following:

(IV) *If f is continuous on [a, b] and α has α continuous derivative on [a, b], the*

$$\int_a^b f\,d\alpha = \int_a^b f\alpha'dx.$$

Illustration. $\int_0^2 x^2\,dx^2 = \int_0^2 x^2 \,.\, 2x dx = 8.$

Now, we state a few properties of RS-Integrals.

3.4 Properties of RS-Integrals

The following are the properties of RS-Integrals

1. If $f_1 \in \Re(\alpha)$ and $f_2 \in \Re(\alpha)$ over [a, b], then
   (i) $f_1 + f_2 \in \Re(\alpha)$ and
   $$\int_a^b (f_1 + f_2)\, d\alpha = \int_a^b f_1\, d\alpha + \int_a^b f_2\, d\alpha.$$
   (ii) $f_1 + f_2 \in \Re(\alpha)$ and $\int_a^b cf_1\, d\alpha = c\int_a^b f_1\, d\alpha$, c is a constant.
2. A function f is RS-integrable w.r.t. α if and only if for each $\varepsilon > 0$, there exists a partition P of [a, b] such that $U(P, f, \alpha) - L(P, f, \alpha) < \varepsilon$
3. If $f \in \Re(\alpha_1)$ and $f \in \Re(\alpha_2)$, then $f \in \Re(\alpha_1 + \alpha_2)$ and
   $$\int_a^b f\, d(\alpha_1 + \alpha_2) = \int_a^b f\, d\alpha_1 + \int_a^b f\, d\alpha_2$$
4. **(Integration by Parts)**
   If f is continuous and α monotone on [a, b], then
   $$\int_a^b f\, d\alpha \;= \mid f(x)\ \alpha\ (x)\ |_\alpha^\beta - \int_a^b \alpha\, df$$
5. If $f \in \Re(\alpha)$ over [a, b] and if $a < c < b$, then $f \in \Re(\alpha)$ on [a, c] and [c, b]. Further $\int_a^b f\, d\alpha = \int_a^c f\, d\alpha + \int_c^b f\, d\alpha$.

Theorem 5:

If f is R-S Integrable on [a, b] and if

f (x) | ≤ M on [a, b]

then, $$\left| \int_a^b f\, d\alpha \right| \le M\ [\alpha\ (b) - \alpha\ (a)].$$

Proof:

We have

$$\left| \int_a^b f\, d\alpha \right| \le \int_a^b |f|\, dx$$

$$\le \int_a^b M\, d\alpha = M \int_a^b d\alpha = M\ [\alpha\ (b) - \alpha\ (a)]$$

Thus $$\left| \int_a^b f\, d\alpha \right| \le M\ [\alpha\ (b) - \alpha\ (a)]$$

Theorem 6:

If $f \in R(\alpha_1)$ and $f \in R(\alpha_2)$, then

$f \in R(\alpha_1 + \alpha_2)$ and

$$\int_a^b f\, d(\alpha_1 + \alpha_2) = \int_a^b f\, d\alpha_1 + \int_a^b f\, d\alpha_2 .$$

Proof:

$f \in R(\alpha_1) \Rightarrow \exists$ a partition P_1 s.t.

$$U(P_1, f, \alpha_1) - L(P_1, f, \alpha_1) < \varepsilon/2 \qquad ...(1)$$

$f \in R(\alpha_2) \Rightarrow \exists$ a partition P_2 s.t.

$$U(P_2, f, \alpha_2) - L(P_2, f, \alpha_2) < \varepsilon/2 \qquad ...(2)$$

where ε is an arbitrary positive number.

If the partitions P_1 and P_2 are replaced by their common refinement P, even the inequalities (1) and (2) hold.

thus $U(P, f, \alpha_1) - L(P, f, \alpha_1) < \varepsilon/2$ ...(3)

$U(P, f, \alpha_2) - L(P, f, \alpha_2) < \varepsilon/2$ ...(4)

Since $\sum_{i=1}^{n} M_i \; [\alpha\, (x_i) - \alpha\, (x_{i=1})]$

$$= \sum_{i=1}^{n} M_i \; [(\alpha_1 + \alpha_2)\,(x_i) - (\alpha_1 + \alpha_2)\,(x_{i=1})]$$

where $\alpha = \alpha_1 + \alpha_2$

$$\sum_{i=1}^{n} M_i \; [(\alpha_1\,(x_i) + \alpha_2\,(x_i) - \alpha_1\,(x_{i=1}) - \alpha_2\,(x_{i=1})]$$

$$= \sum_{i=1}^{n} M_i \; [\{\alpha_1\,(x_i) - \alpha_2\,(x_{i-1}) + \Sigma\,\{\alpha_2\,(x_i) - \alpha_2\,(x_{i-1})\}]$$

$$= \sum_{i=1}^{n} M_i \; [\alpha_1\,(x_i) + \alpha_2\,(x_i)] + \sum_{i=1}^{n} M_i \; [\alpha_1\,(x_i) - \alpha_2\,(x_{i-1})]$$

Thus, $U(P, f, \alpha) = U(P, f, \alpha_1) + U(P, f, \alpha_2)$. . . .(5)

Similarly,

$L(P, f, \alpha) = L(P, f, \alpha_1) + L(P, f, \alpha_2)$. . . . (6)

subtracting (6) from (5) we have

$$U(P, f, \alpha) - L\,(P, f, \alpha) = [U(P, f, \alpha_1) - L(P, f, \alpha_1)] + [U(P, f, \alpha_2) - L(P, f, \alpha_2)]$$

$\in/2 + \in/2 = \in.$

Hence, $f \in R(\alpha)$ where $\alpha = \alpha_1 + \alpha_2$.

Again, from (5) we have

g.*l*.b. $U(P, f, \alpha)$ = g.i.b. $[U(P, f, \alpha_1) + U(P, f, \alpha_2)]$

g.*l*.b. $U(P, f, \alpha) \geq$ g.i.b. $U(P, f, \alpha_1)$ + g.i.b. $U(P, f, \alpha_2)$.

Thus, $\int_a^b fd\alpha \geq \int_a^b fd\alpha_1 + \int_a^b fd\alpha_2$. . . . (7)

Similarly,

l.u.b. $L(P, f, \alpha) \leq$ l.u.b. $L(P, f, \alpha_1)$ + l.u.b. $L(P, f, \alpha_2)$

de $\int_a^b fd\alpha \leq \int_\alpha^b fd\alpha_1 + \int_\alpha^b fd\alpha_2$

Now (7) and (8) $\int_a^b d\alpha = \int_\alpha^b fd\alpha_1 + \int_\alpha^b fd\alpha_2$

Theorem 7:

If $f \in \Re[a, b)$ and a is monotone increasing on [a, b] such that $a' \in \Re[a, b,]$, then $f \in \Re(\alpha)$, and

$$\int_a^b f\, d\alpha = \int_a^b f\alpha'\, dx.$$

Proof:

Let $\in > 0$ be any given number.

Since f is bounded, there exists $M > 0$, such that

$|f(x) \leq M, \forall\, x \in [a, b]$

Again since $f, \alpha' \in \Re\,[a, b]$ therefore $f\alpha' \in \Re\,[a, b]$ and consequently $\exists\, \delta_1 > 0, \delta_2 > 0$ such that

$$\left|\Sigma f(t_i)\, \alpha'(t_i)\, \Delta x_i - \int \alpha'\, dx\right| < \varepsilon/2 \qquad \text{. . . (1)}$$

for $\mu(P) < \delta_1$ and all $t_i \in \Delta x_i$, and

$$\left|\Sigma \alpha'(t_i) \Delta x_i - \int \alpha'\, dx\right| < \varepsilon/4M \qquad \text{. . . (2)}$$

for $\mu(P) < \delta_1$ and all $t_1 \in \Delta x_i$.

Now, for $\mu(P) < \delta_2$ and all $t_i \in \Delta x_i$, $s_i \in \Delta x_i$, (2) gives

$$\Sigma|\alpha'(t_i) - \alpha'(s_i)|\, \Delta x_i < 2.\frac{\varepsilon}{4M} = \frac{\varepsilon}{2M}. \qquad \text{. . . (3)}$$

Again Let $\delta = \min(\delta_1, \delta_2)$, and P any partition with $\mu(P) < \delta$.

Then, for all $t_i \in \Delta x_i$, by Lagrange's Mean Value Theorem, there are points $s_i \in \Delta x_i$ such that

$\Delta\alpha_i = \alpha'(s_i)\, \Delta x_i$ . . . (4)

Thus, we have

$$\left|\Sigma\, f(t_i)\,\Delta\alpha_i - \int f\alpha'\,dx\right| = \left|\Sigma\, f(t_i)\alpha'(s_i)\,\Delta x_i - \int f\alpha' dx\right|$$

$$= \left|\Sigma\, f(t_i)\alpha'(t_i)\,\Delta x_i - \int f\alpha'\,dx\right.$$

$$\left. + \Sigma\, f(t_i)[\alpha'(s_i) - \alpha'(t_i)]\,\Delta x_i\right|$$

$$\leq \left|\Sigma\, f(t_i)\alpha'(t_i)\,\Delta x_i - \int f\alpha'\,dx\right|$$

$$+ \Sigma|f(t_i)|\,|\alpha'(s_i) - \alpha'(t_i)\,\Delta x_i$$

$$< \frac{\varepsilon}{2} + M\frac{\varepsilon}{2M} = \varepsilon. \qquad \ldots (5)$$

Hence for any $\varepsilon > 0$, $\exists\ \delta > 0$ such that for all partitions with $\mu(P) < \delta$, (5) holds

$$\Rightarrow \lim_{\mu(P)\to 0} \Sigma\, f(t_i)\,\Delta\alpha_i \text{ exists and equals } \int_a^b f\alpha'\,dx$$

$$\Rightarrow f \in \Re(\alpha), \text{ and } \int_a^b f\,d\alpha' = \int_a^b f\alpha'\,dx$$

Hence the theorem.

Theorem 8:

(First Mean Value Theorem). If a function f is continuous on [a, b] and α is monotonic increasing on [a, b], then their exists a number ξ in [a, b] such that

$$\int_a^b f\,d\alpha = f(\xi)\,\{\alpha(b) - \alpha(a)\}.$$

Proof:

Since f is continuous and α is monotonic, therefore $f \in \Re(\alpha)$.

Let m, M be the infimum and supremum of f in [a, b]. Then and we know that

$$m\{\alpha(b) - \alpha(a)\} \leq \int_a^b f\,d\alpha \leq M\{\alpha(b) - \alpha(a)\}$$

Hence, there exists a number μ, $m \leq \mu \leq M$ such that

$$\int_a^b f\,d\alpha = \mu\{\alpha(b) - \alpha(a)\}$$

Again, since f is continuous, there exists a number $\xi \in [a, b]$ such that $f(\xi) = \mu$

$$\therefore \int_a^b f\, d\alpha = f(\xi)\{\alpha(b) - \alpha(a)\}$$

Hence the theorem.

Remark: *It is not possible always to choose ξ such that $a < \xi < b$.*

Consider $\alpha(x) = \begin{cases} 0, & x = a \\ 1, & a < x \le b \end{cases}$

For a continuous function f, we have

$$\int_a^b f\, d\alpha = f(a) = f(a)\{\alpha(b) - \alpha(a)\}.$$

Theorem 9:

(Change of variable). If

(i) *f is a continuous function on [a, b], and*

(ii) *ϕ is a continuous and strictly monotonic function on [α, β] where* $a = \phi(\alpha)$, $b = \phi(\beta)$

then

$$\int_a^b f(x)\, dx = \int_\alpha^\beta f(\phi(y))\, d\phi(y).$$

Proof:

Let ϕ be strictly monotonic increasing function.

Since ϕ is strictly monotonic, it is invertible, i.e.,

$$x = \phi(y) \Rightarrow y = \phi^{-1}(x),\ \forall\ x \in [a, b]$$

so that, we can write

$$\alpha = \phi^{-1}(a),\ \beta = \phi^{-1}(b)$$

Again let

$$P = \{a = x_0, x_1, x_2, ..., x_n = b\}$$

be any partition of [a, b], and

$$Q = \{\alpha = y_0, y_1. y_2, ..., y_n = \beta\},\ y_i = \phi^{-1}(x_i)$$

be the corresponding partition of [α, β], so that

$$\Delta x_i = x_i - x_{i-1} = \phi(y_i)\ \phi(y_{i-1}) = \Delta\phi_i \qquad ...(1)$$

Again, for any $x_i \in \Delta x_i$, let $\gamma_i \in \Delta y_i$ where

$$x_i = \phi(\eta_i) \qquad ...(2)$$

Putting $g(y) = \phi(f(y))$, we have

$$S(P, f) = \sum_i f(\xi_i)\, \Delta x_i$$

$$\Rightarrow \sum_i f(\phi(\eta_i)\, \Delta\phi_i \Rightarrow \sum_i g(\eta_i)\, \Delta\phi_i$$

$$= S(Q, g, \phi) \qquad \ldots (3)$$

Continuity of f implies that $S(P, f) \to \int_a^b f\, dx$ as $\mu(P) \to 0$. Also, continuity of g implies that $S(Q, g, \phi) \to \int_\alpha^\beta g(y)\, d\phi$ as $\mu(Q) \to 0$.

Since uniform continuity of ϕ on $[\alpha, \beta]$ implies that $\mu(Q) \to 0$ as $\mu(P) \to 0$, therefore letting $\mu(P) \to 0$ in (3), we get

$$\int_a^b f(x)\, dx = \int_\alpha^\beta g(y)\, d\phi = \int_a^b f(\phi(y))\, d\phi(y).$$

Hence the theorem.

Example 1:

If $f(x) = x$ and $\alpha(x) = \sin(\pi x/2)$, find $U(P, f, \alpha)$ and $L(P, f, \alpha)$, when $P[0, 4] = \{0, 2, 4\}$. **(D.U., 1995)**

Solution:

The partition of inf [0, 4] is given as $P = \{0, 2, 4\}$.

$\therefore m_1 = \inf$ in $[0, 2] = 0$; $m_2 = \inf f$ in $[2, 4] = 2$;

$M_1 = \sup f$ in $[0, 2] = 2$; $M_2 = \sup f$ in $[2, 4] = 4$.

$U(P, f, \alpha) = M_1 \{\alpha(2) - \alpha(0)\} + M_2 \{\alpha(4) - \alpha(2) = 2(0 - 0) + 4(0 - 0) = 0.$

$[\because \alpha(2) = \sin \pi = 0, \alpha(4) = \sin 2\pi = 0]$

Similarly $L(P, f, \alpha) = m_1 \{\alpha(2) - \alpha(0)\} + m_2 \{\alpha(4) - \alpha(2)\} = 0.$

Example 2:

Let $\alpha(x) = f(x) = \begin{cases} 0 \text{ if } 0 \le x < 1 \\ 1 \text{ if } \le x \le 2 \end{cases}$ *and* $g(x) = \begin{cases} 0 \text{ if } 0 \le x \le 1 \\ 1 \text{ if } < x \le 2 \end{cases}$

(i) *Is $f(\alpha)$? If so, compute* $\int_0^2 f d\alpha$.

(ii) *Is $g \in \mathfrak{R}(\alpha)$? If so, compute* $\int_0^2 g d\alpha$.

Solution:

Consider any portion of [0, 2] as follows:

$P = \{0 = x_0, x_1 \ldots, x_{1-1}, x_1 = 1, x_1 = 1, x_{i+1}, \ldots, x_m = 2\}.$

Let M_k = sup f and m_k = inf f over $[x_{k-1}, x_k]$. Then

$m_k = M_k = 0$ for k = 0, l, ..., l – 1;

$m_k = M_k = 1$ for k = l,........, m.

Now, we have

$$U(P, f, \alpha) = \sum_{k=1}^{l-1} M_k \Delta\alpha_k + M_l \Delta\alpha_l + \sum_{k=l+1}^{m} M_k \Delta\alpha_k$$

$$= 0 + 1.\{\alpha(x_{l-1}) - \alpha(x_{l-1})\} + \sum_{k=l+1}^{m} 1.\{\alpha(x_k) - a(x_{k-1})\}$$

$$= 1(1-0) + \sum_{k=l+1}^{m} 1(1-1) = 1.$$

Similarly, L(P, f, a) = 1.

$$\therefore \int_a^b f d\alpha = \inf\{U(P, f, \alpha)\} = 1 \text{ and } \int_a^b f d\alpha = \sup\{L(P, f, \alpha)\} = 1$$

Hence, $f \in \Re(a)$ and $\int_1^2 f d\alpha = 1$.

(ii) Let m_k = inf g and M_k = sup g over $[x_{x-1}, x_k]$.

$\therefore m_k = M_k = 0$ for k = 0, 1....., l

$m_k = M_k = 1$ for k = l + 1...., m

$$\text{Now, } U(P, g, \alpha) = \sum_{k=0}^{l} M_k \Delta\alpha_k + \sum_{k=l+1}^{m} M_k \Delta\alpha_k$$

$$= 0 + \sum_{k=l+1}^{m} 1.\{\alpha(x_k) - \alpha(x_{k-1})\} = \sum_{k=l+1}^{m} 1\{1\text{-}1\} = 0$$

Similarly, L(P, g, a) = 0.

As argued above, $g \in \Re(\alpha)$ and $\int_0^2 g d\alpha = 0$.

Example 3:

Let f (x) = [x] ∀ x ∈ [– 1, 2] and

$$\alpha(x) = \begin{cases} 1, & x < 0 \\ 2, & 0 \le x < 1 \\ 3, & 1 \le x \le 2. \end{cases} \quad \text{Is } f \in \Re(\alpha). \qquad \textbf{(D.U., 1989)}$$

Solution:

Let P = {– 1 = x_0, x_l, ..., x_l = 0, x_{l+1},.....x_m = 1, x_{m+1},...., x_n = 2} be any partition of [– 1, 2].

Let M_k = sup f over $[x_{k-1}, x_k]$ and m_k = inf f over $[x_{k-1}, x_k]$.

$\therefore$ $m_k = -1$ and $M_k = 0$ for $k = 0, 1, ..., l$

$m_k = 0$ and $M_k = 1$ for $k = l + 1..., m$

$m_k = 1$ and $M_k = 2$ for $k = m + 1...., n$

We have

$$\Delta \alpha_k = \alpha(x_k) - a(x_{k-1}) = \begin{cases} 1-1=0 & k=1,2,...,l-1 \\ 2-1=1 & k=l \\ 2-2=0 & k=l+1.....,m-1 \\ 3-2=1 & k=m \\ 3-3=0 & k=m+1.....,n. \end{cases}$$

$$\text{Now } U(P, f, a) = \sum_{k=1}^{l} M_k \,\Delta\alpha_k + \sum_{k=l+1}^{m-1} M_k \,\Delta\alpha_k + M_m \,\Delta\alpha_m + \sum_{k=m+1}^{n} M_k \,\Delta\alpha_k$$

$$= 0 + 1.0 + 1.1 + 2.0 = 1.$$

$$\therefore \quad \overline{\int_{-1}^{2}} f d\alpha = \inf \{U(P, f, \alpha)\} = .$$

$$\text{Now } L(P, f, \alpha) \therefore \sum_{k=1}^{l-1} m_k \,\Delta\alpha_k + m_l \,\Delta\alpha_l + \sum_{k=l+1}^{m} m_k \,\Delta\alpha_k + \sum_{k=m+1}^{n} m_k \,\Delta\alpha_k$$

$$= -1.0 + (-1).1 + 0 + 1.0 = -1.$$

$$\therefore \quad \underline{\int_{-1}^{2}} f d\alpha = \sup \{L(P, f, \alpha)\} = -1.$$

Since $\overline{\int_{-1}^{2}} f d\alpha \neq \underline{\int_{-1}^{2}} f d\alpha$, $f \notin \Re(\alpha)$.

Hence f is not RS-integrable.

Example 4:

Evaluate $\int_0^3 x \, d[x]$. **(D.U., 1985)**

Solution:

The function $\alpha(x) = (x)$ is monotonically increasing on [0, 3].

For $0 \leq x < 1$, $\alpha(x) = 0$,

For $1 \leq x < 2$, $\alpha(x) = 1$,

For $2 \leq x < 3$, $\alpha(x) = 2$ and $\alpha(3) = [3] = 3$.

Consider any partition P of [0, 3] :

$P = \{0 = x_0, x_1, ..., x_{l-1}, x_l = 1, x_{i+1}, ..., x_{m-1}, x_m = 2, x_{m+1}, ..., x_{n-l}, x_n = 3\}$

$$\therefore U(P, f, \alpha) = \sum_{i=1}^{l-1} M_i \,\Delta\alpha_i + M_i (x_i - x_{l-1}) + \sum_{i=l+1}^{m-1} M_i . \Delta\alpha_i$$

$$+ M_m(x_m - x_{m-1}) + \sum_{i=m+1}^{n-1} M_i \Delta\alpha_i + M_n(x_n - x_{n-1}) \ldots(1)$$

$$= \sum_{i=1}^{l-1} M_i (0-0) + 1(1-0) + \sum_{i=l+1}^{m-1} M_i (1-1)$$

$$+ 2(2-1) + \sum_{i=m+1}^{n-1} M_i (2-2) + 3(3-2) = 6.$$

$\therefore\ \overline{\int_0^3} x d[x] = \inf \{U(P, f, \alpha)\} = 6.$

Similarly, $\underline{\int_0^3} x d[x] = \sup \{L(P, f, \alpha)\} = 6$. Hence $\int_0^3 x d[x] = 6$. **Ans.**

Example 5:

Evaluate $\int_0^3 x d([x] - x)$.

Solution:

By R.S.-Integral property.

We have $\int_0^3 x d([x] - x) = \int_0^3 x d[x] - \int_0^3 x dx = 6 - \frac{9}{2} = \frac{3}{2}$ **Ans.**

Example 6:

Give an example of a function f and integrator α such that $f \notin R[a, b]$, *but* $f \in R_\alpha [a, b]$. **(D.U., 1994)**

Solution:

We have to give an example of a function f and integrator α such that f is RS-integrable w.r.t. α but f is not Riemann integrable. Consider a function

$$f(x) = \begin{cases} 0, & \text{if } x \text{ is rational} \\ 1, & \text{if } x \text{ is irrational.} \end{cases}$$

We know f is not Riemann integrable over any interval [a, b] i.e., $f \notin \mathfrak{R}[a, b]$. Obviously f is bounded over [a, b].

Let us take $\alpha(x) = c$ (constant) for all $x \in [a, b]$.

Then $\int_a^b f(x)\, d\alpha = \int_a^b f(x)\, dc = 0.$

Hence $f \in \mathfrak{R}_\alpha [a, b]$ but $f \notin \mathfrak{R}[a, b]$.

Example 7:

Evaluate $\int_0^1 dx^2$ *from the definition of RS-integral.*

Solution:

Consider a partition of [0, 1] as follows:

$$P = P = \{0, \frac{1}{n}, \frac{2}{n}, ..., \frac{k-1}{n}, ..., \frac{n}{n} = 1\}.$$

We take $t_k = \frac{k}{n} \in \left[\frac{k-1}{n}, \frac{k}{n}\right]$. We have f(x) = x and $\alpha(x) = x^2$.

Now $\int_0^1 x\, dx^2 = \lim_{n\to\infty} S(P, f, \alpha)$ (by definition)

$$\Rightarrow \lim_{n\to\infty} \sum_{k=1}^{n} f(t_k) \{\alpha(x_k) - \alpha(x_{k-1})\}$$

$$\Rightarrow \lim_{n\to\infty} \sum_{k=1}^{n} f\left(\frac{k}{n}\right) \left\{\alpha\left(\frac{k}{n}\right) - \alpha\left(\frac{k-1}{n}\right)\right\} \quad \ldots (1)$$

$$\Rightarrow \lim_{n\to\infty} \sum_{k=1}^{n} \frac{k}{n} \left\{\left(\frac{k}{n}\right) - \left(\frac{k-1}{n}\right)^2\right\}$$

$$\Rightarrow \lim_{n\to\infty} \sum_{k=1}^{n} \frac{k}{n} \left(\frac{2k-1}{n^2}\right) = \lim_{n\to\infty} \sum_{k=1}^{n} \frac{1}{n^3} (2k^2 - k) \quad \ldots (2)$$

We know $\sum_{k=1}^{n} k = \frac{n(n+1)}{2}$, $\sum_{k=1}^{n} k^2 = \frac{n(n+1)(2n+1)}{6}$. $\quad \ldots (3)$

From (2) and (3), we get

$$\int_0^1 x\, dx^2 = \lim_{n\to\infty} \left[\frac{n(n+1)(4n-1)}{6n^3}\right], \text{ after simplication}$$

We have $= \frac{1}{6} \lim_{n\to\infty} \left(1 + \frac{1}{n}\right)\left(4 - \frac{1}{n}\right) = \frac{2}{3}$. **Ans.**

Example 8:

Show that $\int_{-1}^{1} x\, d|x| = 1$.

Solution:

Let f(x) = x, $\alpha(x) = |x|$.

Integrating by parts, we get

$$\int_{-1}^{1} x\, d|x| = \{x|x|\}_{-1}^{1} - \int_{-1}^{1} |x|.1\, dx = 2 - 2\int_0^1 x\, dx = 2 - 1 = 1.$$

Example 9:

Evaluate: $\int_0^3 x^2 d[x]$. **(D.U., 1992)**

Solution:

We know that

$$U(P, f, a) = \sum_{i=1}^{l-1} M_i (0-0) + 1^2 (1-0) + \sum_{i=l+1}^{m-1} M_i (1-1)$$

$$+2^2 (2-1) + \sum_{i=m+1}^{n-1} M_i (2-2) + 3^2 (3-2) = 14.$$

$$\therefore \overline{\int_0^3} x^2 d[x] = \inf \{U(P, f, \alpha) = 14.$$

Similarly, $\underline{\int_0^3} x^2 d[x] = 14$. Hence $\int_0^3 x^2 d[x] = 14$. **Ans.**

Example 10:

Let f be bounded on [–1, 1] and the function α be defined as $\alpha(x)$ = 0 if x < 0; $\alpha(0) = \frac{1}{2}$; $\alpha(x)$ = 1 if x > 0. . . . (1)

Prove that is RS-integrable w.r.t. α iff f is continuous at x = 0 and then

$\int_{-1}^{1} f\, d\alpha = f(0)$. **(D.U., 1989)**

Solution:

Let $P = \{-1 = x_0, x_1, ..., x_{i-2}, x_{i-1} = 0, x_i, ..., x_n + 1\}$ be a partition of [–1, 1]. Let $t_i \in \Delta x_i = [x_{i-1}, x_i]$, $x_{i-1} = 0$.

Now know that $S(P, f, \alpha) = \sum_{j=1}^{n} f(t_j) \{\alpha(x_j) - \alpha(x_{j-1})\}$

$= 0 + f(t_{i-1}) \{\alpha(x_{i-1}) - \alpha(x_{i-2})\} + f(t_i) \{\alpha(x_i) - \alpha(x_{i-1})\} + 0$

$= f(t_{i-1}) \left(\frac{1}{2} - 0\right) + f(t_i)\left(1 - \frac{1}{2}\right)$, using (1)

$= \frac{1}{2} [f(t_{i-1}) + f(t_i)] = f(0)$, when $t_{i-1} = 0 = t_i$.

Using variable x instead of t_i and t_{i-1}, we find that $x \to 0$ from the left as well as from the right, when the norm $\mu(P)$ tends to zero. Thus

lim $S(P, f, \alpha)$ exists iff f (0–) and f (0 +) both exist and equal f (0).

Hence f is RS-integrable w.r.t. α if f is continuous at x = 0.

We know that if f is continuous, then $f \in \Re(\alpha)$.

Hence f is RS-integrable iff f is continuous at x = 0 and

$\int_{-1}^{1} f d\alpha = \lim S(P, f, \alpha) = f(0)$. **Ans.**

Example 11:

Evaluate: $\int_0^2 [x]\, dx^2$, *using definition of integral.*

Solution:

It is clear that $\alpha(x) = x^2$ is monotonically increasing in [0, 2]. Also [x] = 0, $0 \le x < 1$; [x] = 1, $1 \le x < 2$ and [x] = 2 for x = 2.

Take $P_n = \left\{0, \frac{1}{n}, \frac{2}{n}, \ldots \frac{2n}{n} = 2\right\}$ as partition of [0, 2].

$$\therefore \Delta \alpha_r = \alpha\left(\frac{r}{n}\right) - \alpha\left(\frac{r-1}{n}\right) = \left(\frac{r}{n}\right)^2 - \left(\frac{r-1}{n}\right)^2 = \frac{2r-1}{n^2} \quad \ldots (1)$$

Now

$$M_r = \text{suo } [x] \text{ in } \left[\frac{r-1}{n}, \frac{r}{n}\right] = \begin{cases} 0, 1 \le r < n \text{ or } 1 \le r \le n-1 \\ 1, r = n \\ 1, n < 2n \text{ or } n+1 \le r \le 2n-1 \\ 2, r = 2n \end{cases} \quad \ldots(2)$$

and

$$m_r = \inf [x] \text{ in } \left[\frac{r-1}{n}, \frac{r}{n}\right] = \begin{cases} 0, 1 \le r < n \text{ or } 1 \le r \le n-1 \\ 0, r = n \\ 1, n < 2n \text{ or } n+1 \le r \le 2n-1 \\ 1, r = 2n \end{cases} \quad \ldots(3)$$

$$\therefore U(P_n, f, \alpha) = \sum_{r=1}^{2n} M_r \Delta\alpha_r = \sum_{r=1}^{2n} M_r \left(\frac{2r-1}{n^2}\right), \text{ by (1)}$$

Using (2), we get

$$\therefore U(P_n, f, \alpha) = \frac{1}{n^2}\sum_{r=1}^{n-1} 0, (2r-1) + \frac{1}{n^2}(2n-1)$$

$$+ \frac{1}{n^2}\sum_{r=n+1}^{2n-1} 1, (2r-1) + \frac{1}{n^2}, 2\,(4n-1)$$

$$= \left(\frac{2}{n} - \frac{1}{n^2}\right) + \frac{1}{n^2}\left[2\sum_{r=n+1}^{n+(n-1)} r - \sum_{r=n+1}^{n+(n-1)} 1\right] + 2\left(\frac{4}{n} - \frac{1}{n^2}\right)$$

$$= \frac{2}{n} - \frac{1}{n^2} + \frac{1}{n^2}\left[2\left\{(n-1)\,n + \frac{(n-1)n}{2}\right\} - (n-1)\right] + \frac{8}{n} - \frac{2}{n^2}$$

$$= \frac{10}{n} - \frac{3}{n^2} + \frac{3(n-1)}{n} - \frac{1}{n} + \frac{1}{n^2}$$

$$= \frac{6}{n} - \frac{2}{n^2} + 3 \to 3, \text{ as } n \to \infty.$$

Similarly, $L(P_m, f, \alpha) = 0 + 0 + \dfrac{1}{n^2}\displaystyle\sum_{r=n+1}^{2n-1} 1.\,(2r-1) + \dfrac{1}{n^2}(4n-1)$

$$= \frac{4}{n} - \frac{1}{n^2} + \frac{3(n-1)}{n} - \frac{1}{n} + \frac{1}{n^2} = \frac{3}{n} + 3 - \frac{3}{n} = 3$$

$\to 3$, as $n \to \infty$.

$\therefore\ \overline{\int_0^2} f\, d\alpha = 3 = \underline{\int_0^2} f\, d\alpha$. Hence $\displaystyle\int_0^2 [x]\, dx^2 = 3.$ **Ans.**

Example 12:

Using only the definition of the integral, show that:

$\int_0^3 x^3 d[x] = 1.$ **(D.U., 1995)**

Solution:

Use know that

$$\int_0^1 x^3 d[x] = \lim_{n\to\infty} \sum_{k=1}^{n} \left(\frac{k}{n}\right)^3 \left\{\left[\frac{k}{n}\right] - \left[\frac{k-1}{n}\right]\right\}$$

By definition, $\alpha(x) = [x] = 0$ for $0 \le x < 1$ and $\alpha(1) = 1$.

$$\int_0^1 x^3 d[x] = \lim_{n\to\infty} \sum_{k=1}^{n-1} \left(\frac{k}{n}\right)^3 \left\{\left[\frac{k}{n}\right] - \left[\frac{k-1}{n}\right]\right\} + \lim_{n\to\infty} \left(\left\{\left[\frac{n}{n}\right] - \left[\frac{n-1}{n}\right]\right\}\right)$$

$$= \lim_{n\to\infty}\left[\sum_{k=1}^{n-1}\left(\frac{k}{n}\right)^3 (0-0) + 1(1-0)\right] = 1.$$ **Ans.**

Example 13:

Let $f(x) = k$ be a constant function on $[a, b]$ and α monotonically increasing function on $[a, b]$. Prove that

$$\int_a^b f\,d\alpha = k[\alpha(b) - \alpha(a)].$$

Solution:

Let $P = \{a = x_0, x_1, ..., xi-1, x_i, ... x_n = b\}$ be any partition of $[a, b]$. Then

$$\sum_{i=1}^{n} f(t_i)\,\Delta\alpha_i = \sum_{i=1}^{n} k[\alpha(x_i) - \alpha(x_{i-1})]$$

$$= k[\{\alpha(x_1) - \alpha(x_0)\} - \{\alpha(x_2) + \{\alpha(x_2) - \alpha(x_1)\} + ...$$
$$+ \{a(x_n) - a(x_{n-1})\}]$$

$$= k[\alpha(x_n) - \alpha(x_0)] = k[\alpha(b) - \alpha(a)].$$

Thus, $\int_a^b k\,d\alpha = \lim\limits_{i=1} t(t_i)\Delta\alpha_i = k[\alpha(b) - \alpha(a)]$ . . . (1)

Note: If $k = 1$, then $\int_a^b d\alpha = \alpha(b) - \alpha(a)$.

$\therefore\ k\int_a^b d\alpha = k[\alpha(b) - \alpha(a)]$.

From (1) and (2), we obtain . . . (2)

$$\int_a^b k\,d\alpha = k\int_a^b d\alpha.$$ **Ans.**

Example 14:

Suppose a function α increases on $[a, b]$, $a \le x_0 \le b$, α is continuous at x_0. Let f be a function defined in $[a, b]$ such that $f(x_0) = 1$ and $f(x) = 0$ for $x \ne x_0$. Prove that $f \in R(\alpha)$ on $[a, b]$ and $\int_a^b f d\alpha = 0$. **[D.U., 1996]**

Solution:

Since α is continuous at x_0, so for any $\varepsilon > 0$, there exists some $\delta > 0$ such that

$$|\alpha(x) - \alpha(x_0)| < \varepsilon/3 \text{ for } |x - x_0| < \delta \text{ or } x_0 - \delta < x < x_0 + \delta.$$

For $y \in]x_0 - \delta, x_0 + \delta[$, $|\alpha(y) - \alpha(x_0)| < \varepsilon/3$. Thus

$$|\alpha(x) - \alpha(y)| = |\{\alpha(x) - \alpha(x_0)\} - \{\alpha(y) - \alpha(x_0)\}[$$
$$\le |\alpha(x) - \alpha(x_0)| + |\alpha(y) - \alpha(x_0)| < \varepsilon/3 + \varepsilon/3$$

$\therefore\ |\alpha(x) - \alpha(y) < 2\,\varepsilon/3$ for all $x, y \in]x_0 - \delta, x_0 + \delta[$. . . . (1)

We are given that $f(x_0) = 1$ and $f(x) = 0$ for $x \ne x_0$. . . . (2)

Consider a partition P of [a, b,] as follows:

$$P = \{a, x_0 - \frac{1}{2}\delta, x_0 + \frac{1}{2}\delta, b\}.$$

From (2), it follows that

sup f = 0 in $]a\, x_0 - \frac{1}{2}\delta$ [and] $]x_0 + \frac{1}{2}\delta, b[$, and

sup f = 1 in $]x_0 - \frac{1}{2}\delta, x_0 + \frac{1}{2}\delta_0[$ $\quad (\because f(x_0) = 1)$

Obviously inf f = 0 in each of the three sub-intervals stated above.

Now,

$$U(P, f, \alpha) = 0.\left[\alpha\left(x_0 - \frac{1}{2}\delta\right) - \alpha(a)\right] + 1.\left[\alpha\left(x_0 + \frac{1}{2}\delta\right) - \alpha\left(x_0 - \frac{1}{2}\delta\right)\right]$$

$$+\left[0.\alpha\ (b) - \alpha\left(x_0 + \frac{1}{2}\ \delta\right)\right] \qquad ...(3)$$

$$= \alpha\left(x_0 + \frac{1}{2}\delta\right) - \alpha\left(x_0 - \frac{1}{2}\delta\right) < \frac{2\varepsilon}{3}, \text{ using (1).}$$

$$\therefore\ U(P, f, \alpha) < \frac{2\varepsilon}{3} < \varepsilon. \qquad ...(4)$$

As argued in (3), we have

$L(P, f, \alpha) = 0\ [....] + 0\ [....] = 0.$

$\therefore\ U(P, f, \alpha) - L(P, f, \alpha) < \varepsilon.$

Hence, $f \in R_\alpha\ [a, b]$.

It follows that

$$\int_a^b f d\alpha = \inf\ \{U(P, f, \alpha)\} < \varepsilon, \text{ using (4)}$$

Now, $f \le 0$ on $[a, b] \Rightarrow \int_a^b f d\alpha \ge 0.$ Also $\int_a^b f d\alpha \le \varepsilon.$

Hence, $\int_a^b f d\alpha = 0,$ since ε is arbitrary. **Ans.**

Example 15:

Let $\alpha(x) = 0$ for $x \le 0$; $\alpha(x) = 1$ for $x > 0$. . . . (1)

Let f be a bounded function on [–1, 1]. Prove that

$f \in \mathfrak{R}(\alpha)$ iff $f(0+) = f(0)$ and then $\int_{-1}^{1} f\, d\alpha = f(0)$. **(D.U., 1986)**

Solution:

We have

$$S(P, f, \alpha) = \sum_{j=1}^{n} f(t_j)\{\alpha(x_j) - \alpha(x_{j-1})\}$$

$$= 0 + f(t_{i-1})\{\alpha(x_{i-1}) - \alpha(x_{i-2})\} + f(t_i)\{\alpha(x_i) - \alpha(x_{i-1})\} + 0$$

$$= 0 + 0 + f(t_i)(1 - 0) + 0 = f(t_i), \text{ by (1)}$$

$\therefore$ $S(P, f, \alpha) = f(0)$, when $t_i = 0$.

Using variable x instead of t_i, we find that $x \to 0$ from the right, when the norm m(P) tends to zero. Thus lim $S(P, f, \alpha)$ exists iff $f(0+)$ exists and equals f(0). Hence $f \in \mathfrak{R}(\alpha)$ iff $f(0+) = f(0)$ and

$$\int_{-1}^{1} f d\alpha = \lim S(P, f, \alpha) = f(0).$$ **Ans.**

Example 16:

Prove that $\int_0^3 x^2 d([x] - x) = 5$.

Solution:

Integrating by parts we obtain

$$\int_0^3 x^2\, d([x] - x) = \left| x^2([x] - x) \right|_0^3 - \int_0^3 2x([x] - x)dx$$

$$= \{9(|3| - 3) - 0\} - 2\int_0^3 x[x]\, dx + 2\int_0^3 x^2\, dx$$

$$= 0 - 2\left[\int_0^1 x.0\,dx + \int_1^2 x.1\,dx + \int_2^3 x.2dx\right] + 2\left|\frac{x^3}{3}\right|_0^3$$

$$= -2\left[0 + \left|\frac{x^2}{2}\right|_1^2 + \left|x^2\right|_2^3\right] + 18$$

$$= -2(2 - \frac{1}{2} + 9 - 4) + 18 = -13 + 18 = 5.$$ **Ans.**

Example 17:

Show $\int_{-1}^{1} x d|x| = 1$, *by definition of RS-integral.*

Solution:

Let f (x) = x and α (x) = | x |. Then

$\alpha(x) = -x, - \le x < 0$ and $\alpha(x) = x, 0 \le x \le 1$.

Let $P_n = \{-1 = x_0, x_1, ..., x_n = 0 = y_0, y_1, ... y_n = 1\}$

be a partition of [–1, 1], where

$$x_r = -1 + \frac{r}{n}, y_r = \frac{r}{n}; \; r = 1, 2, ..., n.$$

Now, $M_r = \sup f(x)$ in $[x_{r-1,} x_r] = x_r = -1 + \frac{r}{n}$,

$M'_r = \sup f(x)$ in $[y_{r-1}, y_r] = y_r = \frac{r}{n}$. $\quad (\therefore f(x) = x)$

The corresponding infima are

$$m_r = -1 + \frac{r-1}{n}, m'_r = \frac{r-1}{n}.$$

We have

$$\Delta \alpha_r = \alpha(x_r) - \alpha(x_{r-1}) = \begin{cases} -x_r + x_{r-1}, & \text{if } 1 \le r < n \\ \alpha(0) - \alpha(x_{n-1}), & \text{if } r = n \end{cases}$$

$$= \begin{cases} 1 - \frac{r}{n} - 1 + \frac{r-1}{n} = -\frac{1}{n}, & \text{if } 1 \le r < n \\ 0 + \left(-1 + \frac{n-1}{n}\right) = -\frac{1}{n}, & \text{if } r = n \end{cases}$$

$$\text{and } \Delta \alpha_r' = \alpha(y_r) - \alpha(y_{r-1}) = \begin{cases} y_r - y_{r-1}, & \text{if } 1 \le r < n \\ 1 - y_n, & \text{if } r = n \end{cases}$$

$$= \begin{cases} \frac{r}{n} - \frac{r-1}{n} = \frac{1}{n}, & \text{if } 1 \le r < n \\ 1 - \frac{n-1}{n} = \frac{1}{n}, & \text{if } r = n. \end{cases}$$

$$\therefore U(P_n, f, \alpha) = \sum_{r=1}^{n} M_r \, \Delta\alpha_r + \sum_{r=1}^{n} M'_r \, \Delta\alpha'_r$$

$$= \sum_{r=1}^{n} \left(-1 + \frac{r}{n}\right)\left(-\frac{1}{n}\right) + \sum_{r=1}^{n} \left(\frac{r}{n}\right)\left(\frac{1}{n}\right)$$

$$= \frac{1}{n}\sum_{r=1}^{n} 1 - \frac{1}{n^2}\sum_{r=1}^{n} r + \frac{1}{n^2}\sum_{r=1}^{n} r = 1 \to 1 \text{ as } n \to \infty,$$

and $L(P_n, f, \alpha) = \sum_{r=1}^{n}\left(-1+\frac{r-1}{n}\right)\left(-\frac{1}{n}\right) + \sum_{r=1}^{n}\left(\frac{r-1}{n}\right)\frac{1}{n}$

$$= \frac{1}{n}\sum_{r=1}^{n} 1 - \frac{1}{n^2}\sum_{r=1}^{n}(r-1) + \frac{1}{n^2}\sum_{r=1}^{n}(r-1)$$

$$= \frac{1}{n}.n = 1 \to 1 \text{ as } n \to \infty.$$

Hence, $\int_{-1}^{1} f d\alpha = \int_{-1}^{1} x d|x| = 1$.

Example 18:

Give an example of a function f and integrator α such that $f \notin R$ [a, b], but $f \in R_\alpha[a, b]$. **(D.U., 1994)**

Solution:

We solve this problem by choosing f and α in a different manner.

We choose

$$f(x) = \begin{cases} 0, & x = 0 \\ 1/x, & 0 < x < 1 \\ 1, & 1 \le x \le 2 \end{cases} \text{ and } a(x) = \begin{cases} 1, & 0 \le x \le 1 \\ x, & 1 \le x \le 2 \end{cases}.$$

It is easy to verify that the functions f and α are both continuous at x = We see that

$$\int_0^2 f dx = \int_0^1 f dx + \int_1^2 f dx = \int_0^1 \frac{1}{x} dx + \int_1^2 1.dx = \infty + 1 = \infty$$

(Notice that $\int_0^1 \frac{1}{x} dx$ is divergent.)

Thus, $f \notin R$ [0, 2].

Now, $\int_0^2 f d\alpha = \int_0^f f(x) d\alpha(x) + \int_1^2 f(x)\, d\alpha\,(x)$

$$= \int_0^1 \frac{1}{x}.0\, dx + \int_1^2 1.1\, dx = 0 + 1 = 1.$$

($\therefore$ $\alpha(x) = 1$ for $0 \le x \le 1 \Rightarrow d\,\alpha(x) = 0$ etc.)

Hence, $f \in R_\alpha$ [0, 2], but $f \notin R[0, 2]$. **Ans.**

Example 19:

Evaluate: $\int_{-1}^{2} x^5 d(|x|^3)$ **(D.U., 1987, 81)**

Solution:

$$\int_{-1}^{2} x^5 d(|x|^3) = \{x^5 |x|^3\}_{-1}^{2} - \int_{-1}^{2} |x|^3 .5x^4 \, dx$$

$$= (32\times 8+1) - 5\left[\int_{-1}^{2} |x| x^6 dx\right] \quad (\because |x|^2 = x^2)$$

$$= 257 - 5\left[-\int_{-1}^{0} x^7 \, dx + \int_{0}^{1} x^7 \, dx + \int_{1}^{2} x^7 \, dx\right]$$

$$= 257 - 5\left[\frac{1}{8}+\frac{1}{8}+\frac{255}{8}\right] = 257 \times \frac{3}{8} = \frac{771}{8}. \textbf{ Ans.}$$

Example 20:

Evaluate: $\int_{0}^{2} [x] \, dx^2$

Solution:

Let $f(x) = [x]$ and $\alpha(x) = x^2$

By the relation between Riemann Integral and RS-Integral,

$$\int_{0}^{2} [x] \, dx^2 = [x] \, 2xdx$$

$$= 2[\int_{0}^{1} [x]xdx + \int_{1}^{2} [x] \, x \, dx]$$

$$= 2\left[\int_{0}^{1} 0.xdx + \int_{1}^{2} 1.xdx\right] = 3.$$ **Ans.**

Example 21:

Using definition of RS-integral, evaluate

$$\int_{0}^{3} x \, d\,([x]-x).$$

Solution:

Let $f(x) = x$ and $\alpha(x) = [x] - x$. Then

$$\alpha(x) = \begin{cases} 0-x, & 0 \le x < 1 \\ 1-x, & 1 \le x < 2 \\ 2-x, & 2 \le x < 3 \\ 0, & x = 3. \end{cases} \quad \ldots (1)$$

We observe that a (x) is piecewise monotonically decreasing function in [0, 1), [1, 2) and [2, 3) respectively.

$$\text{Let } P_n = \left\{0, \frac{1}{n}, \frac{2}{n}, ... \frac{r-1}{n}, \frac{r}{n}, ... \frac{3n}{3} = 3\right\}$$

be a partition of [0, 3]. If M_r and m_r are the sup, and inf. of f (x) = x in [r – 1/n, r/n] respectively. Then

$$M_r = \frac{r}{n} \text{ and } m_r = \frac{r-1}{n}. \qquad ...(2)$$

Using (1), we obtain

$$\Delta\alpha_r = \alpha\left(\frac{r}{n}\right) - \alpha\left(\frac{r-1}{n}\right) = \begin{cases} -\frac{r}{n} + \frac{r-1}{n} = -\frac{1}{n} & 1 \le r < n \\ 0 + \frac{n-1}{n} = \left(1 - \frac{1}{n}\right), & r = n \\ \left(1 - \frac{r}{n}\right) - 1 - \frac{r-1}{n} = -\frac{1}{n}, & n < r < 2n \\ 0 - \left(1 - \frac{2n-1}{n}\right) = 1 - \frac{1}{n}, & r = 2n \\ \left(2 - \frac{r}{n}\right) - \left(2 - \frac{r-1}{n}\right) = -\frac{1}{n}, & 2n < r < 3n \\ 0 - \left(2 - \frac{3n-1}{n}\right) = 1 - \frac{1}{n}, & r = 3n. \end{cases}$$

$$\therefore \ \Delta\alpha_r = -\frac{1}{n}, \text{ except at } r = n, 2n, 3n$$

$$\text{and } \Delta\alpha_r = 1 - \frac{1}{n} \text{ at } r = n, 2n, 3n.$$

Now

$$U(P_n, f, \alpha) = \sum_{r=1}^{3n} M_r \Delta\alpha_r = \sum_{r=1}^{n-1} \frac{r}{n}\Delta\alpha_r + \frac{n}{n}\Delta\alpha_n + \sum_{r=n+1}^{2n-1} \frac{r}{n}\Delta\alpha_r$$
$$+ \frac{2n}{n}\Delta\alpha_{2n} + \sum_{r=2n+1}^{3n-1} \frac{r}{n}\Delta\alpha_r + \frac{3n}{n}\Delta\alpha_{3n}$$

$$= -\frac{1}{n^2}\sum_{r=1}^{n-1} r + \left(1 - \frac{1}{n}\right) - \frac{1}{n^2}\sum_{r=n+1}^{2n-1} r + 2 - \frac{2}{n} - \frac{1}{n^2}\sum_{r=2n+1}^{3n-1} r + 3 - \frac{3}{n}$$

$$= 6 - \frac{6}{n} - \frac{1}{n^2}.\frac{(n-1)n}{2} - \frac{1}{n^2}\left[(n-1)n + \frac{(n-1)n}{2}\right]$$

$$- \frac{1}{n^2}\left[(n-1)2n + \frac{(n-1)n}{2}\right]$$

$$= 6 - \frac{6}{n} - \frac{1}{2n^2}\left\{n^2 - n + 3n^2 - 3n + 5n^2 - 5n\right\}$$

$$= 6 - \frac{6}{n} - \frac{1}{2n^2}\left\{9n^2 - 9n\right\} = 6 - \frac{6}{n} - \frac{9}{2} + \frac{9}{2n}$$

$$= \frac{3}{2} - \frac{3}{2n} \to \frac{3}{2}, \text{ as } n \to \infty.$$

$\therefore \overline{\int_0^3} f\,d\alpha = \frac{3}{2}$. Similarly, $\underline{\int_0^3} f\,d\alpha = \frac{3}{2}$. Hence $\int_0^3 f\,d\alpha = \frac{3}{2}$. **Ans.**

Example 22:

Using only the definition of RS-Integral, evaluate

$$\int_0^2 x^2\, d[x].$$

Solution:

It is clear that f (x) = x^2 is bounded in [0, 2 and

$$\alpha(x) = [x] = \begin{cases} 0, & 0 \le x < 1 \\ 1, & 1 \le x < 2 \\ 2, & x = 2 \end{cases}$$

is monotonically increasing in [0, 2].

Let $P_n = \left\{0, \frac{1}{n}, \frac{2}{n}, \ldots, \frac{r-1}{n}, \frac{r}{n}, \ldots, \frac{2n}{n} = 2\right\}$ be a partition of [0, 2], for any positive integer n.

For each sub-interval [r – 1/n, r/n] (r = 1, 2, ..., n), we have

$$\Delta\alpha_r = \alpha\left(\frac{r}{n}\right) - \alpha\left(\frac{r-1}{n}\right) = \begin{cases} 0-0=0, & \text{if } 1 \le r < n \\ 1-0=1, & \text{if } r = n \\ 1-1=0, & \text{if } n < r < 2n \\ 1-0=1, & \text{if } r = 2n. \end{cases}$$

$\therefore \Delta\alpha_n = \Delta\alpha_{2n} = 1$ and $\Delta\alpha_r = 0$ if $r \ne n, 2n$. . . . (1)

Now M_r = sup f in [r – 1/n, r/n] = $(r/n)^2$, . . . (2)

$$m_r = \inf f \text{ in } [r - 1/n, r/n] = (r - 1/n)^2. \qquad \ldots (3)$$

$$\therefore U(P_n, f, \alpha) = \sum_{r=1}^{n} M_r \, \Delta\alpha_r = 0 + M_n \, \Delta\alpha_n + M_{2n}, \Delta\alpha_{2n}, \quad \text{using (1)}$$

Now $\overline{\int_0^2} f\, d\alpha = \lim_{n\to\infty} U(P_n, f, \alpha) = 5.$

Similarly we have, $L(P_n, f, \alpha) = 0 + m_n \, \Delta\, \alpha_n + m_{2n} \, \Delta\, \alpha_{2n}$

$$= \left(\frac{n-1}{n}\right)^2 + \left(\frac{2n-1}{n}\right)^2, \text{ by (1) and (3)}$$

$$= \left(1 - \frac{1}{n}\right)^2 + \left(2 - \frac{1}{n}\right)^2 \to 5 \text{ as } n \to \infty$$

$$\therefore \underline{\int_0^2} f d\alpha = \lim_{n\to\infty} L(P_n, f, \alpha) = 5. \text{ Hence } \int_0^2 f d\alpha = 5.$$

Example 23:

If $f(x) = x$ and $\alpha(x) = \sin(\pi x/2)$, find $U(P, f, \alpha)$ and $L(P, f, \alpha)$, where $P[0, 2, 4]$. Prove that, refinements of P need not decrease upper sums or increase Lower sums. **(D.U., 1995)**

Solution:

We have

$$\alpha(x) = \sin(\pi x/2) = \begin{cases} \sin(\pi\, x/2),\ 0 \le x \le 2 \text{ or } 0 \le \pi\, x/2 \le \pi \\ \sin(\pi\, x/2),\ 2 \le x \le 4 \text{ or } \pi \le \pi\, x/2 \le 2\pi \end{cases}$$

It follows that α (x) is not monotonic in the subintervals [0, 2] and [2, 4]. Also $\alpha(0) = \alpha(2) = 4$.

$U(P, f, \alpha) = L(P, f, \alpha) = 0$.

We consider a refinement of partition P = [0. 2. 4] as follows:

P^* [0, 4] = {0, 1, 2, 3, 4}. Then $\alpha(1) = 1$, $\alpha(3) = -1$.

Now $U(P^*, f, \alpha) = M_1 \{\alpha(1) - \alpha(0)\} + M_2 \{\alpha(2) - \alpha(1)\}$
$+ M_3 \{\alpha(3) - \alpha(2) + M_4 \{\alpha(4) - \alpha(3)\}$; where

M_1 = sub f in [0, 1] = 1, M_2 = sup f in [1, 2] = etc.

$\therefore U(P^*, f, \alpha) = 1(1 - 0) + 2(0 - 1) + 3(-1 - 0) + 4 \{0 - (-1)$
$= 1 - 2 - 3 + 4 = 0 = U(P, f, \alpha)$.

Similarly, $L(P^*, f, \alpha) = 0(1 - 0) + 1(0 - 1) + 2(-1 - 0) + 3\{0 - (-1)\}$
$= -1 - 2 + 3 = 0 = L(P, f, \alpha)$.

Hence, the refinements of P need not decrease upper sums or increase lower sums.

Example 24:

Prove that $\int_0^1 x\,d(e^{2x}) = \frac{1}{2}(H\ e^2)$. **(D.U., 1994)**

Solution:

Let f(x) = x, α(x) = e^{2x}. Integrating by parts, we obtain

$$\int_0^1 x\,d(e^{2x}) = |xe^{2x}|_0^1 \int_0^1 e^{2x}.1\,dx$$

$$= e^2 - \frac{1}{2}|e^{2x}|_0^1 = e^2 - \frac{1}{2}(e^2 - 1) = \frac{1}{2}(1 + e^2).$$ **Ans.**

EXERCISES

1. Define f on [–1, 1] by f (x) = 0, x < 0 and f (x) = 1, x ≥ 0. Let the integrator α be defined by α(x) = 0, x ≤ 0; α(x) = 1, x > 1 Show that f ∈ R(α) although lim S(P, f, α) does not exist as μ (P) → 0.
2. Define f on [– 1, 1] by f (x) = 0, x < 0 and f (x) = 1, x ≥ 0. Let the integrator α be defined by α(x) = 0, x ≥ 0; α(x) = 1, x > 0. Show that f ∈ $\Re(\alpha)$ although him S(P, f, α) does not exist as μ (P) → 0.
3. $\int_0^4 x\,d[x] = 10$. 2. $\int_0^4 x\,d([x] - x) = 2$.
4. $\int_0^3 x^2 d([x] - x) = 5$. 4. $\int_0^x d[t] = [x]\ \forall\ x \in R$.
5. $\int_0^5 (x^2 + 1)\,d[x] = 60$.
6. $\int_2^5 [5 - x]\,d(\log[x]) = \log 3$. **(D.U. 1994)**
7. $\int_0^4 x\,d[x] = 10.$
8. $\int_0^3 x^2 d([x] - x) = 5.$
9. Show that $\int_0^1 f\,d\alpha = \frac{29}{2}$, where f (x) = $x^3 + 7$

and $\alpha(x) = \begin{cases} 1+x^3, & 0 < x \le 1 \\ 0, & x = 0 \end{cases}$.

10. Let $f(x) = \begin{cases} 1, & 0 \le x \le \frac{1}{2} \\ 2, & \frac{1}{2} < x \le 1 \end{cases}$ and $a(x) = \begin{cases} 3, & 0 \le x \le \frac{1}{2} \\ 4, & \frac{1}{2} < x \le 1. \end{cases}$

Let $P = \left\{0, \frac{1}{3}, \frac{2}{3}, 1\right\}$. Show that $f \in \mathfrak{R}(\alpha)$ although lim $S(P, f, \alpha)$ does not exist as $\mu(P) \to 0$.

Show that

11. $\int_0^n 2x\,d(x[x]) = n^3 + \sum_{r=1} r^2$. **(D.U., 1994)**

12. Show that $\int_0^2 f\,d\alpha = 0$, where $f(x) = \begin{cases} 0, & 0 \le x \le 1 \\ 1, & 1 < x \le 2 \end{cases}$ and $\alpha(x) = \begin{cases} 0, & 0 \le x \le 1 \\ 1, & 1 \le x \le 2 \end{cases}$.

13. $\int_0^5 (x^2+1)\,d[x] = 60$.

14. Show that $\int_0^2 f\,d\alpha = 0$, where $f(x) = \begin{cases} 0, & 0 \le x \le 1 \\ 1, & 1 < x \le 2 \end{cases}$.

and $\alpha(x) = \begin{cases} 0, & 0 \le x < 1 \\ 1, & 1 \le x \le 2. \end{cases}$

15. Evaluate $\int_0^2 x\,d\alpha$, where $a(x) = \begin{cases} x, & 0 \le x \le 1 \\ 2, x, & 1 < x \le 2. \end{cases}$

16. Evaluate $\int_0^2 x\,d\alpha$, $a(x) = \begin{cases} x, & 0 \le x \le 1 \\ 2+x, & 1 < x \le 2. \end{cases}$

17. Show that $\int_0^1 f\,d\alpha = \frac{29}{2}$, where $f(x) = x^3 + 7$ and

$a(x) = \begin{cases} 1+x^3, & 0 < x \le 1 \\ 0, & x = 0 \end{cases}$.

4

INTEGRATION ON R^3

4.1 Line, Surface and Volume Integrals

In this chapter we will discuss the following three types of integrals viz.,

(i) Line Integrals,

(ii) Surface Integrals,

(iii) Volume (Triple) Integrals.

The discussion starts with the consideration of rectifiable curves and rectification, and goes on to consider the line integrals, surface area and the surface integrals, the volumes and the volume (triple) integrals.

4.2 Rectifiable Curves

We are now in a position to consider some suitable definitions of the length of the curve (rectification) and to discuss the conditions under which a curve is rectifiable. *We shall be concerned with continuous curves only* in the present discussion.

A *curve* in space is defined to be a vector-valued function γ with domain a subset of R and range a subset of R^3. The curve is *continuous* if γ is continuous, and is called a *Jordan* arc if γ is one-to-one.

4.3 Length of a Curve

Let $\gamma = (X, Y, Z)$, where X, Y, Z are three real *single-valued, continuous* functions of t defined on an interval [a, b], be a *continuous curve* in R^3, so that $\gamma(t) = [X(t), Y(t), Z(t)]$ is a point on the curve corresponding to $t \in [a, b]$.

Corresponding to any partition

$P = \{a = t_0, t_1, t_2 \ldots\ldots\ t_n = b\}$

of [a, b], we get an ordered set of points given as

$\{\gamma(t_0), \gamma(t_1) \ldots, \gamma(t_n)\}$

on the curve γ.

$|\gamma(t_i) - \gamma(t_{i-1})|$ represents the distance between the points $\gamma(t_{i-1})$ and $\gamma(t_1)$ and so the sum

$$s(P, a, b) = \sum_{i=1}^{n} |\gamma(t_i) - \gamma(t_{i-1})|$$

is the length of the polygon inscribed in the curve with vertices at the points $\gamma(t_0), \gamma(t_1) \ldots, \gamma(t_n)$ in this order. The length the polygon clearly depends upon the particular partition of [a, b]. Let us consider the set of these sums for all possible partitions of [a, b].

As the mesh of the partition tends to zero (or $n \to \infty$) these polygons approach the curve more and more closely. Accordingly.

The curve is said to be *rectifiable* if this set of sums is bounded and then, the supremum of the set is known as the *length of the curve.*

Hence we can conclude that a continuous curve γ is rectifiable when the supremum of the sums $\Sigma\, |\gamma(t_i) - \gamma(t_{i-1})|$, taken over all partitions of [a, b], exists, i.e., when γ is of bounded variation over [a, b]. The length, s(a, b) of the curve, from t = a, to t = b is then the total variation of γ over [a, b]. Thus we have

$s(a, b) = V(\gamma, a, b)$

The length of the curve over the sub-interval [a, t] is generally denoted by s(t), where

$s(t) = s(a, t) = V(\gamma, a, t) = v_\gamma(t)$

a function of t.

Note : $\gamma = (\phi, \psi, \theta)$ being a vector-valued function, the meaning of $|\gamma(t_i) - \gamma(t_{i-1})|$ or $V(\gamma, a, b)$ should be clearly understood. Here

$|\gamma(t_i) - \gamma(t_{i-1})| = \sqrt{\{[\phi(t_i) - \phi(t_{i-1})]^2 + [\psi(t_i) - \psi(t_{i-1})]^2 + [\theta(t_i) - \theta(t_{i-1})]^2\}}$

4.4 Properties of Rectifiable Curves

The following properties of such curves are the direct consequences of the vector-valued functions of bounded variation.

1. Since for a vector-valued function, $\gamma = (X, Y, Z)$, of bounded variation,

   $V(X, a, b) \le V(\gamma, a, b) \le V(X, a, b) + V(Y, a, b) + V(Z, a, b)$

   $V(X, a, b) \le s(a, b);\ V(y, a, b) \le s(a, b);\ V(Z, a, b) \le s(a, b)$

and $S(a, b) \leq V(X, a, b) + V(y, a, b) + V(Z, a, b)$.

Hence we deduce that

$V(X, a, b) \leq s(a, b)$; $V(Y, a, b) \leq s(a, b)$; $V(Z, a, b) \leq s(a, b)$

and

$s(a, b) \leq V(X, a, b) + V(Y, a, b) + V(Z, a, b)$

2. A curve $\gamma = (X, Y, Z)$ is rectifiable if X, Y, Z are derivable and the derivatives are bounded.
3. Since a function of bounded variation over [a, b] is of bounded variation over each sub-interval of [a, b], an arc (part) of a rectifiable curve corresponding to any of its sub-intervals is also rectifiable.
4. Since for a function of bounded variation.

   $V(\gamma, a, b) - V(\gamma, a, c) + V(\gamma, c, b) \quad a \leq c \leq b$

   it follows that

   $s(a, b) = s(a, c) + s(c, b)$.
5. The arc s(t) of a rectifiable curve is monotone increasing function of t. Since the total variation function $v_y(t)$ of a function of bounded variation is monotone increasing over [a, b], the arc s(t) is also monotone increasing on [a, b].
6. The arc s(t) is continuous over [a, b].

Remark:

If γ (or equivalently, X, Y, Z) is not constant over any sub-interval of [a, b], s(t) is a strictly monotone increasing continuous function of t on [a, b], therefore the function s is invertible and s^{-1} is continuous on [a, b].

Thus t can be regarded as a strictly monotone increasing function of s over [s(a), s(b)], so that the arc length s may be used as a parameter in place of t.

Theorem 1:

If g is a smooth curve in R^3 such that γ' exists and is continuous on [a, b], then g is rectifiable and has a length

$$\int_a^b |\gamma'(t)|\, dt.$$

Proof:

Since the function γ has a bounded derivative, it is of bounded variation on [a, b] and is therefore rectifiable on [a, b]. We have to prove the following:

$$\int_a^b |\gamma'(t)|\, dt = V(\gamma, a, b)$$

For any partition $\{a = t_0, t_1, t_2 \ldots\ldots, t_n = b)$ of $[a, b]$,

$$\Sigma|\gamma(t_i)-\gamma(t_{i-1}) = \Sigma\left|\int_{t_{i-1}}^{t_i} \gamma'(u)\, du\right| \le \Sigma \int_{t_{i-1}}^{t_i} |\gamma'(u)|\, du = \int_a^b |\gamma'(t)|\, dt$$

$$\Rightarrow\ V(\gamma, a, b) \le \int_a^b |\gamma'(t)|\, dt \qquad \ldots(1)$$

Let $\varepsilon > 0$ be given.

Since γ' is uniformly continuous on $[a, b]$, there exists $\delta > 0$ such that

$$|\gamma'(v) - \gamma'(u)| < \varepsilon, \text{ if } |v - u| < \delta$$

Let $P = \{a = t_0, t_1, \ldots\ldots, t_n = b\}$ be a partition of $[a, b]$ with mesh $\mu(P) < \delta$.

Now for $t_{i-1} \le u \le t_i$,

$$|\gamma'(t_i) - \gamma'(u) < \varepsilon$$

or

$$|\gamma'(u)| - \varepsilon < |\gamma'(t_i)|$$

so that, integrating from t_{i-1} to t_i,

$$\int_{t_{i-1}}^{t_i} |\gamma'(u)|\, du - \varepsilon\, \Delta t_i < |\gamma'(t_i)|\, \Delta t_i - |\gamma'(t_i)\, \Delta t_i|$$

$$= \left|\int_{t_{i-1}}^{t_i} [\gamma'(u) + \gamma'(t_i) - \gamma'(u)]\, du\right|$$

$$\le \left|\int_{t_{i-1}}^{t_i} \gamma'(u)\, du\right| + \left|\int_{t_{i-1}}^{t_i} [\gamma'(t_i) - \gamma'(u)]\, du\right|$$

$$\le |\gamma(t_i)-\gamma(t_{i-1})| + \int_{t_{i-1}}^{t_i} |\gamma'(t_i) - \gamma'(u)]\, du$$

$$\Rightarrow\ \int_{t_{i-1}}^{t_i} |\gamma'(u)|\, du < |\gamma'(t_i) - \gamma(t_{i-1})| + 2\varepsilon\, \Delta t_i\,.$$

Writing these inequalities for $i = 1, 2, \ldots, n$, and adding,

$$\int_a^b |\gamma'(t)|\, dt < \sum_{i=1}^{n} |\gamma'(t_i) - \gamma(t_{i-1})| + 2\varepsilon(b-a)$$

$$\le V(\gamma, a, b) + 2\varepsilon(b - a)$$

But since ε is arbitrary

$$\int_a^b |\gamma'(t)|\, dt \le V(\gamma, a, b) \qquad \ldots(2)$$

(1) and (2) imply

$$\int_a^b |\gamma'(t)|\, dt = V(\gamma, a, b)$$

i.e., the length of the curve is $\int_a^b |\gamma'(t)|\, dt$. Hence the theorems.

Note : Since γ is a vector-valued function,

$$|\gamma'(t)| = \sqrt{X'^2(t) + Y'^2(t) + Z'^2(t)}$$

Example 1:

Find the length of the curve

$$x = at^2,\ y = 2at,\ z = at,\ 0 \le t \le 1.$$

Solution:

$$s(0, 1) = \int_0^1 \sqrt{(2at)^2 + (2a)^2 + a^2}\, dt$$

$$= a \int_0^1 \sqrt{5+4t^2}\, dt$$

Put $2t = \sqrt{5} \sinh u$.

$$s(0, 1) = \frac{5a}{4} \int_0^{\log\sqrt{5}} (1+\cosh 2u)\, du$$

$$= \frac{5a}{4}\left[\log\sqrt{5} + \frac{6}{5}\right] = \frac{a}{8}[5 \log 5 + 12].$$

4.5 Line Integrals

Definition: Let, $x = X(t)$, $y = Y(t)$, $z = Z(t)$, $a \le 1 \le b$

be a curve C in a space of three dimensions.

Let a bounded vector-valued function $F = (f, g, h)$ be defined at every point of the curve C, so that f, g, h are all bounded and defined at every point (x, y, z) of the curve.

Let $\{a = t_0, t_1, ..., t_n = b\}$ be any partition of [a, b], and ξ_i any point of Δt_i. Let $)x_i, y_i, z_i)$ be a point of the curve corresponding to $t = t_i$, where

$$x_i = X(t_i),\quad y_i = Y(t_i),\quad z_i = Z(t_i)$$

From the sum

$$\sum_{i-1}^{n}[f(X(\xi_i), Y(\xi_i), Z(\xi_i)\,\Delta x_i + g(X(\xi_i), Y(\xi_i), Z(\xi_i))\,\Delta y_i$$
$$+ h(X(\xi_i), Y(\xi_i), Z(\xi_i))\,\Delta z_i]$$

If, as the norm of the partition tends to zero, the sum tends to a finite limit, independent of the choice of ξ_i in Δt_i, the line integral of F along C exists and is denoted by $\int_C F\,dt$ or

$$\left.\begin{aligned}&\int_C f(X, Y, Z)\,dx + g(X, Y, Z)\,dy + h(X, Y, Z)\,dz\\ &\text{or}\\ &\int_C f\,dx + g\,dy + h\,dz\end{aligned}\right\} \quad (1)$$

and call it the line integral of F = (f, g, h) along C.

Remark:

Vectorial Formulation: *Let r be the position vector of a point (x, y, z) on the curve, so that*

$$r = ix + jy + kz$$
$$= iX(t) + jY(t) + kZ(t),\ a \le t \le b$$

where i, j, k are the unit vectors along the coordinate axes x, y, and z. Let a vector function be represented as

$$F(x, y, z) = iP(x, y, z) + jQ(x, y, z) + kR(x, y, z)$$

$$\therefore \quad F.\frac{dr}{dt} = P\frac{dx}{dt} + Q\frac{dy}{dt} + R\frac{dz}{dt}$$

$$\Rightarrow \quad \int_a^b \left(F.\frac{dr}{dt}\right) dt = \int_\alpha^\beta \left(P\frac{dx}{dt} + Q\frac{dy}{dt} + R\frac{dz}{dt}\right) dt$$

Thus the line integral can be expressed as

$$\int_C F.dr = \int_a^b \left(F.\frac{dr}{dt}\right) dt.$$

4.6 Line Integral with Respect to Arc Length

We know that the line integral $\int f\,dx + g\,dy + h\,dz$ of a function

$$F = (f, g, h) \text{ along a curve C,}$$
$$x = X(t),\ y = Y(t),\ z = Z(t),\ a \le t \le b$$

can be expressed as an ordinary integral as follow:

$$\int_a^b \{f(X, Y, Z)X' + g(X, Y, Z)Y' + h(X, Y, Z)Z'\}\,dt$$

If the curve is *smooth*, it can be represented as

$$x = \theta(s),\ y = \phi(s),\ z = \psi(s)$$

where the arc length s varies from 0 to l (l being the length of the curve as t varies from a to b).

Considering the partitions $\{0 = s_0, s_1, ..., s_n = l\}$ and proceeding (with parameter s in place of t) we deduce that the line integral $\int_C f\,dx + g\,dy + h\,dz$ reduces to

$$\int_0^l \left\{ f(\theta, \phi, \psi)\frac{dy}{ds} + g(\theta, \phi, \psi)\frac{dy}{ds} + h(\theta, \phi, \psi)\frac{dz}{ds} \right\} ds$$

which is equivalent to the ordinary integral, or

$$\int_a^b \left(f\frac{dx}{ds} + g\frac{dy}{ds} + h\frac{dz}{ds} + \right) \frac{ds}{dt}\,dt$$

where $\dfrac{ds}{dt} = \sqrt{\left(\dfrac{dX}{dt}\right)^2 + \left(\dfrac{dY}{dt}\right)^2 + \left(\dfrac{dZ}{dt}\right)^2}$.

Example 2:

Prove that

$$\int \frac{x^2 + y^2}{p}\,ds = \frac{\pi ab}{4}\left[4 + (a^2 + b^2)(a^{-2} + b^{-2})\right].$$

Solution:

when the integral is taken round the ellipse $\dfrac{x^2}{a^2} + \dfrac{y^2}{b^2} = 1$, and p is the length of the perpendicular from the centre to the tangent.

The parametric equation of the ellipse is

$$x = a\cos\phi,\quad y = b\sin\phi$$

$$p = \frac{y - x(dy/dx)}{\sqrt{1 + (dy/dx)^2}} = \frac{ab}{\sqrt{a^2\sin^2\phi + b^2\cos^2\phi}}$$

$$\frac{ds}{d\phi} = \sqrt{a^2\sin^2\phi + b^2\cos^2\phi}$$

$$\therefore \quad \int_C \frac{x^2 + y^2}{p}\,ds$$

$$= \int_0^{2\pi} \frac{x^2 + y^2}{p} \frac{ds}{d\phi} d\phi$$

$$= \frac{1}{ab} \int_0^{2\pi} (a^2 \cos^2 \phi + b^2 \sin \phi)(a^2 \sin^2 \phi + b^2 \cos \phi)\, d\phi$$

$$= \frac{4}{ab} \int_0^{\pi/2} [(a^4 + b^4) \cos^2 \phi \sin^2 \phi + a^2 b^2 (\cos^4 \phi + \sin^4 \phi)]\, d\phi$$

$$= \frac{\pi}{4ab} [a^4 + b^4 + 6a^2 b^2) = \frac{\pi ab}{4}[(a^2 + b^2)(a^{-2} + b^{-2}) + 4].$$

Example 3:

Show that

$$\int_C (y^2 + z^2)\, dx + (z^2 + x^2)\, dy + (x^2 + y^2)\, dz = - \ 2\pi ab^2$$

where the curve C is the part for which $z \geq 0$ of the intersection of the surfaces

$$x^2 + y^2 + z^2 = 2ax,\ x^2 + y^2 = 2bx,\ a > b > 0$$

The curve begins at the origin and runs at first in the positive octant.

Solution:

Curve C is the intersection of the two surfaces,

$$(x - b)^2 + y^2 = b^2$$

$$z^2 = 2(a - b)x$$

So the parametric equation of C is

$$x = b(1 + \cos\theta) = 2b \cos^2 \frac{\theta}{2}$$

$$y = b \sin \theta = 2b \sin \frac{\theta}{2} \cos \frac{\theta}{2}$$

$$z = \sqrt{2(a-b)\, b(1+\cos\theta)}$$

$$= 2\sqrt{b(a-b)} \cos \frac{\theta}{2}$$

and θ varies from π to $-\pi$.

$$\therefore \quad y^2 + z^2 = -\ 4b^2 \cos^4 \frac{\theta}{2} + 4ab \cos^2 \frac{\theta}{2}$$

$$z^2 + x^2 = 4ab \cos^2 \frac{\theta}{2} - 4b^2 \cos^2 \frac{\theta}{2} \sin^2 \frac{\theta}{2}$$

$$x^2 + y^2 = 4b^2 \cos^2 \frac{\theta}{2}$$

$$\int_C (y^2 + z^2)\, dx + (z^2 + x^2)\, dy + (x^2 + y^2)\, dz$$

$$= + 4b^2 \int_{-\pi}^{\pi} \left(b \cos^4 \frac{\theta}{2} - a \cos^2 \frac{\theta}{2}\right) \sin\theta \, d\theta$$

$$- 4b^2 \int_{-\pi}^{\pi} \left(a \cos^2 \frac{\theta}{2} - b \cos^2 \frac{\theta}{2} \sin^2 \frac{\theta}{2}\right) \cos\theta \, d\theta$$

$$+ 4b^2 \sqrt{b(a-b)} \int_{-\pi}^{\pi} \cos^2 \frac{\theta}{2} \sin \frac{\theta}{2} \, d\theta$$

As the first and the third integral on the right vanish,

$$= - 8b^2 \int_0^{\pi} \left(a \cos^2 \frac{\theta}{2} - b \cos^2 \frac{\theta}{2} \sin^2 \frac{\theta}{2}\right) \cos\theta \, d\theta$$

$$= - 8b^2 \int_0^{\pi} a \cos^2 \frac{\theta}{2}\left(2 \cos^2 \frac{\theta}{2} - 1\right) d\theta$$

$$= - 16b^2 \int_0^{\pi/2} a \cos^2 (2 \cos^2 t - 1)\, dt, \; t = \frac{\theta}{2}.$$

$$= - 2\pi ab^2.$$

Example 4:

Show that $\int_C (y^2\, dx + z^2\, dy + x^2\, dz) = -\frac{\pi a^2}{4}$,
where C is the curve of intersection of the sphere $x^2 + y^2 + z^2 = a^2$ *and the cylinder* $x^2 + y^2 = ax$ $(a < 0, z < 0)$, *integrated anticlockwise when viewed from the origin.*

Solution:

$$x^2 + y^2 = ax \Rightarrow (x - \frac{1}{2}a)^2 + y^2 = (\frac{1}{2}a)^2.$$

Thus the parametric equations of C are given by

$$x = ½a\,(1 + \cos\theta),\; y = \frac{1}{2} a \sin\theta$$

$z^2 = a^2 - (x^2 + y^2) = a^2 - ax = \frac{1}{2} a^2 (1 - \cos\theta)$. Given integral is

$$I = \frac{a^2}{8} \int_{-\pi}^{\pi} \sin^2\theta\,(-\sin\theta)\, d\theta + \frac{a^2}{4} \int_{-\pi}^{\pi} (1 - \cos\theta) \cos\theta \, d\theta$$

$$+\frac{a^2}{4}\int_{-\pi}^{\pi}(1+\cos\theta)^2\frac{a\sin\theta\,d\theta}{2\sqrt{2}\sqrt{1-\cos\theta}}$$

$$= 0+\frac{a^2}{4}\left[0-4\int_0^{\pi/2}\cos^2\theta\,d\theta\right]+0$$

(integrands of the first and third integrals being odd)

$$= -a^2\cdot\frac{1}{2}\cdot\frac{\pi}{2} = -\frac{\pi a^2}{4}.$$

Example 5:

Show that, $\int_C x\sqrt{x^2-y^2}\,ds = \frac{4\sqrt{2}\,a^3}{3}$, *where C is half lemniscate* $(x^2+y^2)^2 = a^2(x^2-y^2)$.

Solution:

So yourself

Hint. Changing to polars, the lemniscate is $r^2 = a^2\cos 2\theta$.

$$\frac{ds}{d\theta} = \sqrt{[r^2+(dr/d\theta)^2]};\ x = r\cos\theta = a\cos\theta\sqrt{\cos 2\theta} \text{ etc.}$$

θ varies from $-\pi/4$ to $\pi/4$.

Example 6:

Show that

$$I = \int_C (y^2+z^2)\,dx + (z^2+x^2)\,dy + (x^2+y^2)\,dz = -2\pi\,ab,$$

where the curve C is the part for which $z \geq 0$ *of the intersection of the surfaces*

$$x^2+y^2+z^2 = 2ax,\quad x^2+y^2 = 2bx,\ a > b > 0,$$

the curve begins at the origin and runs at first in the positive contant.

Solution:

The given surface can be written as

$$(x-b)^2+y^2 = b^2 \quad \text{and} \quad z^2 = 2(a-b)x.$$

Thus the parametric equations of C are given as

$$x = b(1+\cos\theta),$$

$$y = b\sin\theta,$$

$$z = \sqrt{[2b(a-b)(1+\cos\theta)]}$$

$$x = 2b\cos^2\theta/2,$$

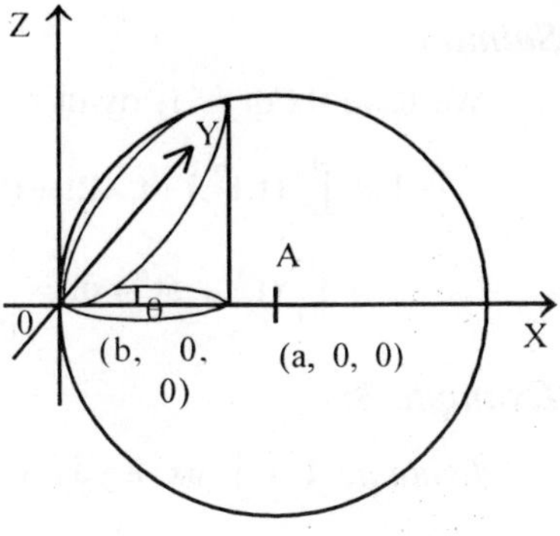

$y = 2b \sin \theta/2 \cos \theta/2,$

$z = 2\sqrt{b(a-b)} \cos \theta/2,$

$-\pi \le \theta \le \pi$

Now we have

$x^2 + y^2 = 2bx = 4b^2 \cos^2 \theta/2,$

$y^2 + z^2 = 2ax - x^2$

$= 4ab \cos^2 \theta/2 - 4b^2 \cos^2 \theta/2,$

$z^2 + x^2 = 2ax - y^2 = 4ab \cos^2 \theta/2 - 4b^2 \sin^2 \theta/2 \cos^2 \theta/2.$

$$\therefore \quad I = \int_{-\pi}^{\pi} (4ab \cos^{\pi} \theta/2 - 4b^2 \cos^2 \theta/2)(-b \sin \theta)\, d\theta$$

$$- \int_{-\pi}^{\pi} (4ab \cos^2 \theta/2 - 4b^2 \sin^2 \theta/2 \cos^2 \theta/2)\, b \cos \theta\, d\theta$$

$$- 4b^2 \sqrt{2(a-b)b} \int_{-\pi} \cos^2 \theta/2 \sin\theta\, d\theta.$$

Using $\sin \theta = 2 \sin \theta/2 \cos \theta/2$ in the first and the third integrals, their values are zero. Thus

$$I = -8b^2 \int_0^{\pi} (a \cos^2 \theta/2 - b \sin^2 \theta/2 \cos^2 \theta/2) \cos \theta\, d\theta$$

$$= -8b^2 \int_0^{\pi} [\tfrac{1}{2} a(1+\cos\theta) - \tfrac{1}{4} b \sin^2 \theta] \cos \theta\, d\theta$$

$$= -8b^2 \left[0 + a \int_0^{\pi/2} \cos^2 \theta\, d\theta - 0 \right]$$

$$= -8ab^2 . \frac{1}{2} . \frac{1}{2} = -2\pi ab^2.$$ **Ans.**

Example 7:

Evaluate, $I = \int_C (xy\, dx + yz dy + zx\, dz),$

where the curve C is given by

$x = t,\ y = t^2,\ z = t^3$; *t varying from –1 to +1.*

Solution

We have dx/dt = 1, dy/dt = 2t, dz/dt = $3t^2$.

$$I = \int_{-1}^{1} (t.t^2.1+t^3.2t+t^3.t.3t^2)\,dt$$

$$= \int_{-1}^{1} (t^3 + 5t^6)\,dt = \left|\frac{1}{4}t^4 + \frac{5}{7}t^7\right|_{-1}^{1} = \frac{10}{7}.$$ **Ans.**

Example 8:

Evaluate $I = \int x\,dx + y\,dy + (x + y - 1)\,dz$

along the line segment joining the points (1, 1, 1) and (2, 3, 4).

Solution:

The equations of the line segment joining the points (1, 1, 1) and (2, 3, 4) are

$$\frac{x-1}{2-1} = \frac{y-1}{3-1} = \frac{z-1}{4-1}\ (=t, \text{say})$$

$\therefore$ x = t + 1, y = 2t + 1, z = 3t+ 1

$\Rightarrow$ dx/dt = 1, dy/dt = 2, dx/dt = 3.

The given points correspond to t = 0 and t = 1 respectively.

$$\therefore\quad I = \int_0^1 [(t+1).1 + (2t + 1).2 + (t + 1 + 2t + 1 - 1).3]\,dt$$

$$= \int_0^1 (14t+6)\,dt = 7 + 6 = 13.$$ **Ans.**

Example 9:

Show that $\int_C (x+y)\,ds = \sqrt{2}a^2$, *where C is the quarter circle* $x^2 + y^2 + z^2 = a^2$, *y = x lying in the first octant.*

Solution:

The parametric equations of $x^2 + y^2 + z^2 = a^2$ are x = a sin θ cos ϕ, y = a sin θ sin ϕ, z = a cos θ.

Now y = x $\Rightarrow$ tan ϕ = 1 $\Rightarrow$ ϕ = π/4.

$$\therefore\quad x = \frac{a}{\sqrt{2}}\sin\theta,\ y = \frac{a}{\sqrt{2}}\sin\theta,\ z = a\cos\theta$$

$$\Rightarrow\quad \frac{dx}{d\theta} = \frac{a}{\sqrt{2}}\cos\theta = \frac{dy}{d\theta},\ \frac{dz}{d\theta} = -a\sin\theta.$$

$$\therefore \quad \left(\frac{ds}{d\theta}\right)^2 = \left(\frac{dx}{d\theta}\right)^2 + \left(\frac{dy}{d\theta}\right)^2 + \left(\frac{dz}{d\theta}\right)^2 = a^2.$$

Thus $\int_C (x + y)\, ds = \frac{a}{\sqrt{2}} \int_0^{\pi/2} (\sin\theta + \sin\theta).\, a\, d\theta$

$$= \frac{2a^2}{\sqrt{2}} \left|-\cos\theta\right|_0^{\pi/2} = \sqrt{2a^2}.$$ **Ans.**

Example 10:

Show that

$$I = \int_C (yz\, dx + zx\, dy + xy\, dz) = 0,$$

where C is the arc of the curve $x = b \cos t$, $y = b \sin t$, $z = at/2\pi$ from the point it intersects $z = 0$ to the point it intersects $z = a$.

Solution:

We have

$$dx/dt = -b \sin t,\ dy/dt = b \cos t,\ dz/dt = a/2\pi.$$

Now $z = 0 \quad \Rightarrow \quad z = 0$ and $z = a \quad \Rightarrow \quad t = 2\pi$. Thus

$$I = \int_0^{2\pi} \left(yz \frac{dx}{dt} + zx \frac{dy}{dt} + xy \frac{dz}{dt}\right) dt$$

$$= \frac{ab^2}{2\pi} \int_0^{2\pi} [(t \sin t)(-\sin t) + (t \cos t)(\cos t) + (\cos t \sin t)]\, dt$$

$$= \frac{ab^2}{2\pi} \int_0^{2\pi} t(\cos^2 t - \sin^2 t)\, dt + \frac{ab^2}{2\pi} \left|\frac{\sin^2 t}{2}\right|_0^{2\pi}$$

$$= \frac{ab^2}{2\pi} \int_0^{2\pi} t \cos 2t + dt + 0$$

$$= \frac{ab^2}{2\pi} \left[\left| t\left(-\frac{\sin 2t}{2}\right)\right|_0^{2\pi} + \frac{1}{2} \int_0^{2\pi} \sin 2t\, dt\right]$$

$$= \frac{ab^2}{2\pi} \left[0 - \frac{1}{4} \left|\cos 2t\right|_0^{2\pi}\right] = 0.$$ Hence the result.

Example 11:

Evaluate $I = \int_C (x^2y^3\, dx + dy + z dz),$

where C is the circle $x^2 + y^2 = R^2$, $z = 0$.

Solution:

The parametric equations of the given curve C are $x = R\cos t$, $y = R\sin t$ and $z = 0$.

$\Rightarrow$ $dx/dt = -R\sin t$, $dy/dt = R\cos t$, $dz/dt = 0$.

$$\therefore \quad I = \int_0^{2\pi} \left[R^5 \cos^2 t \sin^3 t\,(-R\sin t) + R\cos t + 0\right] dt$$

$$= -R^6 \int_0^{2\pi} \sin^4 t \cos^2 t\, dt + R\int_0^{2\pi} \cos t\, dt$$

$$= -4R^6 \int_0^{2\pi} \sin^4 t \cos^2 t\, dt + 0$$

$$= -4R^6 \cdot \frac{3.1}{6.4.2} \cdot \frac{\pi}{2} = -\frac{\pi R^6}{8}.$$

Ans.

Example 12:

Evaluate $\int (x-y)^{-1}\, ds$ *along the line* $2y = x - 4$ *between the points (0, –2) and (4, 0).*

Solution:

We have $dy/dx = 1/2$ and

$$\frac{ds}{dx} = \sqrt{\left[1 + \left(\frac{dy}{dx}\right)^2\right]} = \frac{\sqrt{5}}{2}.$$

$$\therefore \quad \int \frac{dx}{x-y} = \sqrt{5}\int_0^4 \frac{dx}{4+x} \qquad (\because y = \tfrac{1}{2}x - 2)$$

$$= \sqrt{5}\left|\log(4+x)\right|_0^4 = \sqrt{5}\,(\log 8 - \log 4)$$

$$= \sqrt{5}\log 2.$$

Ans.

Example 13:

Prove that

$$\int \frac{x^2 + y^2}{p}\, ds = \frac{\pi ab}{4}\left[a + (a^2 + b^2)(a^{-2} + b^{-2})\right],$$

when the integral is taken round the ellipse $x^2/a^2 + y^2/b^2 = 1$ *and p is length of the perpendicular from the centre to the tangent.*

(D.U., 1993)

Solution:

The parametric equations of the ellipse are

$$x = a\cos\phi,\ y = b\sin\phi. \qquad ...(1)$$

The length 'p' of the perpendicular from the centre to any tangent on the ellipse is given by

$$p = \frac{y - x(dy/dx)}{\sqrt{[1+(dy/dx)^2]}}. \qquad ...(2)$$

From (1), $dx/d\phi = -a\sin\phi$, $dy/d\phi = b\cos\phi$.

and $\dfrac{dy}{dx} = -\dfrac{b}{a}\cot\phi$.

$$\therefore \quad p = \frac{b\sin\phi + b\cos^2\phi/\sin\phi}{\sqrt{[1+(b^2/a^2)\cot^2\phi]}} = \frac{ab}{\sqrt{[a^2\sin^2\phi + b^2\cos^2\phi}}.$$

Now, $\dfrac{ds}{d\phi} = \sqrt{\left[\left(\dfrac{dx}{d\phi}\right)^2 + \left(\dfrac{dy}{d\phi}\right)^2\right]} = \sqrt{[a^2\sin^2\phi + b^2\cos^2\phi]}$

$$\int \frac{x^2+y^2}{p}\,ds = \int \frac{x^2+y^2}{p}\frac{ds}{d\phi}\,d\phi$$

$$= \frac{1}{ab}\int_0^{2\pi} (a^2\cos^2\phi + b^2\sin^2\phi)(a^2\sin^2\phi + b^2\cos^2\phi)\,d\phi$$

$$= \frac{4}{ab}\int_0^{2\pi} [(a^4+b^4)\cos^2\phi\sin^2\phi + a^2b^2(\cos^4\phi + \sin^4\phi)]\,d\phi$$

$$= \frac{4}{ab}\left[(a^4+b^4).\frac{1}{4.2}.\frac{\pi}{2} + 2a^2b^2\left(\frac{3}{4}.\frac{1}{2}.\frac{\pi}{2}\right)\right]$$

$$= \frac{\pi}{4ab}(a^4+b^4+6a^2b^2) = \frac{\pi ab}{4}\left[\frac{a^4+b^4}{a^2b^2} + 6\right]$$

$$= \frac{\pi ab}{4}[(a^2+b^2)(a^{-2}+b^{-2}) + 4].$$ **Ans.**

Example 14:

(i) Evaluate $\int x\,ds$, along the ellipse $x^2/a^2 + y^2/b^2 = 1$ lying in the first quadrant.

(ii) Evaluate $\int y\,ds$, along the arc of the parabola $y^2 = 2px$ cut off by $x^2 = 2py$.

Solution:

(i) By previous example we have

$x = a\cos\phi.\ y = b\sin\phi.\ ds/d\phi = \sqrt{[a^2\sin^2\phi + b^2\cos^2\phi]}.$

$$\therefore \quad \int xy\,ds = \int_0^{\pi/2} ab\sin\phi\cos\phi\sqrt{[a^2\sin^2\phi + b^2\cos^2\phi]}\,d\phi.$$

$$= \frac{ab}{2(a^2-b^2)}\int_{b^2}^{a^2}\sqrt{t}\,dt,\ a^2\sin^2\phi + b^2\cos^2\phi = t$$

$$= \frac{ab}{3(a^2-b^2)}(a^2-b^2) = \frac{ab(a^2+ab+b^2)}{3(a+b)}.$$

(ii) The two parabolas intersect at the points given by

$$(x^2/2p)^2 = 2px \quad\Rightarrow\quad x(x^3 - 8p^2) = 0 \quad\Rightarrow\quad x = 0,\ x = 2p.$$

Now $y^2 = 2px \quad\Rightarrow\quad y = \sqrt{2px} \quad\Rightarrow\quad \dfrac{dy}{dx} = \sqrt{\dfrac{p}{2x}}.$

$$\therefore \quad \frac{ds}{dx} = \sqrt{1+\left(\frac{dy}{dx}\right)^2} = \sqrt{\frac{2x+p}{2x}}.$$

$$\therefore \quad \int y\,ds = \int_{x=0}^{2p}\sqrt{2px}\sqrt{\frac{2x+p}{2x}}\,dx$$

$$= \sqrt{p}\,\frac{1}{3}\left|(2x+p)^{3/2}\right|_0^{2p} = \frac{1}{3}\sqrt{p}[(5p)^{3/2} - p^{3/2}]$$

$$= \frac{1}{3}p^2\,(5\sqrt{5-1}).$$

Ans.

Example 15:

Evaluate

$$\int\frac{xdx + ydy + zdz}{\sqrt{(x^2+y^2+z^2-x-y+2z)}}$$

along the straight line joint the points (1, 1, 1) and (4, 4, 4).

Solution:

The equations of the line joining (1, 1, 1) and (4, 4, 4) are

$$\frac{x-1}{4-1} = \frac{y-1}{4-1} = \frac{z-1}{4-1} = (t,\ \text{say})$$

$$\therefore \quad x = 3t+1,\ y = 3t+1,\ z = 3t+1$$

and $dx/dt = dy/dt = dz/dt = 3$.

Now $(x^2 + y^2 - x - y + 2z) = 3(3t + 1)^2$.

$$\therefore \quad I = \int_0^1 \frac{9(3t + 1)\,dt}{\sqrt{3(3t+1}} = 3\sqrt{3}. \qquad \textbf{Ans.}$$

Example 16:

Show that

$$\int_C xyz\,ds = (\sqrt{3/32})\,R^4.$$

where C is the quarter of the circle

$x^2 + y^2 + z^2 = R^2$, $x^2 + y^2 = R^2/4$ *lying in the first octant.*

Solution:

The parametric equations of the curve C are

$$x = \tfrac{1}{2} R \cos t,\ y = \tfrac{1}{2} R \sin t,\ z = \tfrac{1}{2}\sqrt{3}\,R.$$

$$\text{Now} \quad \frac{ds}{dt} = \sqrt{\left[\left(\frac{dx}{dt}\right)^2 + \left(\frac{dy}{dt}\right)^2 + \left(\frac{dz}{dt}\right)^2\right]} = \frac{1}{2}R.$$

$$\therefore \quad \int xyz\,ds = \int_0^{\pi/2}\left(\frac{1}{2}R\cos t\right)\left(\frac{1}{2}R\sin t\right).\frac{\sqrt{3}}{2}R \times \frac{1}{2}R\,dt$$

$$= \frac{\sqrt{3R^4}}{16}\int_0^{\pi/2} \sin t \cos t\,dt = \frac{\sqrt{3R^4}}{32}. \qquad \textbf{Ans.}$$

4.7 Surface

Definition: A surface in R^3 is a vector-valued function with domain a subset of R^2 and range a subset of R^3.

If X, Y, Z are three real-valued functions defined on a domain $E \subset R^2$, where

$$x = X(u, v),\ y = Y(u, v),\ z = Z(u, v),\ (u, v) \in E$$

then the set

$$\{(X(u, v), Y(u, v), Z(u, v) : (u, v) \in E\} \qquad ...(1)$$

is a surface in R^3.

A surface of the form

$$z = \psi(x, y) \quad x \in R, y \in R \qquad ...(2)$$

which is met in not more than one point by any line parallel in the axis of z, is known as quadratic regular with respect to z-axis. Such a surface is projectible in a one-to-one manner on the xy-plane.

$x = \theta(y, z)$ and $y = \phi(z, x)$ are surfaces quadratic (regular) with respect to x-axis and y-axis, respectively.

A surface which can be divided by a finite number of smooth curves into a finite number of portions each of which is quadratic (regular) with respect to the axes, is called a *piecewise quadratic surface.*

A surface is said to be smooth if X, Y, Z possess continuous first order partial derivatives at each point of E and

$$\frac{\partial(y, z)}{\partial(u, v)}, \frac{\partial(z, x)}{\partial(u, v)}, \frac{\partial(x, y)}{\partial(u, v)}$$

do not all vanish simultaneously at any point.

4.8 The Surface Area

Let it be required to compute the area S of a surface bounded by a curve C. The surface is defined by the eqn.

$$z = \psi(x, y)$$

where the function ψ is continuous and possesses continuous partial derivatives. Denote the projection of C on the xy-plane bounded by Γ, and let D be the domain on the xy-plane bounded by Γ. Let σ denote the area of D.

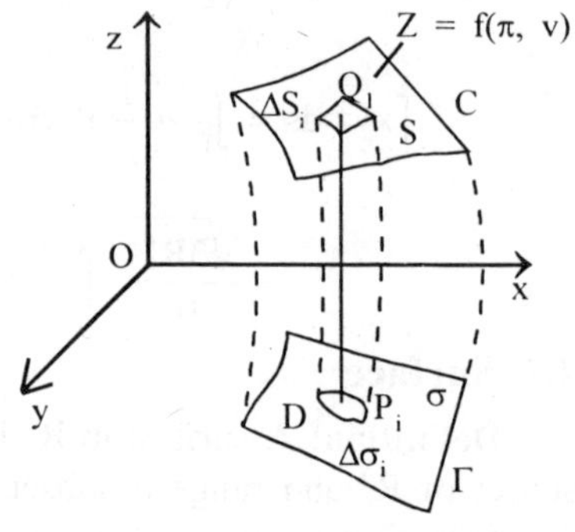

Now let an arbitrary partition P of D give rise to sub-domains of areas $\Delta\sigma_i$, $\Delta\sigma_2$, ..., $\Delta\sigma_n$. In each sub-domain take a point $P_1(\xi_1, \eta_i)$. To the point P_i, there will correspond, on the surface, a point $Q_i(\xi_i, \eta_i)\ \psi(\xi_i, \eta_i))$. Through Q_i draw a tangent plane to the surface, and on this plane pick out a sub-domain ΔS_i which projects onto $\Delta\sigma_i$ on the xy-plane.

Consider the sum $\sum_{i=1}^{n} \Delta S_i$. Now let us

If the limit of this sum exists, when the norm $\mu(P)$ of the partition (the greatest of the diameters of the sub-domains) approaches zero, the surface is said to be *Squarable* and its area S is given by the equation.

$$S = \lim_{\mu(P)\to 0} \sum_{i=1}^{n} \Delta S_i .$$

Now, we have to calculate the values of S.

If γ' denote the angle between the tangent plane (at Q_i) and the xy-plane, we have from analytical geometry that

$$\Delta\sigma_i = \Delta S_i \cos \gamma_1$$

and

$$\cos \gamma' = \frac{1}{\sqrt{1 + \psi_x^2(\xi_i, \eta_i) + \psi_y^2(\xi_i, \eta_i)}}$$

$$\therefore \quad \Delta S_i = \sqrt{1 + \psi_x^2(\xi_i, \eta_i) + \psi_y^2(\xi_i, \eta_i)}\ \Delta\sigma_i$$

We know that

$$\therefore \quad S = \lim_{\mu(P)\to 0} \sum_i \Delta S_i$$

$$= \lim_{\mu(P)\to 0} \sum_i \sqrt{1 + \psi_x^2(\xi_i, \eta_i) + \psi_y^2(\xi_i, \eta_i)}\ \Delta\sigma_i$$

$$= \iint_D \sqrt{1 + \left(\frac{\partial z}{\partial x}\right)^2 + \left(\frac{\partial z}{\partial y}\right)^2}\ dx\, dy \qquad \text{...(1)}$$

is a formula to compute the area of the surface $z = \psi(x, y)$.

If the equation of the surface is given in the form given below:

$$x = \theta(y, z), \quad \text{or} \quad y = \phi(z, x)$$

then the corresponding formulas for calculating the surface area are of the form

$$S = \iint_{D_1} \sqrt{1 + \left(\frac{\partial x}{\partial y}\right)^2 + \left(\frac{\partial x}{\partial z}\right)^2}\ dy\, dz \qquad \text{...(2)}$$

$$S = \iint_{D_2} \sqrt{1 + \left(\frac{\partial y}{\partial z}\right)^2 + \left(\frac{\partial y}{\partial x}\right)^2}\ dz\, dx \qquad \text{...(3)}$$

where D_1, D_2 are the domains in the yz-plane and zx-plane in which the given surface is projected.

To find the area of a smooth surface represented parametrically as

$$x = X(u, v), \quad y = Y(u, v), \quad z = Z(u, v), \quad (u, v) \in E$$

we have only to change the variables in the above formula, where the equation of surface is given by $z = \phi\ (x, y)$.

Now, put $x = X(u, v)$, $y = Y(u, v)$ so that

$$z = \psi[X(u, v), Y(u, v)] = Z(u, v)$$

Also we know that

$$\frac{\partial z}{\partial x} = -\frac{\partial(Y, Z)}{\partial(u,v)} / \frac{\partial(X, Y)}{\partial(u,v)}; \frac{\partial z}{\partial y} = -\frac{\partial(Z, X)}{\partial(u,v)} / \frac{\partial(X, Y)}{\partial(u,v)}$$

and then

$$\sqrt{1+\left(\frac{\partial z}{\partial x}\right)^2 + \left(\frac{\partial z}{\partial y}\right)^2} = \sqrt{\left[\frac{\partial(X, Y)}{\partial(u,v)}\right]^2 + \left[\frac{\partial(Y, Z)}{\partial(u,v)}\right]^2 + \left[\frac{\partial(Z, X)}{\partial(u,v)}\right]^2} / \frac{\partial(X, Y)}{\partial(u,v)}$$

and the Jacobian of transformation $J = \frac{\partial(X, Y)}{\partial(u,v)}$.

Thus formula (I) yields

$$S = \iint_D \sqrt{\left[\frac{\partial(X, Y)}{\partial(u,v)}\right]^2 + \left[\frac{\partial(Y, Z)}{\partial(u,v)}\right]^2 + \left[\frac{\partial(Z, X)}{\partial(u,v)}\right]^2}\, du\, dv \quad ...(4)$$

which is the required formula to calculated the surface area.

Again

$$\left.\begin{aligned} &\left[\frac{\partial(X, Y)}{\partial(u,v)}\right]^2 + \left[\frac{\partial(Y, Z)}{\partial(u,v)}\right]^2 + \left[\frac{\partial(Z, X)}{\partial(u,v)}\right]^2 = AC - B^2 \\ &\text{where } \left(\frac{\partial X}{\partial u}\right)^2 + \left(\frac{\partial Y}{\partial u}\right)^2 + \left(\frac{\partial Z}{\partial u}\right)^2 = A \\ &\frac{\partial X}{\partial u}\frac{\partial X}{\partial v} + \frac{\partial Y}{\partial u}\frac{\partial Y}{\partial v} + \frac{\partial Z}{\partial u}\frac{\partial Z}{\partial v} = B \\ &\left(\frac{\partial X}{\partial v}\right)^2 + \left(\frac{\partial Y}{\partial v}\right)^2 + \left(\frac{\partial Z}{\partial v}\right)^2 = C \end{aligned}\right\} \quad ...(5)$$

Thus (4) may b expressed as

$$S = \iint_D \sqrt{AC - B^2}\, du\, dv \quad ...(6)$$

Remarks:

The following results of differential geometry are very useful.

1. (a) For the surface x = X(u, v), y = Y(u, v), z = Z(u, v) the direction consines of the normal to the surface at any point (x, y, z) are

$$\frac{\partial(y, z)}{\partial(u, v)} / \sqrt{K},\ \frac{\partial(z, x)}{\partial(u, v)} / \sqrt{K},\ \frac{\partial(x, y)}{\partial(u, v)} / \sqrt{K},$$

where $K = \left[\frac{\partial(y, z)}{\partial(u, v)}\right]^2 + \left[\frac{\partial(z, x)}{\partial(u, v)}\right]^2 + \left[\frac{\partial(x, y)}{\partial(u, v)}\right]^2$

$= AC - B^2$

(b) The elementary surface area dS is given by

$DS = \sqrt{K}\ du\ dv$

2. (a) For the surface $z = \psi(x, y)$, the direction cosines of the normal at any point (x, y, z) of the surface area are

$$-\frac{\partial z}{\partial x}/\sqrt{K},\quad -\frac{\partial z}{\partial y}/\sqrt{K},\quad 1/\sqrt{K}$$

where $K = 1 + \left(\frac{\partial z}{\partial x}\right)^2 + \left(\frac{\partial z}{\partial y}\right)^2$

(b) The elementary surface area dS is given by

$$dS = \sqrt{K}\ dx\ dy = \sqrt{1 + \left(\frac{\partial z}{\partial x}\right)^2 + \left(\frac{\partial z}{\partial y}\right)^2}\ dx\ dy.$$

Example 17:

Compute the surface area S of the sphere

$x^2 + y^2 + z^2 = a^2.$

Solution:

The surface area of the sphere is twice the surface area of the upper half-sphere, $z = \sqrt{a^2 - x^2 - y^2}$.

Now we have

$$\frac{\partial z}{\partial x} = -\frac{x}{\sqrt{a^2 - x^2 - y^2}},\quad \frac{\partial z}{\partial y} = -\frac{y}{\sqrt{a^2 - x^2 - y^2}}$$

so that

$$\sqrt{1 + \left(\frac{\partial z}{\partial x}\right)^2 + \left(\frac{\partial z}{\partial y}\right)^2} = \frac{a}{\sqrt{a^2 - x^2 - y^2}}$$

The domain of integration is the circle $x^2 + y^2 = a^2$ on the xy-plane. Thus by formula (1), we have the following expression

$$\frac{1}{2}S = \int_{-a}^{a} dx \int_{-\sqrt{(a^2 - x^2)}}^{\sqrt{a^2 - x^2}} \frac{a}{\sqrt{a^2 - x^2 - y^2}}\, dy$$

On passing to polars, we have

$$S = 2\int_0^{2\pi}\left[\int_0^a \frac{a}{\sqrt{a^2 - r^2}}\, r\, dr\right] d\theta = 4\pi a^2$$

Ans.

Example 18:

Find the area of the part of the surface of the cylinder $x^2 + y^2 = a^2$ which is cut out by the cylinder $x^2 + z^2 = a^2$.

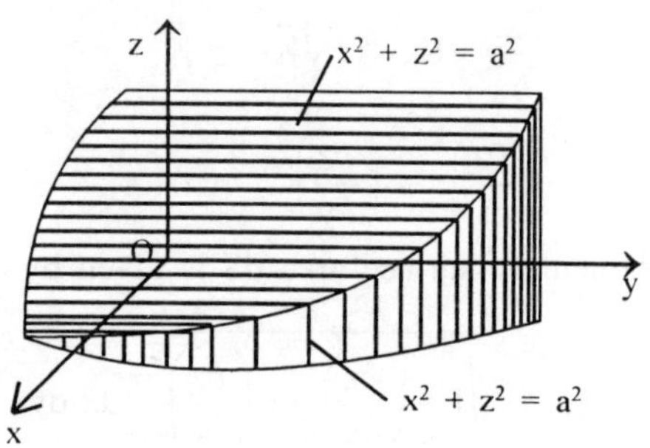

Solution:

The figure show ⅛th of the desired surface.

The equation of the surface has the form

$$y = \sqrt{a^2 - x^2}$$

so we have

$$\frac{\partial y}{\partial x} = -\frac{x}{\sqrt{a^2 - x^2}},\quad \frac{\partial y}{\partial z} = 0$$

$$\sqrt{1 + \left(\frac{\partial y}{\partial x}\right)^2 + \left(\frac{\partial y}{\partial z}\right)^2} = \frac{a}{\sqrt{a^2 - x^2}}$$

The domain of integration is a quarter circle $x^2 + z^2 = a^2$, $x \geq 0$, $z \geq 0$, on the xz-plane. Thus by formula (3), we have the following result.

$$\frac{1}{8}S = \int_0^a dx \int_0^{\sqrt{(a^2 - x^2)}} \frac{a}{\sqrt{a^2 - x^2}}\, dz = a^2$$

$\therefore \quad S = 8a^2.$

Ans.

Example 19:

The x and y coordinates of a point on the paraboloid $2z = x^2/a + y^2/b$ are expressed in the form

$$x = a \tan \theta \cos \phi, \; y = b \tan \theta \sin \phi$$

where θ is the angle of inclination of the normal at any point to the axis of z. Show that the area of the cap of the surface cut off by the curve $\theta = \lambda$ is $2\pi ab(\sec^3 \lambda - 1)/3$.

Solution:

Here we have

$$1 + \left(\frac{\partial z}{\partial x}\right)^2 + \left(\frac{\partial z}{\partial y}\right)^2 = 1 + \frac{x^2}{a^2} + \frac{y^2}{b^2}$$

so that by formula (1), surface areas S is given by the formula

$$S = \iint_D \sqrt{1 + \frac{x^2}{a^2} + \frac{y^2}{b^2}}\, dx\, dy$$

Let us put $x = a \tan \theta \cos \phi$, $y = b \tan \theta \sin \phi$.

$$\therefore \quad \sqrt{1 + \frac{x^2}{a^2} + \frac{y^2}{b^2}} = \sec\theta$$

and we know that

$$J = \frac{\partial(x, y)}{\partial(\theta, \phi)} = ab \sec^2 \theta \tan \theta$$

$$\therefore \quad S = \int_0^{2\pi} d\phi \int_0^{\lambda} \sec \theta \; ab \sec^2 \theta \tan \theta \, d\theta$$

$$= 2\pi \int_0^{\lambda} ab \sec^3 \theta \tan \theta \, d\theta = \frac{2\pi ab}{3}(\sec^3 \lambda - 1).$$ **Ans.**

Example 20:

A surface is given by the equations

$$x = c \sin u, \quad y = c \cos v, \quad z = c(\cos u + \cos v)$$

prove that its area bounded by

$$u = 0, \; u = \pi/2; \; v = 0, \; v = \pi/2$$

is

$$\frac{\pi c^2}{2}\left[1 - \sum_{n=1}^{\infty}(-1)^n \frac{p_2^2 n}{2n-1}\right]$$

$$\text{where} \quad p_{2n} = \frac{(2n-1)(2n-3)\ldots.3.1}{2n(2n-2)\ldots 4.2}.$$

Solution:

Here we have

$$A = c^2 \cos^2 u + c^2 \sin^2 u = c^2$$

$$B = c^2 \sin u \sin v$$

$$C = c^2 \sin^2 v + c^2 \sin^2 v = 2 c^2 \sin^2 v$$

$$\therefore \quad S = \iint_D c^2 \sqrt{2 \sin^2 v - \sin^2 u \sin^2 v}\, du\, dv$$

$$= c^2 \int_0^{\pi/2} \sin v\, dv \int_0^{\pi/2} \sqrt{1 + \cos^2 u}\, du$$

$$= c^2 \int_0^{\pi/2} (1+\cos^2 u)^{1/2}\, du$$

$$= c^2 \int_0^{\pi/2} \left[1 + \tfrac{1}{2} \cos^2 u + \frac{\frac{1}{2}\left(\frac{1}{2}-1\right)}{2!} (\cos^2 u)^2 + \ldots. \right.$$

$$\left. + \frac{\frac{1}{2}\left(\frac{1}{2}-1\right)\ldots\left(\frac{1}{2}-n+1\right)}{n!} (\cos^2 u)^n + \ldots..\right] du$$

$$= c^2 \int_0^{\pi/2} \left[1 + \sum_{n=1}^{\infty} \frac{(-1)^{n-1}}{2n-1} \frac{(2n-1)(2n-3)\ldots..3.1}{2n(2n-2)\ldots 4.2} \cos^{2n} u\right] du$$

$$= \frac{\pi}{2} c^2 \left[1 - \sum_{n=1}^{\infty} \frac{(-1)^n p_{2n}^2}{2n-1}\right]. \qquad \textbf{Ans.}$$

Example 21:

Prove that the surface of the cylinder $x^2 + y^2 = y$, contained between $z = 0$ and the surface $z = y^2/(x^2 + y^2)$ is $\pi/2$.

Solution:

From $x^2 = y - y^2$ or $x = \sqrt{y-y^2}$, we get

$$\frac{\partial x}{\partial y} = \frac{1-2y}{2\sqrt{y-y^2}}, \quad \frac{\partial x}{\partial z} = 0. \text{ Thus,}$$

$$1+\left(\frac{\partial x}{\partial y}\right)^2+\left(\frac{\partial x}{\partial z}\right)^2=1+\frac{(1-2y)^2}{4(y-y^2)}=\frac{1}{4y(1-y)}$$

Since half of the given surface lies above the plane x = 0 and half of it below x = 0, so the surface area is given by

$$S=2\int\int_0^1\frac{dy\,dz}{2\sqrt{y(1-y)}}=\int_0^1\frac{dy}{\sqrt{y(1-y)}}\int_0^1 dz,$$

where $z_1 = y^2/(x^2 + y^2) = y^2/y = y$.

$$\therefore\quad S=\int_0^1\frac{y\,dy}{\sqrt{(1-y)}}=2\int_0^{\pi/2}\sin^2\theta\,d\theta,\; y=\sin^2\theta$$

Hence $S = 2.\frac{1}{2}.\frac{\pi}{2}=\frac{\pi}{2}.$ **Ans.**

Example 22:

Find the area of the part of the surface of the paraboloid $y^2 + z^2 = 4ax$ enclosed by the cylinder $y^2 = ax$ and the plane $x = 3a$. **(D.U., 1985)**

Solution:

We have

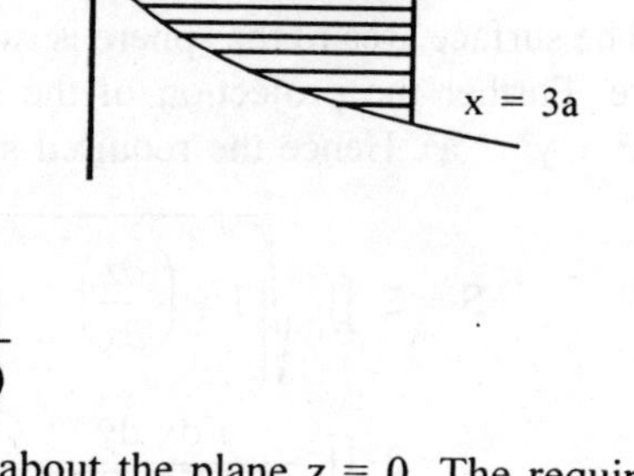

$$z=\sqrt{4ax-y^2}.$$

$$\therefore\quad \frac{\partial z}{\partial x}=\frac{2a}{z},$$

$$\frac{\partial z}{\partial y}=-\frac{y}{z}\text{ and so}$$

$$1+\left(\frac{\partial z}{\partial x}\right)^2+\left(\frac{\partial z}{\partial y}\right)^2$$

$$=\frac{z^2+4a^2+y^2}{z^2}=\frac{4a(a+x)}{(4ax-y^2)}$$

The given surface is symmetrical about the plane z = 0. The required surface area is

$$S=2\iint_E\sqrt{1+\left(\frac{\partial z}{\partial x}\right)^2+\left(\frac{\partial z}{\partial y}\right)^2}\,dx\,dy$$

$$= 4\sqrt{a} \iint_E \frac{\sqrt{a+x}}{\sqrt{4ax-y^2}} \, dx \, dy$$

$$= 8\sqrt{a} \int_0^{3a} \sqrt{a+x} \, dx \int_0^{\sqrt{ax}} \frac{dy}{\sqrt{4ax-y^2}}$$

$$= 8\sqrt{a} \int_0^{3a} \sqrt{a+x} \left| \sin^{-1} \frac{y}{\sqrt{4ax}} \right|_0^{\sqrt{ax}} dx$$

$$= 8\sqrt{a} \sin^{-1}\left(\frac{1}{2}\right) . \frac{2}{3} \left| (a+x)^{3/2} \right|_0^{3a}$$

$$= \frac{16\sqrt{a}}{3} . \frac{\pi}{6} \left[(4a)^{3/2} - a^{3/2} \right] = \frac{56\pi a^2}{9}. \qquad \textbf{Ans.}$$

Example 23:

Find the surface area of the sphere $x^2 + y^2 + z^2 = a^2$.

Solution:

We have $z = \sqrt{a^2 - x^2 - y^2}$ and so

$$\frac{\partial z}{\partial x} = -\frac{x}{\sqrt{a^2-x^2-y^2}}, \quad \frac{\partial z}{\partial y} = -\frac{y}{\sqrt{a^2-x^2-y^2}}$$

$$\therefore \quad 1 + \left(\frac{\partial z}{\partial x}\right)^2 + \left(\frac{\partial z}{\partial y}\right)^2 = \frac{a^2}{(a^2-x^2-y^2)}.$$

The surface area of the sphere is twice the surface area of the upper half-sphere. Further the projection of the sphere on the xy-plane is the circle $R : x^2 + y^2 = a^2$. Hence the required surface area is

$$S = 2 \iint_R \sqrt{\left[1 + \left(\frac{\partial z}{\partial x}\right)^2 + \left(\frac{\partial z}{\partial y}\right)^2 \right]} \, dx \, dy$$

$$= 2 \iint_R \frac{a \, dx \, dy}{\sqrt{a^2-x^2-y^2}}$$

Putting $x = r\cos\theta$, $y = r\sin\theta$ so that $x^2 + y^2 = r^2$ and $|J| = r \, dr \, d\theta$.

$$\therefore \quad S = 2a \int_0^{2\pi} d\theta \int_0^a \frac{r\,dr}{\sqrt{a^2 - r^2}} = 2a \int_0^{2\pi} \left| -\sqrt{a^2 - r^2} \right|_0^a d\theta$$

$$= 2a^2 \int_0^{2\pi} d\theta = 4\pi a^2.$$ **Ans.**

Example 24:

Find the surface area of the part of the cone $z^2 = x^2 + y^2$ inside the cylinder $x^2 + y^2 = 2ax$. **(D.U., 1989)**

Solution:

We have $z = \sqrt{x^2 + y^2}$ so that

$$\frac{\partial z}{\partial x} = -\frac{-x}{\sqrt{x^2 + y^2}}, \quad \frac{\partial z}{\partial y} = \frac{-y}{\sqrt{x^2 + y^2}}$$

$$\therefore \quad 1 + \left(\frac{\partial z}{\partial x}\right)^2 + \left(\frac{\partial z}{\partial y}\right)^2 = 2.$$

Since the surface is symmetrical about the plane $z = 0$, we have

$$S = 2 \iint_R \sqrt{\left[1 + \left(\frac{\partial z}{\partial x}\right)^2 + \left(\frac{\partial z}{\partial y}\right)^2\right]}\, dx\, dy = 2\sqrt{2} \iint_R dx\, dy$$

where $R : (x - a)^2 + y^2 = a^2$ is a circle of radius a

$$\therefore \quad \iint_R dx\, dy = \text{Area of a circle f radius } a = \pi a^2.$$

Hence, $S = 2\sqrt{2}\pi a^2$. **Ans.**

Example 25:

Find the surface area of the part of the cone $z^2 = x^2 + y^2$ inside the cylinder $x^2 + y^2 = 2x$.

Solution:

Do yourself. **Ans. $2\sqrt{2}\pi$**

Example 26:

Find the surface area of the part of $y^2 + z^2 = x^2$ inside the cylinder $x^2 + y^2 = a^2$.

Solution:

Do your self. **Ans.** $2\sqrt{2}\pi a^2$

[**Hint.** Project on yz-plane.]

$$S = 2 \iint_R \sqrt{\left[1 + \left(\frac{\partial z}{\partial x}\right)^2 + \left(\frac{\partial z}{\partial y}\right)^2 dy\, dz\right]}$$

Example 27:

Find the area of the surface of the part of $y^2 + z^2 = x^2$ *cut off by thee cylinder* $x^2 - y^2 = a^2$ *and the planes* $y = b$, $y = -b$.

Solution:

The solids $y^2 + z^2 = x^2$ and $x^2 - y^2 = a^2$ intersect in the planes $z = \pm a$. The required surface area is

$$S = 2 \int_{-b}^{b} \int_{-a}^{a} \sqrt{\left[1 + \left(\frac{\partial z}{\partial x}\right)^2 + \left(\frac{\partial z}{\partial y}\right)^2\right]}\, dy\, dz$$

$$= 2 \int_{-b}^{b} \int_{-a}^{a} \sqrt{2}\, dy\, dz$$

$$= 2\sqrt{2}\,(2a \,.\, 2b) = 8\sqrt{2}ab.$$ **Ans.**

Example 27:

Show that the surface area of $z = 2 - x^2 - y$, *where* (x, y) *ranges over the triangle* $x = 0$, $y = 1$, $y = x$ *is*

$$\frac{1}{2} \log(\sqrt{2} + \sqrt{3}) + (\sqrt{2}/6).$$

Solution:

The vertices of the given triangle are (0, 0), (0, 1) and (1, 1). From $z = 2 - x^2 - y$, we obtain the following

$\partial z/\partial x = -2x$, $\partial z/\partial y = -1$. Thus

$$1 + \left(\frac{\partial z}{\partial x}\right)^2 + \left(\frac{\partial z}{\partial y}\right)^2 = 2 + 4x^2.$$

The required surface area is given by

$$S = \int_{x=0}^{1} \int_{y=x}^{1} \sqrt{2 + 4x^2}\, dx\, dy$$

$$= \int_0^1 \sqrt{2+4x^2}\,(1-x)\,dx$$

$$= 2\int_0^1 \sqrt{\tfrac{1}{2}+x^2}\,dx - \int_0^1 x\sqrt{2+4x^2}\,dx$$

$$= \left[x\sqrt{\tfrac{1}{2}+x^2} + \tfrac{1}{2}\log\{x+\sqrt{\tfrac{1}{2}+x^2}\}\right]_0^1 - \frac{2}{3.8}\left|(2+4x^2)^{3/2}\right|_0^1$$

$$= \sqrt{3/2} + \tfrac{1}{2}\{\log(1+\sqrt{3/2}) - \log\sqrt{\tfrac{1}{2}}\} - \tfrac{1}{12}(6\sqrt{6}-2\sqrt{2})$$

$$= \frac{1}{2}\log(\sqrt{2}+\sqrt{3}) + (\sqrt{2}/6).$$ **Ans.**

Example 28:

Find the surface area of the part of $z^2 = x^2 + y^2$ cut off by the cylinder $z^2 = 2py$.

Solution:

We have $z = \sqrt{x^2+y^2}$. Then

$$1 + \left(\frac{\partial z}{\partial x}\right)^2 + \left(\frac{\partial z}{\partial y}\right)^2 = 1 + \frac{x^2}{x^2+y^2} + \frac{y^2}{x^2+y^2} = 2.$$

The projection of the required surface on xy-plane is R : $x^2 + y^2 = 2py$ or $x^2 + (y-p)^2 = p^2$, which is circle of radius p.

Since the surface is symmetrical about the plane $z = 0$, we have

$$S = \iint_R \sqrt{2}\,dx\,dy$$

$= 2\sqrt{2}$ (Area of a circle of radius p) $= 2\sqrt{2}\pi p^2$. **Ans.**

Example 29:

Show that the area of the surface of the sphere

$x^2 + y^2 + z^2 = a^2$ cut off by $x^2 + y^2 = ax$ is $2(\pi - 2)a^2$.

Solution:

We have $z = \sqrt{a^2 - x^2 - y^2}$. The required surface area is

$$S = 4\iint_R \sqrt{\left[1 + \left(\frac{\partial z}{\partial x}\right)^2 + \left(\frac{\partial z}{\partial y}\right)^2\right]}\,dx\,dy$$

where R : $x^2 + y^2 = a^2$ is the projection of sphere on the xy-plane. Changing to polars, we get

$$S = 4a \int_0^{\pi/2} d\theta \int_0^{a\cos\theta} \frac{r\,dr}{\sqrt{a^2 - r^2}}$$

$(\because x^2 + y^2 = ax \Rightarrow r^2 = ar\cos\theta \Rightarrow r = a\cos\theta)$

$$= 4a \int_0^{\pi/2} \left| -\sqrt{a^2 - r^2} \right|_0^{a\cos\theta} d\theta = 4a^2 \int_0^{\pi/2} (1 - \sin\theta)\, d\theta$$

$$= 4a^2 \left[\frac{\pi}{2} - 1 \right] = 2a^2 (\pi - 2).$$ **Ans.**

Example 30:

Show that the surface area of the torus

$$x = (a - \cos v)\cos u,\ y = (a - \cos v)\sin u,\ z = \sin v$$

where $-\pi < u, v < \pi,\ a > 1$, *is* $4\pi^3 a$. **(D.U., 1986)**

Solution:

We have

$$\frac{\partial(x,y)}{\partial(u,v)} = \begin{vmatrix} \partial x/\partial u & \partial x/\partial v \\ \partial y/\partial u & \partial y/\partial v \end{vmatrix} = \begin{vmatrix} -(a-\cos v)\sin u & \sin v\cos u \\ (a-\cos v)\sin u & \sin v\cos u \end{vmatrix}$$

$$= -(a - \cos v)\sin v$$

$$\frac{\partial(y,z)}{\partial(u,v)} = \begin{vmatrix} \partial y/\partial u & \partial y/\partial v \\ \partial z/\partial u & \partial z/\partial v \end{vmatrix}$$

$$= \begin{vmatrix} (a-\cos v)\cos u & \sin v\cos u \\ 0 & \cos v \end{vmatrix} = (a - \cos v)\cos u\cos v$$

$$\frac{\partial(z,x)}{\partial(u,v)} = \begin{vmatrix} \partial z/\partial u & \partial z/\partial v \\ \partial x/\partial u & \partial x/\partial v \end{vmatrix}$$

$$= \begin{vmatrix} 0 & \cos v \\ -(a-\cos v)\sin v & \sin v\cos u \end{vmatrix} = (a - \cos v)\cos u\cos v$$

$$\therefore \quad \sqrt{\left[\left(\frac{\partial(x,y)}{\partial(u,v)}\right)^2 + \left(\frac{\partial(y,z)}{\partial(u,v)}\right)^2 + \left(\frac{\partial(z,x)}{\partial(u,v)}\right)^2 \right]}$$

$= (a - \cos v)\,[\sin^2 v + \cos^2 u\cos^2 v + \sin^2 u\cos^2 v] = (a - \cos v)$.

The required surface area is

$$S = \int_{-\pi}^{\pi}\int_{-\pi}^{\pi}(a - \cos v)\, du\, dv = 2\pi \int_{-\pi}^{\pi}(a - \cos v)\, dv$$

$$= 2\pi\,(2a\pi + 0) = 4a^2\pi.$$

Example 31:

Find the area of that part of the surface of the cylinder $x^2 + y^2 = a^2$ which is cut out by the cylinder $x^2 + z^2 = a^2$.

Solution:

The equation of the given surface is $y = \sqrt{a^2 - x^2}$. Then

$$1 + (\partial y/\partial x)^2 + (\partial y/\partial x)^2 = a^2/(a^2 - x^2)$$

The given surface is symmetrical about the plane $y = 0$.

Hence the required surface area is

$$S = 2\iint_R \sqrt{\left[1 + \left(\frac{\partial y}{\partial x}\right)^2 + \left(\frac{\partial y}{\partial z}\right)^2\right]}\, dx\, dz$$

$$= 2a \iint_R \frac{dx\, dz}{\sqrt{a^2 - x^2}},\ R : x^2 + z^2 = a^2$$

$$= 8a \int_0^a \frac{dx}{\sqrt{a^2 - x^2}} \int_0^{\sqrt{a^2 - x^2}} dz \quad \text{(+ive quadrant used)}$$

$$= 8a \int_0^a dx = 8a^2.$$

Example 32:

Find the area of the surface out from the cylinder $x^2 + z^2 = a^2$ by the cylinder $x^2 + y^2 = a^2$.

Solution:

Do yourself. **Ans. $8a^2$**

Example 33:

Find the surface area of the part of the sphere $x^2 + y^2 + z^2 = a^2$ above $z = 0$ plane cut off by the vertical cylinder erected on one loop of the curve whose equation in polar coordinates is $r = a \cos 2\theta$.

Solution:

The surface area consists of two parts, one above and one below the xy-plane.

We have

$$z = \sqrt{a^2 - x^2 - y^2}$$

The required surface area

$$= \iint \frac{a\,dx\,dy}{\sqrt{a^2 - x^2 - y^2}}$$

$$= \int_{-\pi/4}^{\pi/4} \int_0^{a\cos 2\theta} \frac{a\,r\,dr\,d\theta}{\sqrt{a^2 - r^2}}$$

$(x = r\cos\theta,\ y = r\sin\theta)$

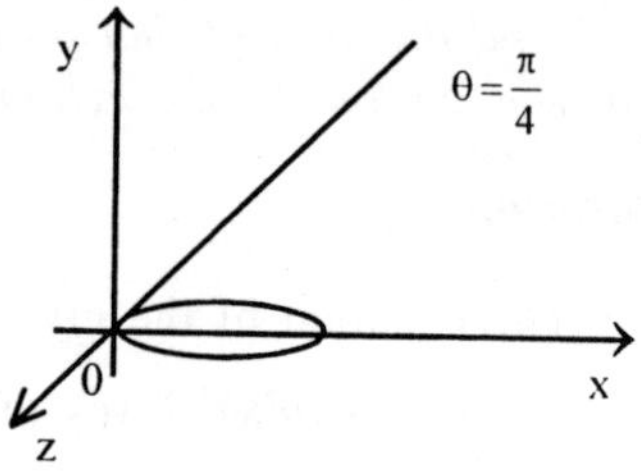

$$= 2a \int_0^{\pi/4} \left| -\sqrt{a^2 - r^2} \right|_0^{a\cos 2\theta} d\theta$$

$$= 2a^2 \int_0^{\pi/4} (1 - \sin 2\theta)\,d\theta$$

$$= 2a^2 \left| \theta + \tfrac{1}{2}\cos\theta \right|_0^{\pi/4} = \tfrac{1}{2} a^2 (\pi - 2).$$ **Ans.**

y, $\theta = \frac{\pi}{4}$, 0, x, z

Example 34:

Find the surface area of the part of the sphere $x^2 + y^2 + z^2 = a^2$ inside the cylinder erected on loop of the curve $r = a\cos 3\theta$.

Solution:

Do yourself. **Ans.** $\frac{2}{3}a^2(\pi - 2)$

Example 35:

Find the area of the surface of paraboloid $x^2 + y^2 = az$, which lies between the planes $z = 0$ and $z = a$.

Solution:

$$\frac{\partial z}{\partial x} = \frac{2x}{a},\ \frac{\partial z}{\partial y} = \frac{2y}{a}.$$ Thus

$$1 + \left(\frac{\partial z}{\partial x}\right)^2 + \left(\frac{\partial z}{\partial y}\right)^2 = \frac{1}{a^2}[a^2 + 4(x^2 + y^2)].$$

Projecting on the plane $z = a$ gives $E : x^2 + y^2 = a^2$. Thus

$$S = \frac{1}{2} \iint_E \sqrt{[a^2 + 4(x^2 + y^2)]}\,dx\,dy$$

$$= \frac{1}{2} \int_0^{2\pi} \int_0^{\pi} \sqrt{a^2 + 4r^2}\ r\, dr\, d\theta \quad (x = r \cos \theta,\ y = r \sin \theta)$$

$$= \frac{2\pi}{a} \frac{2}{3.8} \left| (a^2 + 4r^2)^{2/2} \right|_0^a = \frac{\pi}{6a} [(5a^2)^{3/2} - a^2]$$

$= \pi a^2 (5\sqrt{5} - 1)/6.$ **Ans.**

Example 36:

Find the area of the surface of the part of $x^2 = y^2 + z^2$ between the cylinder $y^2 = z$ and the plane $y = z - 2$.

Solution:

$$x = \sqrt{y^2 + z^2} \Rightarrow \partial x/\partial y = y/x,\ \partial x/\partial z = z/x.$$

$$\therefore \quad 1 + \left(\frac{\partial x}{\partial y}\right)^2 + \left(\frac{\partial x}{\partial z}\right)^2 = \frac{x^2 + y^2 + z^2}{x^2} = \frac{x^2 + x^2}{x^2} = 2. \qquad \text{...(1)}$$

The cylinder $y^2 = z$ and the planeey $= z - 2$ intersect in the points given by $y^2 = y + z$ or $(y - 2)(y + 1) = 0$.

Also $(z - 2)^2 = z \Rightarrow z^2 - 5z + 4 = 0$ Þ $(z - 1)(z - 4) = 0$.

The required surface area is

$$S = \iint \sqrt{\left[1 + \left(\frac{\partial x}{\partial y}\right)^2 + \left(\frac{\partial x}{\partial z}\right)^2\right]}\ dy\, dz$$

$$= \sqrt{2} \int_{z=1}^{4} \int_{y=-1}^{2} dy\, dz = 9\sqrt{2}.$$ **Ans.**

Example 37:

Show that the area of the surface of the paraboloid $x^2/a + y^2/b = 2z$ inside the cylinder $x^2/a^2 + y^2/b^2 = k$ is

$$\frac{2}{4} \pi ab \left[(1+k)^{3/2} - 1\right].$$ **(D.U., 1993, 87)**

Solution:

From $2z = x^2/a + y^2/b$, we have

$$\frac{\partial z}{\partial x} = \frac{x}{a},\ \frac{\partial z}{\partial y} = \frac{y}{b} \Rightarrow 1 + \left(\frac{\partial z}{\partial x}\right)^2 + \left(\frac{\partial z}{\partial y}\right)^2 = 1 + \frac{x^2}{a^2} + \frac{y^2}{b^2}$$

The surface area is

$$S = \iint_D \sqrt{1 + \frac{x^2}{a^2} + \frac{y^2}{b^2}}\ dx\,dy.$$

Now, we put x = a tan θ cos ϕ, y = b tan θ sin ϕ.

$$\therefore \quad 1 + \frac{x^2}{a^2} + \frac{y^2}{b^2} = \sqrt{1 + \tan^2\theta} = \sec\theta,$$

$$\frac{x^2}{a^2} + \frac{y^2}{b^2} = k \Rightarrow \tan^2\theta = k \Rightarrow \theta = \tan^{-1}\sqrt{k}.$$

and $J = \dfrac{\partial(x, y)}{\partial(\theta, \phi)} = \begin{vmatrix} \partial x/\partial\theta & \partial x/\partial\phi \\ \partial y/\partial\theta & \partial y/\partial\phi \end{vmatrix}$

i.., $J = ab \sec^2\theta \tan\theta$.

$$\therefore \quad S = \int_0^{2\pi} d\theta \int_0^{\theta} \sec\theta .\, ab \sec^2\theta \tan\theta\, d\theta$$

$$= 2\pi\, ab \int_0^{\theta} \sec^2\theta .\, (\sec\theta \tan\theta)\, d\theta \qquad ...(1)$$

$$= \frac{2}{3}\pi\, ab \left|\sec^2\theta\right|_0^{\tan^{-1}\sqrt{k}} = \frac{2}{3}\pi ab \left[(1 + \tan^2\theta)^{3/2}\right]_0^{\tan^{-1}\sqrt{k}}$$

$$= \frac{2}{3}\pi\, ab \left[(1 + k)^{3/2} - 1\right].$$ **Ans.**

Example 38:

The x and y coordinates of a point on the paraboloid

$2z = x^2/a + y^2/b$ are expressed in the form

$x = a \tan\theta \cos\phi,\ y = b \tan\theta \sin\phi,$

where θ is the inclination of the normal at any point to the axis of z. Show that the area of the cap of the surface cut off the by the curve $\theta = \lambda$ is $\frac{2}{3}\pi ab(\sec^3\lambda - 1)$.

Solution:

Do yourself.

EXERCISES

1. Find the surface area of the part of $z = xy$ cut off by the cylinder $x^2 + y^2 = a^2$. **(Ans.** $\frac{2}{3}\pi[(1 + a^2)^{3/2} - 1]$)

   Ex. Find the area of the portion of the surface $az = xy$ that lies inside the cylinder $(x^2 + y^2)^2 = 2a^2 xy$. **(D.U., 1996)**

   [**Hint.** Changing to polars, we get]

   $$S = 4a \iint \frac{dx\,dy}{\sqrt{a^2 + x^2 + y^2}} = 4a \int_0^{\pi/4} d\theta \int_0^{a\sqrt{\sin 2\theta}} \frac{r\,dr}{\sqrt{a^2 + r^2}}$$

2. Find the surface area of the part of $2x = x^2 + y^2$ cut off by the cylinder $x^2 + y^2 = 1$. **(Ans.** $\frac{2}{3}\pi(\sqrt{8} - 1)$

   [**Hint.** use formula (B) and change to polars to get.]

   $$S = \int_0^{2\pi} d\theta \int_0^1 \sqrt{1 + r^2}\; r\,dr.$$

3. The sphere $x^2 + y^2 + z^2 = a^2$ is pierced by the cylinder $(x^2 + y^2)^2 = a^2(x^2 - y^2)$. Find the area of the spherical surface inside the cylinder. **(Ans.** $8a^2\,[(\pi/4) + 1 - \sqrt{2})$

   Hint. $$S = 2a \iint_E \frac{dx\,dy}{\sqrt{a^2 - x^2 - y^2}} = 8a \int_0^{\pi/4} d\theta \int_0^{a\sqrt{\cos 2\theta}} \frac{r\,dr}{\sqrt{a^2 - r^2}}$$

4. Show that the area of the surface of the solid
   $x = a \sin\theta \cos\phi$, $y = (a + w)\sin\theta \sin\phi$, $z = (a + \varepsilon)\cos\theta$
   is $4\pi a^2 + 8\pi a\,(w + \varepsilon)/3$, where ε^2, w^2 and we are neglected. **(D.U., 1994)**

5. Calculate the area of the spherical surface given by
   $x = a\cos\theta\cos\phi$, $y = a\cos\theta\sin\phi$, $z = a\sin\theta$
   where $0 < \phi < 2\pi$, $-\frac{1}{2}\pi < \theta < \frac{1}{2}\pi$. **(Ans.** $4\pi a^2$)

4.9 Surface Integrals of the First Type

Definition : Let S be a smooth (or piecewise smooth) surface bounded by a smooth (or piecewise smooth) contour C. Let a bounded function f be defined at all points (x, y, z) of the surface.

Let the surface be partitined into sub-surfaces of areas $\Delta S_1, \Delta S_2, \ldots, \Delta S_n$ by means of smooth (or piecewise smooth) curves. Let $P(\xi_1, \eta_1, \zeta_1)$ be a point of ΔS_1. Now we have the sum

$$\sum_i f(\xi_1, \eta_1, \zeta_1)\,\Delta S_i \qquad \ldots(1)$$

It the sum tends to a finite limit as the norm of the partition (the maximal of the diameters of the sub-areas ΔS_i) tends to zero, and for all positions of (ξ_1, η_1, ζ_1) in ΔS_1, the limit is called the *surface integral of the first type* of the function f over the surface S and is denoted by the symbol

$$\iint_S f(x, y, z)\, dS. \quad ...(2)$$

Remark:

Let a plane surface D bounded by contour Γ be the projection of the surface S on the xy-plane and $\Delta\sigma_i$ the corresponding subareas of D. The we have

$$\Delta\sigma_i = \Delta S_i \cos\gamma_i$$

where γ_i is the angle of in clination of the normal at (ξ_1, η_1, ζ_1) to the surface ΔS_i with the z-axis. Sum (1) then becomes as

$$\sum_i f(\xi_i, \eta_i, \zeta_i)\, \Delta S_i = \sum_i f(\xi_i, \eta_i, \zeta_i) \frac{\Delta\sigma_i}{\cos\gamma_i}$$

which, in the limit, yields

$$\iint_S f(x, y, z)\, dS = \iint_D f(x, y, z) \frac{dx\, dy}{\cos\gamma} \quad ...(3)$$

where z, on the right hand side, is expressed in terms of x, y with the help of the equation of the surface.

Relation (3) expresses a surface integral in terms of a double integral.

4.10 Reducing a Surface Integral of First Type to a Double Integral

(1) If the surface is represented by the equ. $z = \phi(x, y)$ then we know

$$\Delta S_i = \sqrt{1+\psi_x^2 + \psi_y^2}\, \Delta x_i\, \Delta y_i$$

so the in the limit, (1) takes following form

$$\iint_S f(x, y, z)\, dS = \iint_D f(x, y, z) \sqrt{1 + \left(\frac{\partial z}{\partial x}\right) + \left(\frac{\partial z}{\partial y}\right)^2}\, dx\, dy \quad ...(4)$$

where z, in the double integral, is expressed in terms of x, y and D is the projection of S on the xy-plane.

If the surface is represented by equation of the following form

$$x = \theta(y, z) \text{ or } y = \phi(z, x)$$

then interchanging the roles of the variables x, y, z, the formula correspond as

$$\iint_S f(x, y, z)\, dS = \iint_{D_1} f(\theta(y, z), y, z) \sqrt{1 + \left(\frac{\partial x}{\partial y}\right)^2 + \left(\frac{\partial x}{\partial z}\right)^2}\, dy\, dz \quad ...(5)$$

$$\iint_S f(x, y, z)\, dS = \iint_{D_2} f(x, \phi(z, x), z) \sqrt{1 + \left(\frac{\partial y}{\partial z}\right)^2 + \left(\frac{\partial y}{\partial x}\right)^2}\, dz\, dx \quad ...(6)$$

which D_1 and D_2 are respectively the projection of S on the yz-plane and zx-plane.

Remark 1:

If the surface S is composed of several parts, each of which can be represented by an equation of the form

$$x = \theta(y, z),\ \ y = \phi(z, x) \text{ or } z = \phi(x, y).$$

Then by definition, the surface integral over S is equal to the sum of the integrals over the parts of S, applying formulas (4), (5) or (6) to these integrals separately, we reduce the integral over S to the sum of the double integrals.

Remark 2:

In case of a smooth surface S is represented by a parametric equation as

$$x = X(u, v)\ \ y = Y(u, v),\ \ z = Z(u, v)$$

then by applying arguments essentially the same as above, the surface integral of a bounded function f over S can be expressed in terms of a double integral, as follows:

$$\left.\begin{aligned} \iint_S f(x, y, z)\, dS &= \iint_D f(X, Y, Z)\sqrt{k}\, du\, dv \\ \text{or} \qquad &= \iint_D f(X, Y, Z)\sqrt{AC - B^2}\, du\, dv \end{aligned}\right\} \quad ...(7)$$

where the domain D in uv-plane corresponds to S, and

$$K = \left[\frac{\partial(x, y)}{\partial(u, v)}\right]^2 + \left[\frac{\partial(y, z)}{\partial(u, v)}\right]^2 + \left[\frac{\partial(z, x)}{\partial(u, v)}\right]^2$$

$$A = \left(\frac{\partial x}{\partial u}\right)^2 + \left(\frac{\partial y}{\partial u}\right)^2 + \left(\frac{\partial z}{\partial u}\right)^2$$

$$B = \frac{\partial x}{\partial u}\frac{\partial x}{\partial v} + \frac{\partial y}{\partial u}\frac{\partial y}{\partial v} + \frac{\partial z}{\partial u}\frac{\partial z}{\partial v}$$

$$C = \left(\frac{\partial x}{\partial v}\right)^2 + \left(\frac{\partial y}{\partial v}\right)^2 + \left(\frac{\partial z}{\partial v}\right)^2$$

Remark 3:

We have considered surface integrals of scalar function f. This notice can be easily generalized to a vector function F defined on S. Let

$$F(x, y, z) = iP(x, y, z) + jQ(x, y, z) + kR(x, y, z)$$

where P, Q, R are scalar functions. We define the surface integral of F over S by the relation as

$$\iint_S F\, dS = i\iint_S P\, dS + j\iint_S Q\, dS + k\iint_S R\, dS \quad ...(8)$$

where the surface integrals of scalar functions on the right side can be expressed as double integrals, and call it the surface integral of the first type of the vector function F over the surface S.

4.11 Oriented Surface, Positive and Negative Sides

A surface is said to be two sided (*bilateral*) if it is possible to distinguish one of its sides from the other. We assume that a surface has two distinct sides in such a way that it is impossible to pass from one side to he other along a continuous path which lies on the surface and which does not cross one of the bounding curves.

The concept of a side of a surface is closely related to the orientation of its bounding curve. A side of a surface (a region) is said to be positive or positively oriented if the orientation of its boundary is positive.

Let S be a two-sided smooth surface and C its bounding curve. Take a small area σ bounded by a curve Γ on the surface. Take a point P in σ and the normal PNP' to the surface at P. The half lines PN and PN', which do not pierce the surface, are drawn in opposite directions from the surface; one of these directions, say that of PN, is chosen as the positive direction of the normal at P, and PN may be called the positive normal, PN' the negative normal. That side of the surface σ which faces the positive direction of the normal at P will be called the positive (or upper) side of the surface; the other side will be the negative (or lower) side.

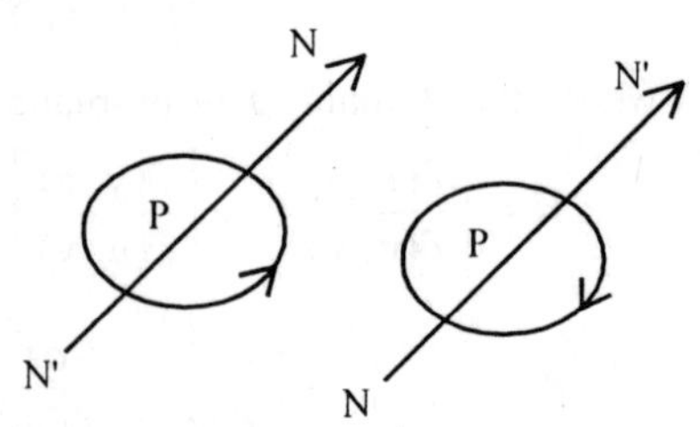

The concept of a side of a surface is closely related to the orientation of its contour. A side of a surface is said to be positive or positively oriented if the orientation of its boundary is positive.

To correlate these two ideas, we introduce for each contour entering into the boundary its orientation according to the following rule:

'A direction in which the contour C is described is considered to be positive if the surface S is always kept on the left of a person who is on the surface and walking round the contour in this direction, so that the normal PN goes from his feet to his head. The opposite side is referred to as the negative one.'

Thus the side of the surface with this orientation of the boundary is called the positive side and the other the negative side. The two sides of the surface are then called *orientable* and the process of choosing a certain side of a two-sided surface is referred to as the *orientation* of the surface. The one-sided surfaces are thus non-orientable.

4.12 Surface Integral of the Second Type: Flux Across a Surface

Definition: Let S be a smooth (or piecewise smooth) two-sided (oriented) surface, bounded by a smooth (or piecewise smooth) contour C. Let a bounded vector-valued function F = (f, g, h) be defined for all points of a certain side, say positive side of the surface. The projection (or resolved part) F_n of F in the direction of the normal to the surface at arbitrary point (x, y, z) can be written as

$$F_n = f \cos \alpha + g \cos \beta + h \cos \gamma$$

where $\cos \alpha$, $\cos \beta$, $\cos \gamma$ are the direction cosines of the normal to the surface at (x, y, z).

The surface integral

$$\iint_S (f \cos \alpha + g \cos \beta + h \cos \gamma)\, dS \qquad ...(1)$$

is called the *Surface integral of the second type* of the vector valued function F = (f, g, h) over the surface S and will be denoted as

$$\iint_S (f\, dy\, dz + g\, dz\, dx + h\, dx\, dy) \qquad ...(2)$$

Thus by definition, we have the following relation

$$\iint_S (f \cos \alpha + g \cos \beta + h \cos \gamma)\, dS$$

$$= \iint_S f\, dy\, dz + g\, dz\, dx + h\, dx\, dy. \qquad ...(3)$$

Remark:

We have defined the surface integral of the second type on the basis of the notion of the surface integral of the first type. But it can be defined directly also as follows:

4.13 Reducing a Surface Integral of the Second Type to a Double Integral

The definition of the surface integral of the second type, we have the following results:

(i) Let S be a smooth surface determined by an equation

$Z = \phi(x, y)$

and let h(x, y, z) be a bounded function defined on S. Then for the surface integral of the second type taken over the positive side of S we have the following relation.

$$\iint_S h(x, y, z)\, dx\, dy = \iint_D h(x, y, \psi(x, y))\, dx\, dy \qquad ...(8)$$

where D is the projection of S on the z = 0 plane, and since S is positively oriented, the sign before the double integral is to be positive if D is positively oriented, otherwise negative.

If the integral is taken over the other side (negative) of S, the sign before the double integral is to be negative if D is positively oriented, otherwise positive.

We, similarly derive the formula as

$$\iint_S f(x, y, z)\, dy\, dz = \iint_{D_1} f(\theta(y, z), y, z)\, dy\, dz \qquad ...(9)$$

$$\iint_S g(x, y, z)\, dz\, dx = \iint_{D_2} g(x, \phi(z, x), z)\, dz\, dx \qquad ...(10)$$

where D_1, D_2 are the projections of S on yz and zx planes respectively. So finally we have

$$\iint_S f\, dy\, dz + g\, dz\, dx + h\, dx\, dy = \iint_{D_1} f(\theta(y, z), y, z)\, dy\, dz$$

$$+ \iint_{D_2} g(x, \phi(z, x), z)\, dz\, dx$$

$$+ \iint_D h(x, y, \psi(x, y))\, dx\, dy \qquad ...(11)$$

(ii) Let S be a smooth surface represented as

$x = X(u, v),\ \ y = Y(u, v),\ \ z = Z(u, v),\ (u, v) \in D$

where D is the surface in uv-plane corresponding to the surface S in the xy-plane.

Let F = (f, g, h) be a bounded vector valued function defined on S. Then for the surface integral of the second type taken over the positive side of S, we have the following relation

$$\iint_S h(x, y, z)\, dx\, dy = \iint_D h(x, y, z)\, dx\, dy$$

$$= \iint_D h(X, Y, Z) \frac{\partial(X, Y)}{\partial(u, v)} du\, dv$$

$$\iint_S f(x, y, z)\, dy\, dz = \iint_D f(X, Y, Z) \frac{\partial(Y, Z)}{\partial(u, v)} du\, dv$$

$$\iint_S g(x, y, z)\, dz\, dx = \iint_D g(X, Y, Z) \frac{\partial(Z, X)}{\partial(u, v)} du\, dv$$

where D is also oriented in the same sense as S.

Thus, finally we have

$$\iint_S f\, dy\, dz + g\, dz\, dx + h\, dx\, dy$$

$$\iint_D \left[f \frac{\partial(Y, Z)}{\partial(u, v)} + g \frac{\partial(Z, X)}{\partial(u, v)} + h \frac{\partial(X, Y)}{\partial(u, v)} \right] du\, dv. \qquad ...(12)$$

4.14 Vectorial Formation

Let

$$r = ix + jy + kz$$

be the position vector of any point on the surface S, and

$$F(x, y, z) = iP(x, y, z) + jQ(x, y, z) + kR(x, y, z)$$

be a vector function, with P, Q, R as its components.

Let n denote the unit vector along the normal at any point on the side of the surface under consideration, so that

$$a = i \cos\alpha + j \cos\beta + k \cos\gamma$$

$$\therefore \quad \iint_S F.n\, dS = \iint_S (P \cos\alpha + Q \cos\beta + R \cos\gamma)\, dS \qquad ...(13)$$

Thus the surface integral of the second type of F (or the surface integral of the first type of F.n) over S is $\iint_S F.n\, dS$.

4.15 Properties of Surface Integrals

The surface integrals (of the second type) taken over the opposite sides of a surface have different signs, i.e.,

$$\iint_S F.n\, dS = -\iint_{S'} F.n\, dS$$

where S, S' are the two sides of the surface.

If a surface S is broken up into m part S_1, S_2 ..., S_m, the integral over the whole surface S (say, over its positive side is equal to the sum of the integrals taken over the corresponding (i.e. positive sides of the surfaces S_1, S_2, ... S_m;

$$\iint_S F.n\, dS = \iint_{S_1} F.n\, dS + \iint_{S_2} F.n\, dS + ... + \iint_{S_m} F.n \quad ...(14)$$

4.16 Relation Between the Two Types of Surface Integrals

Let S be a smooth (or piecewise smooth) oriented surface and C its bounding curve. Then the surface integral of the second type of a function P, taken over a certain side (say, positive) of the surface is $\iint_S f\, dy\, dz$, which is equivalent to $\iint_S f \cos\alpha\, dS$. The letter being the surface integral of the first type of the function f cos α.

Similarly, the surface integral of the second type of the function F = (f, g, h) is written as

$$\iint_S (f\, dy\, dz + g\, dz\, dx + h\, dx\, dy) \quad ...(1)$$

which is equivalent to the following result.

$$\iint_S (f \cos\alpha + g \cos\beta + h \cos\gamma)\, dS.$$

The letter being a surface integral of the first type of the function (f cos α + g cos β + h cos γ). Also (f cos α + g cos β + h cos γ) is the projection or the component of the function F along the normal to the surface.

Thus, the surface integral of the second type of a function is same as the surface integral of the first type of the normal component of the function.

In symbolic representation of the surface integral,

$$\iint_S f\, dS, \ \iint_S f \cos\alpha,\, dS \text{ or } \iint_S f\, dy\, dz.$$

Example 39:

Find the value of the surface integral $\iint_S \frac{dS}{r}$, *where S is the portion of the surface of the hyperbolic paraboloid* $z = xy$ *cut off by the cylinder* $x^2 + y^2 = a^2$, *and r is the distance from a point on the surface to the z-axis.*

Solution:

For any point (x, y) on the surface, $r = \sqrt{x^2 + y^2}$.

Now,

$$\iint_S \frac{dS}{\sqrt{x^2 + y^2}} = \iint_D \frac{1}{\sqrt{x^2 + y^2}} \sqrt{1 + \left(\frac{\partial z}{\partial x}\right)^2 + \left(\frac{\partial z}{\partial y}\right)^2}\, dx\, dy$$

where D is the projection of E on xy-plane

$$= \iint_D \frac{\sqrt{1 + x^2 + y^2}}{\sqrt{x^2 + y^2}}\, dx\, dy = \int_0^{2\pi} d\theta \int_0^a \frac{\sqrt{1 + r^2}}{r}\, r\, dr$$

$$= \pi\left[a\sqrt{a^2 + 1} + \log\left(a + \sqrt{a^2 + 1}\right)\right]$$

Example 40:

Evaluate the surface integral

$$\iint_S p(x^4 + y^4 + z^4)\, dS$$

where S is the surface $x^2/a^2 + y^2/b^2 + z^2/c^2 = 1$, and p is the perpendicular from the origin to the tangent plane at the point (x, y, z).

Symmetry of the integrand and the surface S show that its value is 8 times the value of the integral over the portion of the surface in the first octant.

Solution:

Let

$$\left.\begin{aligned} x &= a\cos\theta\cos\phi \\ y &= b\cos\theta\sin\phi \\ z &= c\sin\theta \end{aligned}\right\}, \quad \begin{aligned} 0 \le \phi \le \pi/2 \\ 0 \le \theta \le \pi/2 \end{aligned}$$

We have

$$\frac{\partial(x, y)}{\partial(\theta, \phi)} = -\, ab\cos\theta\sin\theta$$

$$\frac{\partial(y, z)}{\partial(\theta, \phi)} = -\, bc\cos^2\theta\cos\phi$$

$$\frac{\partial(z, x)}{\partial(\theta, \phi)} = -\, ac\cos^2\theta\sin\phi$$

So that

$$\left[\frac{\partial(x, y)}{\partial(u, v)}\right]^2 + \left[\frac{\partial(y, z)}{\partial(u, v)}\right]^2 + \left[\frac{\partial(z, x)}{\partial(u, v)}\right]^2$$

$$= a^2b^2c^2 \cos^2\theta \left\{ \frac{\cos^2\theta\cos^2\phi}{a^2} + \frac{\cos^2\theta\sin^2\phi}{b^2} + \frac{\sin^2\theta}{c^2} \right\}$$

and

$$p = 1/\sqrt{\frac{\cos^2\theta\cos^2\phi}{a^2} + \frac{\cos^2\theta\sin^2\phi}{b^2} + \frac{\sin^2\theta}{c^2}}$$

Now the surface integral becomes as

$$\iint_S p(x^4 + y^4 + z^4)\, dS$$

$$= 8\int_0^{\pi/2} d\theta \int_0^{\pi/2} abc\cos\theta[a^4\cos^4\theta\cos^4\phi + b^4\cos^4\theta\sin^4\phi + c^4\sin^4\theta]\, d\phi$$

$$= 8abc\int_0^{\pi/2} \cos\theta\left[a^4\cos^4\theta.\frac{3.1}{4.2}\frac{\pi}{2} + b^4\cos^4\theta.\frac{3.1}{4.2}\frac{\pi}{2} + c^4\sin^4\theta.\frac{\pi}{2}\right] d\theta$$

$$= 8abc\,.\,\frac{\pi}{2}\left[\frac{3}{8}(a^4 + b^4)\frac{4.2}{5.3} + \frac{c^4}{5}\right] = \frac{4\pi abc\,(a^4 + b^4 + c^4)}{5}. \qquad \textbf{Ans.}$$

Example 41:

Evaluate: $I = \iint_S (x\, dy\, dz + dz\, dx + xz^2\, dx\, dy)$

where S is the outer side of the part of the sphere $x^2 + y^2 + z^2 = 1$ in the first octant.

Solution:

Let us denote the projections of the surface S on the coordinate planes yz, zx and xy by D_1, D_2, D_3 respectively. These are quarter circles of radius 1. Then, we have the following results

$$I_1 = \iint_S x\, dy\, dz = \iint_{D_1} \sqrt{1 - y^2 - z^2}\, dy\, dz$$

$$I_2 = \iint_S dz\, dx = \iint_{D_2} dz\, dx$$

$$I_3 = \iint_S xz^2\, dx\, dy = \iint_{D_2} x(1 - x^2 - y^2)\, dx\, dy.$$

The second integral I_2 is simply the area of the domain D_2, (quarter circle $x^2 + z^2 = 1$), i.e., equal to $\pi/4$.

$$I_1 = \int_0^{\pi/2} d\theta \int_0^1 \sqrt{1-r^2}\ r\ dr = \frac{\pi}{6}$$

$$I_3 = \int_0^{\pi/2} \cos\theta\ d\theta \int_0^1 r(1-r^2)\ r\ dr = \frac{2}{15}$$

Thus, we have

$$I = I_1 + I_2 + I_3 = \frac{\pi}{6} + \frac{\pi}{4} + \frac{2}{15} = \frac{5\pi}{12} + \frac{2}{15} \qquad \textbf{Ans.}$$

Example 42:

Evaluate the surface integral $\iint z \cos\gamma\ dS$, *over the outer side of the sphere* $x^2 + y^2 + z^2 = 1$, *where* γ *is the inclination of the normal at any point of the sphere with the z-axis.*

Solution:

Here the z-coordinate can be expressed as a single-valued function of x and y for the whole surface. Let us break it into two parts; the upper hemisphere S_1, above the xy-plane, and the lower hemisphere S_2 below it. We have the following equations

$$z = \sqrt{1-x^2-y^2},\ z = -\sqrt{1-x^2-y^2}$$

We can write

$$\iint_S z\cos\gamma\ dS = \iint_{S_1} z\cos\gamma\ dS + \iint_{S_2} z\cos\gamma\ dS\ .$$

Here, of the two integrals on the right hand side, the firsts over the upper side of the upper hemisphere in the upward direction and the second is over the lower side of the lower hemisphere in the downward direction. The second integral will be negative and will therefore have to be taken with a negative sign.

$$\therefore \quad \iint_{S_1} z\cos\gamma\ dS = \iint_{S_1} z\ dx\ dy = \iint_D \sqrt{1-x^2-y^2}\ dx\ dy$$

$$\iint_{S_2} z\cos\gamma\ dS = \iint_{S_1} z\ dx\ dy = -\iint_D -\sqrt{1-x^2-y^2}\ dx\ dy$$

where D is the circle $x^2 + y^2 - 1$.

Thus, we have

$$\iint_S z\cos\gamma\ dS = 2\iint_D \sqrt{1-x^2-y^2}\ dx\ dy$$

$$= 2 \int_0^{2\pi} d\theta \int_0^1 \sqrt{1-r^2}\ r\, dr = \frac{4\pi}{3}.$$ **Ans.**

Example 43:

Show that

$$I = \iint_S (yz\, dy\, dz + zx\, dz\, dx + xy\, dx\, dy) = \tfrac{3}{8}$$

where S is the outer surface of the sphere $x^2 + y^2 + z^2 = 1$ in the first octant.

Solution:

Let we have

$$\left.\begin{aligned} x &= \cos\theta \cos\phi \\ y &= \cos\theta \sin\phi \\ z &= \sin\theta \end{aligned}\right\}, \quad \begin{aligned} 0 \le \theta \le \pi/2 \\ 0 \le \phi \le \pi/2 \end{aligned}$$

So that

$$\frac{\partial(y, z)}{\partial(\theta, \phi)} = -\cos^2\theta \cos\phi, \quad \frac{\partial(yz, x)}{\partial(\theta, \phi)} = -\cos^2\theta \sin\phi,$$

$$\frac{\partial(x, y)}{\partial(\theta, \phi)} = -\sin\theta \cos\theta.$$

The negative signs show that the correspondence is inverse and so the double integrals are to be taken with negative signs.

Thus, we get

$$I = 3 \int_0^{\pi/2} \cos^3\theta \sin\theta\, d\theta \int_0^{\pi/2} \sin\phi \cos\phi\, d\phi = \tfrac{3}{8}.$$ **Ans.**

Example 44:

Evaluate $\iint_S x\, dS$, *where S is the entire surface of the solid bounded by the cylinder $x^2 + y^2 = 1$ and the planes $z = 0$, $z = x + 2$.*

Solution:

By the Fig. the surface consists of three parts: S_1, the circular base in the xy-plane, S_2, the elliptic plane section, i.e., part of the plane $z = x + 2$ inside the cylinder $x^2 + y^2 = 1$, and S_3, the lateral surface of the cylinder.

On S_1, we have

$z = 0, x^2 + y^2 = 1$

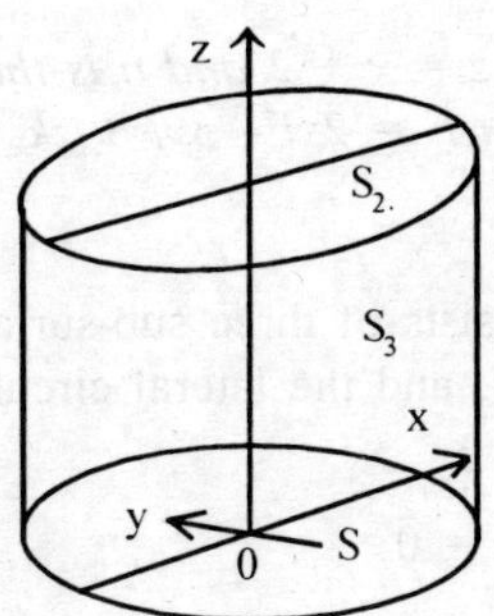

$$\therefore \quad \iint_{S_1} x\, dS = \iint_{S_1} x\, dx\, dy = \int_{-1}^{1} x\, dx \int_{-\sqrt{v(1-x^2)}}^{\sqrt{(1-x^2)}} dy = 0$$

On S_2, we have

$z = x + 2, x^2 + y^2 = 1$

$$\therefore \quad \iint_{S_2} x\, dS = \iint_{S_1} x\sqrt{1+1}\, dx\, dy = \sqrt{2} \iint_{S_1} x\, dx\, dy = 0$$

On S_2, we have

$x^2 + y^2 = 1, 0 \le z \le x + 2$

put

$x = \cos\theta, y = \sin\theta, z = z$

$\pi \le \theta \le \pi, 0 \le z \le \cos\theta$

So that

$$\frac{\partial(y, z)}{\partial(\theta, z)} = \cos\theta, \frac{\partial(z, x)}{\partial(\theta, z)} = \sin\theta, \frac{\partial(x, y)}{\partial(\theta, z)} = 0$$

$$\therefore \quad \iint_S x\, dS = \iint_D \cos\theta \sqrt{\cos^2\theta + \sin^2\theta}\, d\theta\, dz$$

where D is the corresponding dominion the θz-plane.

$$= \int_{-\pi}^{\pi} \cos\theta\, d\theta \int_{\theta}^{2+\cos\theta} dz = \int_{-\pi}^{\pi} \cos\theta\,(2+\cos\theta)\, d\theta = \pi$$

$$\iint_S x\, dS = \iint_{S_1} x\, dS + \iint_{S_2} x\, dS + \iint_{S_3} x\, dS = \pi.$$ **Ans.**

Example 45:

Evaluate $\iint_S$ F.n dS, *where S is the entire surface of the solid formed by* $x^2 + y^2 = 1$, $z = 0$, $z = x + 2$ *and n is the outward drawn unit normal and the vector function* $F = 2xi - 3yj + zk$.

Solution:

The surface S consists of three sub-surfaces, the circular base S_1, the elliptic plane section S_2 and the lateral circular section S_2.

On S_1, we have

$$x^2 + y^2 = 1,\ z = 0$$

$$n = -k,\ F.n = -z$$

$$\therefore \quad \iint_{S_1} F.n\ dS = \iint_{S_1} -z\ dS = \iint_{S_1} -z\ dx\ dy = 0$$

On S_2, we have

$$z = x + 2,\ x^2 + y^2 = 1$$

$$n = \frac{1}{\sqrt{2}}\ i + 0j + \frac{1}{\sqrt{2}} k$$

$$F.n = \frac{1}{\sqrt{2}}(-2x + z) = \frac{1}{\sqrt{2}}(-x + 2)$$

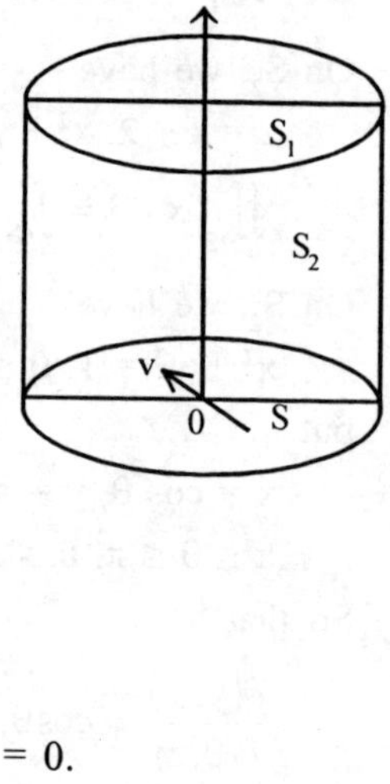

$$\therefore \quad \iint_{S_2} F.n\ dS = \frac{1}{\sqrt{2}} \iint_{S_2} (-2x+z)\ dS$$

$$= \frac{1}{\sqrt{2}} \iint_{S_1} (-2x+x+2)\sqrt{1+1}\ dx\ dy$$

$$= \iint_{S_1} (2-x)\ dx\ dy$$

S₁ S₂ V 0 S

where S_1 is the circular base, $x^2 + y^2 = 1$, $z = 0$.

$$\iint_{S_2} F.n\ dS = \int_{-\pi}^{\pi} d\theta \int_0^1 (2 - r\cos\theta)\ r\ dr = \int_{-\pi}^{\pi} \left(1 - \tfrac{1}{3}\cos\theta\right) d\theta = 2\pi$$

On S_3, we have

$x^2 + y^2 = 1,\ 0 \le z \le x + 2$

Now, let

$$x = \cos\theta,\ y = \sin\theta,\ z = z$$

$$-\pi \le \theta \le \pi,\quad 0 \le 2 + \cos\theta$$

$$F.n = (i2\cos\theta - j3\sin\theta + kz) . (i\cos\theta + j\sin\theta)$$
$$= 2\cos^2\theta - 3\sin^2\theta$$

$$\therefore \quad \iint_{S_2} F.n\, dS$$

$$= \int_{-\pi}^{\pi}\int_{0}^{2+\cos\theta} (2\cos^2\theta - 3\sin^2\theta)\sqrt{\cos^2\theta + \sin^2\theta}\, d\theta\, dz$$

$$= \int_{-\pi}^{\pi} (2-5\sin^2\theta)(2+\cos\theta)\, d\theta = -2\pi$$

$$\therefore \quad \iint_S F.n\, dS = \iint_{S_1} F.n\, dS + \iint_{S_2} F.n\, dS + \iint_{S_3} F.n\, dS$$

$$= 0 + 2p - 2p = 0.$$ **Ans.**

EXERCISES

1. Evaluate the integrals:
   (i) $\iint_S (z+2x+\frac{4}{5}y)\, dS$, over the plane $x/2 + y/3 + z/4 = 1$, lying in the first octant.
   (ii) $\iint_S xyz\, dS$, over the portion of $x + y + z = 1$, lying in the first octant.
   (iii) $\iint_{S'} x\, dS$ where S is the portion of the sphere $x^2 + y^2 + z^2 = a^2$ lying in the first octant.
2. Evaluate the surface integrals:
   (i) $\iint_S x^2y^2 z\, dS$, where S is the positive side of the lower half of the sphere $x^2 + y^2 + z^2 = a^2$.
   (ii) $\iint_S z^2\, dx\, dy$, where S is the outer side of the ellipsoid $\frac{x^2}{a^2} + \frac{y^2}{b^2} + \frac{z^2}{c^2} = 1$.
   (iii) $\iint_S (x^2\, dy\, dz + y^3\, dz\, dx + z^2\, dx\, dy)$, where S is the outer surface of the sphere $x^2 + y^2 + z^2 = 1$.
   (iv) $\iint_S (xz\, dx\, dy + xy\, dy\, dz + yz\, dz\, dx)$, where S is the outer side of the pyramid formed by the planes $x = 0$, $y = 0$, $z = 0$ and $x + y + z = 1$.

3. Compute the integrals:

 (i) $\iint_S \sqrt{a^2 - x^2 - y^2}\ dS$, where S is the hemisphere ,

 $z = \sqrt{a^2 - x^2 - y^2}$.

 (ii) $\iint_{S'} x^2 y^2\ dS$ where S is the hemisphere $z = \sqrt{R^2 - x^2 - y^2}$.

 (iii) $\iint_{S'} \frac{dS}{r^2}$, where S is the cylinder $x^2 + y^2 = a^2$, bounded by the planes $z = 0$, $z = h$, and r is the distance between a point on the surface and the origin.

4. Evaluate the surface integral:

 $\iint_S (x \cos \alpha + y \cos \beta + z \cos \gamma)\ dS$

 $\cos \alpha$, $\cos \beta$, $\cos \gamma$ being the direction cosines of the outward drawn normal of the surface S, where

 (i) S is the positive side of the cube formed by the planes $x = 0$, $y = 0$, $z = 0$, and $x = 1$, $y = 1$, $z = 1$.

 (ii) S is the outer surface of the ellipsoid $\frac{x^2}{a^2} + \frac{y^2}{b^2} + \frac{z^2}{c^2} = 1$ lying above the xy-plane

5. Evaluate $\iint_S z^2\ dS$, where S is the part of the outer surface of the cylinder $x^2 + y^2 = 4$ between the planes $z = 0$, $x + 3$.

6. Find the value of the surface integral:

 $\iint_S (yz\ dx\ dy + xz\ dy\ dz + xy\ dx\ dz)$,

 where S is the outer side of the surface situated in the first octant and formed by the cylinder $x^2 + y^2 = a^2$ and the planes $x = 0$, $y = 0$, $z = 0$, $z = h$.

7. Evaluate $\iint_E (x + y + z)(lx + my + nz)\ dS$, where E is the surface of the region $x^2 + y^2 \leq a^2$, $0 \leq z \leq h$.

8. Compute $\iint_S F.n\ dS$, when n is the outward drawn unit normal of the surface, and the vector function F and the oriented surface S are given in each case.

(i) $F = x^2i + y^2j + z^2k$, S is part cone $z^2 = x^2 + y^2$, $1 \le z \le 2$, $n.k > 0$, $n.i > 0$.

(ii) $F = y^2i + zj - xk$, S is part of cylinder $y^2 = 1 - x$ between the planes $z = 0$, $z = x$; $x \ge 0$ with $n.i > 0$.

9. Compute the integral:

$$\iint_S (y^2z\,dx\,dy + xz\,dy\,dz + x^2y\,dx\,dz),$$

where S is the outer side of the surface situated in the first octant and formed by the paraboloid of revolution $z = x^2 + y^2$, cylinder $x^2 + y^2 = 1$ and the coordinate planes.

ANSWERS

1. (i) $4\sqrt{16}$	(ii) $\sqrt{3}/120$	(iii) $\pi a^3/4$
2. (i) πa^3	(ii) $2\pi R^6/15$	(iii) $2\pi \tan^{-1}(h/a)$ (iv) 1/8
3. (i) $2\pi a^7/105$	(ii) 0	(iii) $56\pi a^2/9$
4. (i) 3	(ii) $2\pi abc$	7. π
6. $a^2h(2a/3 + \pi h/8)$	5. $\frac{1}{3}\pi ah[3(1 + m)a^2 + 3nah + 2nh^2]$	
9. (i) $15\pi/4$	(ii) 4/15	8. $\pi/8$.

Example 46:

Evaluate $\iint_S xyz\,dS$, *over the portion of* $x + y + z = 1$, *lying in the first octant.*

Solution:

$z = 1 - x - y \Rightarrow \partial z/\partial z = -1, \partial z/\partial y = -1.$

$\therefore \sqrt{[1 + (\partial z/\partial x)^2 + (\partial z/\partial y)]^2} = \sqrt{3}$. Thus

$$I = \iint_S xyz\,dS = \sqrt{3}\iint_D xyz\,dx\,dy,$$

where $D : x + y = 1$ is the projection of $S : x + y + z = 1$ on xy-plane.

$$\therefore \quad I = \sqrt{3}\int_0^1\int_0^{1-x} xy(1 - x - y)\,dx\,dy$$

$$= \sqrt{3}\int_0^1 \left|\tfrac{1}{2}x(1 - x)\,y^2 - \tfrac{1}{3}\,xy^3\right|_0^{1-x} dx$$

$$= \sqrt{3}\int_0^1 x(1 - x)^3\left(\tfrac{1}{2} - \tfrac{1}{3}\right) dx$$

$$= \frac{\sqrt{3}}{6} \int_0^1 x(1 - x^3 - 3x + 3x^2)\, dx$$

$$= \frac{\sqrt{3}}{6} \left(\frac{1}{2} - \frac{1}{5} - 1 + \frac{3}{4} \right) = \frac{\sqrt{3}}{120}.$$ **Ans.**

Example 47:

Evaluate $I = \iint_S \left(z + 2x + \frac{4}{3}y\right) dS$, *over the plane* $x/2 + y/3 + z/4 = 1$, *lying in the first octant.*

Solution:

The equation of the plane is

$S : 6x + 4y + 3z = 12 \Rightarrow \partial z/\partial x = -2,\ \partial z/\partial y = -4/3.$

$\therefore\ 1 + (\partial z/\partial x)^2 + (\partial z/\partial y)^2 = 61/9$. Thus

$$I = \iint_D \left(z + 2x + \tfrac{4}{3}y\right) \sqrt{\frac{61}{9}}\, dx\, dy,$$

where $D : 3x + 2y = 6$ is the projection of S on xy-plane. Now we have

$$I = \frac{\sqrt{61}}{3} \int_{x=0}^{2} \int_{y=0}^{\left(3-\frac{3}{2}x\right)} 4\, dx\, dy \qquad (\because 3z + 6x + 4y = 12)$$

$$= \frac{4}{3}\sqrt{61} \int_0^2 \left(3 - \tfrac{3}{2}x\right) dx = \tfrac{4}{5}\sqrt{61} \left| 3x - \tfrac{3}{4}x^2 \right|_0^2 = 4\sqrt{61}.$$

Example 48:

Evaluate $\iint_S x\, dS$, *where S* is the portion of the sphere $x^2 + y^2 + z^2 = a^2$ lying in the first octant.

Solution:

Hence we have $z = \sqrt{a^2 - x^2 - y^2}$

$$\text{So } \frac{\partial z}{\partial x} = \frac{-x}{\sqrt{a^2 - x^2 - y^2}},\ \frac{\partial z}{\partial y} = \frac{-y}{\sqrt{a^2 - x^2 - y^2}}$$

$$\therefore\ 1 + \left(\frac{\partial z}{\partial x}\right)^2 + \left(\frac{\partial z}{\partial y}\right)^2 = \frac{a^2}{\sqrt{a^2 - x^2 - y^2}}.$$ Thus

$$\iint_S x\, dS = \iint_D x\sqrt{1+\left(\frac{\partial z}{\partial x}\right)^2+\left(\frac{\partial z}{\partial y}\right)^2} = dx\ dy,$$

where D is the projection of S on xy-plane.

$$\therefore \quad \iint_S x\, dS = \iint_D \frac{x\, dx\, dy}{\sqrt{a^2 - x^2 - y^2}}$$

$$= a\int_0^{\pi/2}\int_0^a \frac{r\cos\theta . r\, dr\, d\theta}{\sqrt{a^2 - r^2}},$$

changing the above equation into polar form.

$$= a\int_0^{\pi/2} \cos\theta\, d\theta \int_0^a \frac{r^2\, dr}{\sqrt{a^2 - r^2}}$$

$$= a\int_0^{\pi/2} a^2 \sin^2 t\, dt, \quad r = a\sin t$$

$$= a^2 . \frac{1}{2} . \frac{\pi}{2} = \frac{\pi a^2}{4}.$$ **Ans.**

Example 49:

Evaluate $\iint_S \sqrt{a^2 - x^2 - y^2}\, dS$, *where S is the hemisphere* $z = \sqrt{a^2 - x^2 - y^2}$.

Solution:

By Example 48 we have

$$\iint_S \sqrt{a^2 - x^2 - y^2}\, dS = a\iint_D \frac{\sqrt{a^2 - x^2 - y^2}}{\sqrt{a^2 - x^2 - y^2}}\, dx\, dy,$$

$$= a\int_0^{2\pi}\int_0^a r\, dr\, d\theta = \pi a^2 .$$ **Ans.**

Example 50:

Show that $I = \iint_D (yz\, dy\, dz + zx\, dz\, dx + xy\, dx\, dy) = \frac{3}{8}$,

where S is the surface of the sphere $x^2 + y^2 + z^2 = 1$ *in the first quadrant.*

Solution:

By the symmetry of the integrand and the region of integration, we know that

$$I = 3\iint_D xy\, dx\, dy,\ D: x^2 + y^2 = 1\ (x \geq 0,\ y \geq 0)$$

$$= 3\int_0^{\pi/2}\int_0^1 r^2 \sin\theta \cos\theta \,.\, r\, dr\, d\theta \text{ (changing to polars)}$$

$$= 3\int_0^1 r^3\, dr \int_0^{\pi/2} \sin\theta \cos\theta\, d\theta = \frac{3}{4}\left|\frac{\sin^2\theta}{2}\right|_0^{\pi/2} = \frac{3}{8}. \qquad \textbf{Ans.}$$

Example 51:

Evaluate the surface integral $\iint z\cos\gamma\, dS$, *over the outer side of the sphere* $x^2 + y^2 + z^2 = 1$, *where* γ *is the inclination of the normal at any point of the sphere with the z-axis.*

Solution:

We have given that

$$\iint_S z\cos\gamma\, dS = \iint_{S_1} z\, dx\, dy + \iint_{S_2} z\, dx\, dy$$

where S_1 is the upper hemisphere above the xy-plane, and the lower hemisphere S_2 below it, also

$$\iint_S z\cos\gamma\, dS = \iint_D \sqrt{1 - x^2 - y^2}\, dx\, dy - \iint_D \sqrt{1 - x^2 - y^2}\, dx\, dy$$

[$\because z = \pm\sqrt{1 - x^2 - y^2}$. Here the first integral is over the upper side of the upper hemisphere in the upward direction, and the second integral is over the lower side of the lower hemisphere in downward direction. So the second integral will be negative.]

Thus we have

$$\iint_S z\cos\gamma\, dS = \iint_D \sqrt{1 - x^2 - y^2}\, dx\, dy,$$

where D is the circle $x^2 + y^2 = 1$.

$$\iint_S z\cos\gamma\, dS = 2\int_0^{2\pi} d\theta \int_0^1 \sqrt{1 - r^2}\,.\, r\, dr$$

$$= 4\pi\left|-\tfrac{1}{3}(1 - r^3)^{3/2}\right|_0^1 = \frac{4\pi}{3}. \qquad \textbf{Ans.}$$

Example 52:

Evaluate $\iint_S x^2y^2\, dS$, *where S is the hemisphere* $z = \sqrt{a^2 - x^2 - y^2}$.

Solution:

It is given that

$$\iint_S x^2y^2\ dS = a\iint_D \frac{x^2y^2\ dx\ dy}{\sqrt{a^2-x^2-y^2}},$$

(D is the projection of S on xy-plane)

$$= a\int_0^{2\pi}\int_0^a \frac{r^4 \sin^2\theta \cos^2\theta . r\ dr\ d\theta}{\sqrt{a^2-r^2}},\ \text{changing to polars}$$

$$= 4a\int_0^{2\pi} \sin^2\theta\cos^2\theta\ d\theta \int_0^a \frac{r^2\ dr}{\sqrt{a^2-r^2}}$$

$$= 4a.\frac{1}{4.2}.\frac{\pi}{2}\int_0^{2\pi} a^5 \sin^5 t\ dt,\ r = a\sin t$$

$$= \frac{\pi a^6}{4}.\frac{4}{5}.\frac{2}{3} = \frac{2\pi a^6}{15}.$$ **Ans.**

Example 53:

Evaluate $\iint x^2y^2 z\ dS$, *where S is the positive side of the lower half of the sphere* $x^2 + y^2 + z^2 = a^2$.

Solution:

Do your self. **(Ans.** $2\pi a^7/105$)

Example 54:

Evaluate $\iint_S \frac{dS}{r}$, *where S is the portion of the surface of the hyperbolic* $z = xy$ *cut off by the cylinder* $x^2 + y^2 = a^2$, *and r is the distance from a point on the surface to the z-axis.*

Solution:

We have $r^2 = x^2 + y^2$. Now we have

$$\iint_S \frac{dS}{r} = \iint_D \frac{1}{\sqrt{x^2+y^2}}\sqrt{\left[1+\left(\frac{\partial z}{\partial x}\right)^2+\left(\frac{\partial z}{\partial y}\right)^2\right]}\ dx\ dy$$

$$\iint_D \frac{\sqrt{1+x^2+y^2}}{\sqrt{x^2+y^2}}\ dx\ dy,$$

where D is the projection of the surface on xy-plane.

Changing the above equation into polar form we have

$$\iint_S \frac{dS}{r} = \int_0^{2\pi}\int_0^a \frac{\sqrt{1+r^2}}{r}\,.\,r dr\, d\theta$$

$$= 2\pi \int_0^a \sqrt{1+r^2}\; dr$$

$$= \pi\left[a\sqrt{a^2+1} + \log\left(a + \sqrt{a^2+1}\right)\right].$$ **Ans.**

Example 55:

Evaluate $\iint_S x^2\, dS$ *where S is the part of the outer surface of the cylinder* $x^2 + y^2 = 4$ *between the planes* $z = 0$ *and* $x = x + 3$.

Solution:

Refer to the adjoining Figure. The given surface S corresponds to the lateral surface F_2 of the cylinder. Changing to cylindrical co-ordinates:

We have S : $x = 2\cos\theta$, $y = 2\sin\theta$,

$z = z$; where

$-\pi \le \theta \le \pi$, $0 \le z \le 3 + 2\cos\theta$.

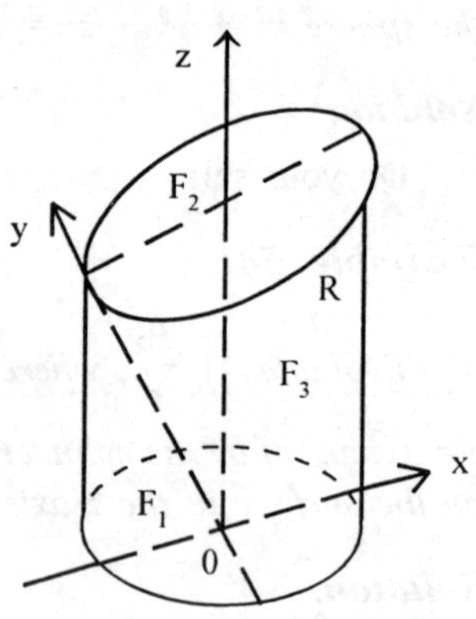

Now

$$\frac{\partial(y, z)}{\partial(\theta, z)} = 2\cos\theta,$$

$$\frac{\partial(z, z)}{\partial(\theta, z)} = 2\sin\theta,$$

$$\frac{\partial(x, y)}{\partial(\theta, z)} = 0. \text{ Now,}$$

$$\iint_S z^2\, dS = \iint s^2\sqrt{K\,\theta^2}\; d\theta = \iint x^2\sqrt{4\cos^2\theta + 4\sin^2\theta + 0}\; dz\, d\theta$$

$$= 2\int_{-\pi}^{\pi}\int_0^{3+2\cos\theta} z^6\; dx\, d\theta$$

$$= \frac{2}{3}\int_{-\pi}^{\pi} (3+2\cos)^2\; d\theta$$

$$= \frac{2}{3} \int_{-\pi}^{\pi} (27 + 54 \cos\theta + 36 \cos^2\theta + 8 \cos^3\theta)\, d\theta$$

$$= \frac{2}{3} \left[54\pi + 0 + 36.4.\tfrac{1}{2}.\frac{\pi}{2} + 0 \right] = 60\pi. \qquad \textbf{Ans.}$$

Example 56:

Evaluate $\iint_S \frac{dS}{r^2}$ *where S is the cylinder $x^2 + y^2 = a^2$, bounded by the planes $z = 0$, $z = h$ and r is the distance between a point on the surface and the origin.*

Solution:

We have $r^2 = x^2 + y^2 + z^2$.

Changing to cylinder co-ordinates we have

$S : x = a \cos\theta,\ y = a \sin\theta,\ z = z$; where

$-\pi \le \theta \le \pi,\quad 0 \le z \le h.$ Also $r^2 = a^2 + z^2$.

Now $\frac{\partial(y, z)}{\partial(\theta, z)} = a \cos\theta,\ \frac{\partial(z, x)}{\partial(\theta, z)} = a \sin\theta,\ \frac{\partial(x, y)}{\partial(\theta, z)} = 0.$

$$\therefore \quad \iint_S \frac{dS}{r^2} = \int_{-\pi}^{\pi} \int_0^h \frac{\sqrt{a^2 \cos^2\theta\, a^2 \sin^2\theta + 0}}{a^2 + z^2}\, dz\, d\theta$$

$$= 2\pi a \int_0^h \frac{dz}{a^2 + z^2} = 2\pi \left| \tan^{-1} \frac{z}{a} \right|_0^h$$

$$= 2\pi \tan^{-1} \frac{h}{a}. \qquad \textbf{Ans.}$$

Example 57:

Evaluate $I = \iint_E (x + y + z)(lx + my + nz)\, dS$, *where E is the surface of the region $x^2 + y^2 \le a^2$, $0 \le z \le h$.*

Solution:

The given surface is similar to that of Fig. 4.4, p. 370 with the only difference that the face F_2 $(z = h)$ is parallel to the face $F_1(\alpha = 0)$. The total surface F is given by

$E = F_1 \cup F_2 \cup F_3$.

On F_3 : $x = a \cos\theta,\ y = a \sin\theta,\ z = z$, where

$-\pi \le \theta \le \pi,\ \ 0 \le z \le h$. Now

$$I = \int_{-\pi}^{\pi}\int_{0}^{h}[a(\cos\theta + \sin\theta) + z]\,[a(l\cos\theta + m\sin\theta) + nz]\,a\,dz\,d\theta$$

$$= a\int_{-\pi}^{\pi}\int_{0}^{h}(nz^2 + a^2\,l\cos^2\theta + a^2\,m\sin^2\theta)\,d\theta\,dz.$$

other terms vanish.

$$\therefore\ \ I_2 = \tfrac{2}{3}\pi anh^3 + \pi a^3 h\,(l + m) \qquad ...(1)$$

On $F_1 : x^2 + y^2 = a^2,\ z = 0$. We have

$$I_1 = \iint_{F_1}(x + y)\ (lx + my)\,dS$$

$$= \int_{-\pi}^{\pi}\int_{0}^{a} r^2\,(\cos\theta + \sin\theta)\ l\cos\theta + m\sin\theta)\ r\,dr\,d\theta,$$

changing to polars

$$= \frac{a^4}{4}\int_{-\pi}^{\pi}(l\cos^2\theta + m\sin^2\theta)\,d\theta, \text{ other terms vanish.}$$

$$\therefore\ \ I_1 = 1/4\,(l + m)\,\pi a^4. \qquad ...(2)$$

On $F_2 : x^2 + y^2 = a^2,\ z = h$. We have

$$I_1 = \iint_{F_2}(x + y + h)\,(lx + my + nh)\,dS$$

$$= \iint(x + y)\,(lx + my)\,dS + nh^2\iint dS$$

$$+ (n + l)h\iint x\,dS + (n + m)h\iint y\,dS\,.$$

Using (2), we get

$$I_2 = \tfrac{1}{4}\,(l + m)\,\pi a^4 + nh^2\,(\pi a^2) + 0 + 0. \qquad ...(3)$$

Since unit normal to S_1 is negative, therefore

$$I = -\,I_1 + I_2 + I_3$$

$$= nh^2\pi a^2 + \tfrac{2}{3}\pi anh^3 + \pi a^3 h\,(l + m), \text{ by (1), (2), (3)}$$

$$= \tfrac{1}{3}\pi ah\,[\,3\,(l + m)\,a^3 + 3\,nah + 2nh^2].$$ **Ans.**

4.17 Stokes' Theorem

We have seen that the Green's theorem expresses a double integral over a plane region in terms of a line integral over the boundary curve of the plane. The *Stokes' theorem* is one generallization of the Green's theorem in as much as it expresses a surface integral over a surface in terms of a line integral over the boundary curve of the surface. The second generalization of the Green's theorem is *Gauss's theorem* .

Theorem 2:

State and Prove Stokes' Theorem

If S is a smooth oriented surface bounded by a curve C oriented in the same sense of f, g, h are three functions of x, y, z which are continuous and which posses continuous first order partial derivatives at each point of a domain containing S, then

$$\int_C (f dx + g dy + h dz)$$

$$= \iint \left[\left(\frac{\partial h}{\partial y} - \frac{\partial g}{\partial z} \right) dy\, dz + \left(\frac{\partial f}{\partial z} - \frac{\partial h}{\partial x} \right) dz\, dx + \left(\frac{\partial g}{\partial x} - \frac{\partial f}{\partial y} \right) dx\, dy \right].$$

(D.U., 1995, 94)

Proof:

Suppose that the oriented surface is given as

$$x = x(u, v),\ y = y(u, v),\ z = z(u, v); \quad (u, v) \in D$$

where D is an oriented surface in u, v plane.

Let Γ denote the oriented boundary curve of the domain D, where Γ is given as

$$u = u(t),\ v = v(t);\ a \le t \le b.$$

The proof the theorem will be given according to the following steps:

1. The line integral along C is expressed as an ordinary integral.
2. The ordinary integral is expressed as a line integral along Γ.
3. The line integral along Γ is expressed as a double integral over D (by Green's theorem).
4. The double integral over D is expressed as a surface integral over S.

We have

$$\int_C (f dx + g dy + h dz)$$

$$= \int_a^b \left[f \left(\frac{\partial x}{\partial u} \frac{\partial u}{\partial t} + \frac{\partial x}{\partial v} \frac{\partial v}{\partial t} \right) dt + g \left(\frac{\partial y}{\partial u} \frac{\partial u}{\partial t} + \frac{\partial y}{\partial v} \frac{\partial v}{\partial t} \right) dt + h \left(\frac{\partial z}{\partial u} \frac{\partial u}{\partial t} + \frac{\partial z}{\partial v} \frac{\partial v}{\partial t} \right) dt \right]$$

$$= \int_\Gamma \left(f \frac{\partial x}{\partial u} + g \frac{\partial y}{\partial u} + h \frac{\partial z}{\partial u} \right) du + \left(f \frac{\partial x}{\partial v} + g \frac{\partial y}{\partial v} + h \frac{\partial z}{\partial v} \right) dv$$

$$= \iint_D \left[\frac{\partial}{\partial u} \left(f \frac{\partial x}{\partial v} + g \frac{\partial y}{\partial v} + h \frac{\partial z}{\partial v} \right) \right.$$

$$\left. - \frac{\partial}{\partial v} \left(f \frac{\partial x}{\partial u} + g \frac{\partial y}{\partial u} + h \frac{\partial z}{\partial u} \right) \right] du\, dv. \quad \text{...(1)}$$

Now,

$$\frac{\partial}{\partial v} \left(f \frac{\partial x}{\partial v} \right) = \left(\frac{\partial f}{\partial x} \frac{\partial x}{\partial u} + \frac{\partial f}{\partial y} \frac{\partial y}{\partial u} + \frac{\partial f}{\partial z} \frac{\partial z}{\partial u} \right) \frac{\partial x}{\partial v} + f \frac{\partial^2 x}{\partial u \partial v},$$

$$\frac{\partial}{\partial v} \left(f \frac{\partial x}{\partial u} \right) = \left(\frac{\partial f}{\partial x} \frac{\partial x}{\partial v} + \frac{\partial f}{\partial y} \frac{\partial y}{\partial v} + \frac{\partial f}{\partial z} \frac{\partial z}{\partial v} \right) \frac{\partial x}{\partial v} + f \frac{\partial^2 x}{\partial v \partial u},$$

$$\therefore \quad \frac{\partial}{\partial u} \left(f \frac{\partial x}{\partial v} \right) - \frac{\partial}{\partial v} \left(f \frac{\partial x}{\partial u} \right) = \frac{\partial f}{\partial z} \frac{\partial (z, x)}{\partial (u, v)} - \frac{\partial f}{\partial y} \frac{\partial (z, x)}{\partial (u, v)} \text{ etc. etc.}$$

Thus, the double integral on the R.H.S. of (1) is

$$= \iint_D \left(\frac{\partial h}{\partial y} - \frac{\partial g}{\partial z} \right) \frac{\partial (z, x)}{\partial (u, v)} + \left(\frac{\partial f}{\partial z} - \frac{\partial h}{\partial x} \right) \frac{\partial (z, x)}{\partial (u, v)}$$

$$\left. + \left(\frac{\partial g}{\partial x} - \frac{\partial f}{\partial x} \right) \frac{\partial (z, x)}{\partial (u, v)} \right] du\, dv$$

$$= \iint_S \left[\left(\frac{\partial h}{\partial y} - \frac{\partial g}{\partial z} \right) dy\, dz + \left(\frac{\partial f}{\partial z} - \frac{\partial h}{\partial x} \right) dz\, dx + \left(\frac{\partial g}{\partial x} - \frac{\partial f}{\partial x} \right) dx\, dy \right].$$

Hence the theorem.

Cor. 1:

By the definition of surface integral, the expression of Stokes' theorem is equivalent to

$$\int_C (f\, dx + g dy + h\, dz = \iint_S \left[\left(\frac{\partial h}{\partial y} - \frac{\partial g}{\partial z} \right) \cos \alpha \right.$$

$$\left. + \left(\frac{\partial f}{\partial z} - \frac{\partial h}{\partial x} \right) \cos \beta + \left(\frac{\partial g}{\partial x} - \frac{\partial f}{\partial y} \right) \cos \gamma \right] : dS,$$

where $\cos \alpha$, $\cos \beta$, $\cos \gamma$ are the direction cosines of the normal at any point of thee surface.

Cor 2:

Vectorial Form of Stokes Theorem

Let r = xi + yj + zk be the position vector of any point on the surface so that the unit normal at any point on the surface is

n cos αi + cos βj + cos γk. We write

$F(x, y, z) = f(x, y, z)i + g(x, y, z)\, j + h(x, y, z)\, k.$

$$\text{Then curl } F = \left(\frac{\partial h}{\partial y} - \frac{\partial g}{\partial z}\right) i + \left(\frac{\partial f}{\partial z} - \frac{\partial h}{\partial x}\right) j + \left(\frac{\partial g}{\partial x} - \frac{\partial f}{\partial y}\right) k,$$

$$\text{curl } F.\, n = \left(\frac{\partial h}{\partial y} - \frac{\partial g}{\partial z}\right) \cos\alpha + \left(\frac{\partial f}{\partial z} - \frac{\partial h}{\partial x}\right) \cos\beta + \left(\frac{\partial g}{\partial x} - \frac{\partial f}{\partial y}\right) \cos\gamma,$$

and F. dr = j Ix + g dy + h dz.

By Cor, 1, Stokes theorem can be written as

$$\int_C F.dr = \iint_S \text{curl } F.n \, dS.$$

This is the vectorial form of Stocke's Theorem.

Example 58:

Use Stokes' theorem to find the line integral

$$\int_C (x^2 y^3 \, dx + dy + z \, dz),$$

where C is the circle $x^2 + y^2 = a^2$, $z = 0$.

Solution:

By Stokes' theorem we have

$$\int_C (x^2 y^3 \, dx + dy + z \, dz) = \int_C (f dx + g + dy + h \, dz)$$

$$= \iint_S \left[\left(\frac{\partial h}{\partial y} - \frac{\partial g}{\partial z}\right) dy\, dz + \left(\frac{\partial f}{\partial z} - \frac{\partial h}{\partial x}\right) dz\, dx + \left(\frac{\partial g}{\partial x} - \frac{\partial f}{\partial y}\right) dx\, dy\right]$$

$$= \iint_S [(0 - 0)\, dy\, dz + (0 - 0)\, dz\, dx + (0 - 3x^2y^2)\, dx\, dy]$$

$$= -3 \iint_S x^2 y^2 \, dx\, dy, \; S : x^2 + y^2 = a^2$$

$$= -3 \int_{-\pi}^{\pi} \int_0^a r^2 \cos^2\theta \, r^2 \sin^2\theta \, r \, dr \, d\theta \quad \text{(change to polars)}$$

$$= -\frac{3a^2}{6} \times 4 \int_0^{\pi/2} \cos^2\theta \sin^2\theta \, d\theta = -2a^6 . \frac{1.1}{4.2} . \frac{\pi}{2} = -\frac{\pi a^6}{8}.$$

Ans.

Example 59:

Verify Stokes' theorem when

$$F = (x^2y^3,\ 1,\ z),\quad C : x^2 + y^2 = a^2,\ z = 0.$$

Solution:

To evaluate the line integral over C directly, we put $x = a\cos\theta$, $y = a\sin\theta$, $z = 0$. Then we have

$$\int_C (x^2y^3\,dx + dy + dz)$$

$$= \int_{-\pi}^{\pi} [a^2\cos^2\theta \,.\, a^3\sin^3\theta\,(-a\sin\theta) + 1.a\cos\theta + 0]\,d\theta$$

$$= -4a^6 \int_0^{\pi/2} \sin^4\theta\cos^2\theta\,d\theta + 0$$

$$= -4a^6 \cdot \frac{3.1.1}{6.4.2} \cdot \frac{\pi}{2} = -\frac{\pi a^6}{8}.$$

Stokes' theorem is verified.

Example 60:

Using Stokes' theorem, prove that

$$\int_C (y\,dx + z\,dy + x\,dz) = -2\pi a^2\sqrt{2},$$

where C is the curve $x^2 + y^2 + z^2 - 2ax - 2ay = 0,\ x + y = 2a.$

(D.U., 1996)

Solution:

The given curve can be written as

$$C : (x-a)^2 + (y-a)^2 + z^2 = 2a^2,\ (x-a) + (y-a) = 0$$

or $\quad C : X^2 + Y^2 + Z^2 = 2a^2,\ X + Y = 0\ (X = x - a,\ Y = y - a,\ Z = z).$

$$\therefore \quad \int_C (y\,dx + z\,dy + x\,dz) = \int_C [(Y + a)\,dX + Z\,dY + (X + a)\,dZ]$$

Let $F = (Y + a, Z, X + a)$ so that

$$\text{curl } F = \begin{vmatrix} i & j & k \\ \partial/\partial X & \partial/\partial Y & \partial/\partial Z \\ Y+a & Z & X+a \end{vmatrix} = -(i + j + k).$$

By Stokes' theorem,

$$\int_C F.dr = \iint_S \text{curl } F.n\,dS$$

$$= - \iint_S (i+ j+k).n \, dS \qquad ...(1)$$

where S is any surface whose boundary is C. We take S as the surfacc of the plane X + Y = 0, bounded by the circle

$$X^2 + Y^2 + Z^2 = 2a^2, \; X + Y = 0.$$

The unit normal to the plane X + Y = 0 is given as

$n = \frac{1}{\sqrt{2}} i + \frac{1}{\sqrt{2}} j + 0k$ and so by (1),

$$\int_C (y\,dx + z\,dy + x\,dz) = - \iint_S \left(\frac{1}{\sqrt{2}} + \frac{1}{\sqrt{2}} + 0 \right) dS$$

$$= - \sqrt{2} \iint_S dS$$

$= - \sqrt{2}$(area of great circle of radius $\sqrt{a}$)

$= - \sqrt{2}(\pi.2a^2) = -2\sqrt{2}\pi a^2.$ **Ans.**

Example 61:

Verify Stokes' theorem when

$$F = (2x - y, - yz^2, - y^2z); \; S : x^2 + y^2 + z^2 = a^2, \; z \geq 0.$$

(D.U., 1994)

Solution:

The boundary C os S is a circle with equations

$x = a \cos \theta, \; y = a \sin \theta, \; z = 0.$ Now

$$\Rightarrow \int_C (f\,dx + g\,dy + h\,dz) \;\; \Rightarrow \int_C [(2x - y)\,dx - yz^2\,dy - y^2z\,dz]$$

$$\Rightarrow \int_{-\pi}^{\pi} [a\,(2 \cos \theta - \sin \theta)\,(-a \sin \theta) - 0 - 0]\,d\theta$$

$$\Rightarrow a^2 \left[\tfrac{1}{2} \left| \cos 2\theta \right|_{-\pi}^{\pi} + 4 \int_0^{\pi/2} \sin^2 \theta \, d\theta \right] = 4a^2 . \tfrac{1}{2} . \frac{\pi}{2} = \pi a^2.$$

We have

$$\iint_S \left[\left(\frac{\partial h}{\partial y} - \frac{\partial g}{\partial z} \right) dy\,dz + \left(\frac{\partial f}{\partial z} - \frac{\partial h}{\partial x} \right) dz\,dx + \left(\frac{\partial g}{\partial x} - \frac{\partial f}{\partial y} \right) dx\,dy \right]$$

$$\Rightarrow \iint_S [(-2yz + 2yz)\,dy\,dz + (0 - 0)\,dz\,dx + (0 + 1)\,dx\,dy]$$

$= \iint_S dx\,dy = \pi a^3$. Hence Stokes' theorem is verified.

Example 62:

Using Stokes' theorem, show that

$$\int_C (ydx + zdy + xdz) = -\iint_S (\cos\alpha + \cos\beta + \cos\gamma)\, dS.$$

Solution:

Let f = y, g = z, h = x so that

$$\frac{\partial h}{\partial y} - \frac{\partial g}{\partial z} = \frac{\partial f}{\partial z} - \frac{\partial h}{\partial x} = \frac{\partial g}{\partial x} - \frac{\partial f}{\partial y} = 0 - 1 = -1. \quad ...(1)$$

Let F = (y, z, x). By Stokes' theorem, we have

$$\int_C F.dr = \iint_S \text{curl } F.n\, dS$$

$$= \iint_S \left[\left(\frac{\partial h}{\partial y} - \frac{\partial g}{\partial z}\right)\cos\alpha + \left(\frac{\partial f}{\partial z} - \frac{\partial h}{\partial x}\right)\cos\beta + \left(\frac{\partial g}{\partial x} - \frac{\partial f}{\partial y}\right)\cos\gamma\right] dS$$

$$= -\iint_C (\cos\alpha + \cos\beta + \cos\gamma)\, dS, \text{ using (1).}$$

Example 63:

Show that

$$I = \iint_S [(y - z)\, dy\, dz + (z - x)dz\, dx + (x - y)dx\, dy] = a^3\pi,$$

where S is the portion of the surface $x^2 + y^2 - 2ax + az = 0,\ z \geq 0.$

Solution:

Let F = $(y^2 + z^2, z^2 + x^2, x^2 + y^2)$ so that

$$\text{curl } F = \begin{vmatrix} i & j & k \\ \partial/\partial X & \partial/\partial Y & \partial/\partial Z \\ y^2 + z^2 & z^2 + x^2 & x^2 + y^2 \end{vmatrix}$$

$= 2(y - z, z - x, x - y)$. By Stokes' theorem,

$$\int_C F.dr = \iint_S \text{curl } F.n\, dS.$$

Thus, $\quad I = \frac{1}{2}\int_C [(y^2 + z^2)\, dx + (z^2 + x^2)\, dy + (x^2 + y^2)\, dz],$

where C is the curve $(x - a)^2 + y^2 = a^2,\ z = 0,$

We take $x - a = a\cos\theta$ [i.e., $x = a(1 + \cos\theta)$], $y = a\sin\theta,\ z = 0.$

$$\therefore \quad I = \frac{1}{2}\int_{-\pi}^{\pi} [(a^2 \sin^2\theta\ (-a\sin\theta) + a^2(1 + \cos\theta)^2 a\cos\theta + 0]$$

$$= \frac{1}{2}a^2 \int_{-\pi}^{\pi} (-\sin^3 \theta + \cos \theta + 2 \cos^2 \theta + \cos^3 \theta)\, d\theta$$

$$= \frac{1}{2}a^3 \left[0 + 0 + \int_{-\pi}^{\pi} (2\cos^2 \theta + \cos^3 \theta)\, d\theta\right]$$

(We know that $\sin^3 \theta$ is an odd function)

$$= a^3 \int_0^{\pi} (2 \cos^2 \theta + \cos^3 \theta)\, d\theta,$$

(integrand being an even function)

$$= 4a^3 \int_0^{\pi/2} \cos^2 \theta\, d\theta + \frac{1}{4} a^3 \int_0^{\pi} (\cos 3\theta + 3\cos \theta)\, d\theta$$

$$= 4a^3 . \frac{1}{2} . \frac{\pi}{2} + 0 = \pi a^3.$$ **Hence Proved.**

Example 63:

Show that $\iint_S \text{curl } F . n\, dS = 12\pi$

where $F = (x - z, x^3 + yz, - 3xy^2)$ *and S is the surface of the cone* $z = 2 - \sqrt{x^2 + y^2}$ *above the xy-plane.*

Solution:

By Stokes' theorem, we have

$$\iint_S \text{curl } F . n\, dS = \int_C F.dr, \qquad ...(1)$$

where the boundary curve C is given by the equation

$x^2 + y^2 = 4$, $z = 0$ or $x = 2 \cos \theta$, $y = 2 \sin \theta$, $z = 0$.

Now, we have

$$\int_C F.dr = \int_C [(x-z)\, dx + (x^3 + yz)\, dy - 3xy^2\, dz]$$

$$\int_C (dx + x^3\, dy) \quad (z = 0 \text{ along C})$$

$$= \int_{-\pi}^{\pi} [2\cos\theta\, (-2 \sin\theta) + 8 \cos^3 . 2 \cos \theta]\, d\theta$$

$$= 0 + 16 \times 4 \int_0^{\pi/2} \cos^4 \theta\, d\theta = 64 . \frac{3}{4} . \frac{1}{2} . \frac{\pi}{2} = 12\pi. \qquad ...(2)$$

From (1) and (2), $\iint_S \text{curl } F . n\, dS = 12\pi.$ **Ans.**

Example 64:

Verify Stokes' theorem for $F = zi + xj + yk$, where S is the part of the paraboloid $z = 1 - x^2 - y^2$ for which $z \geq 0$, n, k > 0.

Solution:

Do your self **(Ans. π)**

[**Hint.** The boundary C of S is $x = \cos\theta$, $y = \sin\theta$, $z = 0$.]

Example 65:

Verify Stokes' theorem for $F = 2yi + 3x\,j - z^2\,k$, where S is the upper half surface of the sphere $x^2 + y^2 + z^2 = 9$ and C is its boundary.

(Ans. 9π)

Solution:

Do your self

Example 66:

Using Stokes' theorem, prove that

$$\int_C [(y+z)\,dx + (z+x)dy + (x+y)\,dz] = 0,$$

where C is the circle $x^2 + y^2 + z^2 = a^2$, $x + y + z = 0$.

Solution:

Let f = y + z, g = z + x, h = x + y.

Then $\dfrac{\partial h}{\partial y} - \dfrac{\partial g}{\partial z} = 1 - 1 = 0$. Similarly,

$$\frac{\partial f}{\partial z} - \frac{\partial h}{\partial x} = 0,\ \frac{\partial g}{\partial x} - \frac{\partial f}{\partial y} = 0.$$

By Stokes' theorem,

$$\int_C (f\,dx + g\,dy + h\,dz) = \iint_S \left[\underbrace{\left(\frac{\partial h}{\partial y} - \frac{\partial g}{\partial z}\right)}_{0} dy\,dz + \ldots + \ldots\right].$$

Example 67:

Using Stokes' theorem, show that

$$\int_C (c^x\,dx + 2y\,dy - dz) = 0$$

where C is the curve $x^2 + y^2 = 4$, $z = 0$.

Solution:

Do your self.

Example 68:

$F = 3y\, i - xz\, j - yx^2\, k$,

where S is the surface of the paraboloid $2z = x^2 + y^2$ bounded by $z = 2$ and C is its boundary.

Solution:

Do your self. **(Ans.** Common value = 20π)

Hint. The boundary C of S is a circle with equations $x^2 + y^2 = 4$, $z = 2$ with parametric equations $x = 2\cos\theta$, $y = 2\sin\theta$, $z = 2$, $0 \le \theta \le 2\pi$).

Example 69:

Show that $\iint_S \text{curl } F.n\, dS = -\pi$,

where $F = (2y(1 - x),\ x - x^2 + y^2,\ x^2 + y^2 + z^2)$ *and*

$S : x^2 + y^2 + z^2 = 1,\ z \ge 0.$

Solution:

Do your self.

4.18 Gauss' Divergence Theorem

Theorem 3:

State and Prove Gauss' Divergence Theorem Statements : *If a three-dimensional regular domain E is bounded by a smooth oriented surface S; and f, g, h are three functions which along with their partial derivatives f_x, g_y, h_z are continuous at each point of E and S, then*

$$\iiint_E (f_x + g_y + h_z)\, dx\, dy\, dz = \iint_S (f\, dy\, dz + g\, dz\, dx + h\, dx\, dy),$$

where the surface integral is taken over the exterior of the surface S.

Proof:

Let the given surface S is such that a line parallel to each axis cuts it in at the most two points. Then it is possible to split the external surface S into two sub-surfaces S_1 and S_2 both of which project into the same region of the xy-plane; where S_1, S_2 are given by the equations

$z = \phi(x, y), z = \psi(x, y)$

(x, y) varies over the same region of the xy-plane.

These two sub-surfaces S_1 and S_2 have the same boundary curve, the boundary curve of S_1 differs from that of S_2 in orientation only. Thus if D is the region of the xy-plane on which S_1 projects, then –D is the region on which S_2 projects.

Thus we have

$$\iiint_E h_z \, dx\, dy\, dz = \iint_D \left[\int_{\psi(x,y)}^{\phi(x,y)} h_z \, dz \right] dx\, dy$$

$$= \iint_D \left| h(x, y, z) \right|_{\psi(x,y)}^{\phi(x,y)} dx\, dy$$

$$= \iint_D [h(x, y, \phi) - h(x, y, \psi)] \, dx\, dy$$

$$= \iint_D h(x, y, \phi) \, dx\, dy - \iint_D h(x, y, \phi) \, dx\, dy$$

$$= \iint_D h(x, y, z) \, dx\, dy + \iint_{-D} h(x, y, z) \, dx\, dy$$

$$= \iint_{S_1} h(x, y, z) \, dx\, dy + \iint_{S_2} h(x, y, z) \, dx\, dy$$

$$= \iint_S h(x, y, z) \, dx\, dy$$

$$\therefore \quad \iiint_E h_z \, dx\, dy\, dz = \iint_S h(x, y, z) \, dx\, dy \qquad \text{...(1)}$$

Similarly, we have

$$\iiint_E g_y \, dx\, dy\, dz = \iint_S g(x, y, z) \, dx\, dz. \qquad \text{...(2)}$$

$$\iiint_E f_x \, dx\, dy\, dz = \iint_S f(x, y, z) \, dy\, dz. \qquad \text{...(3)}$$

Adding (1), (2) and (3), we get

$$\iiint_E (f_x + g_y + h_z) \, dx\, dy\, dz = \iint_S (f\, dy\, dz + g\, dz\, dx + h\, dx\, dy).$$

Cor. 1:

The above expression can also be written as

$$\iiint_E (f_x + g_y + h_z) \, dx\, dy\, dz = \iint_S (f \cos\alpha + g \cos\beta + h \cos\gamma) \, dS.$$

where $\cos\alpha$, $\cos\beta$, $\cos\gamma$ are the direction cosines of the outward drawn unit normal of S.

Cor. 2:

Vectorial Form of Gauss's Theorem

Let F = fi + gj + hk

$$\therefore \quad \text{div } F = \frac{\partial f}{\partial x} + \frac{\partial g}{\partial y} + \frac{\partial h}{\partial z} = (f_x + g_y + h_z)$$

Let n = i cos α + j cos β + k cos γ, so that

F. n = f cos α + g cos β + h cos γ, By Cor. 1,

$$\iiint_E \text{div } F \, dV = \iint_S F.n \, dS,$$

where dV denotes element of volume.

This is the vector form of Gauss's Theorem.

Example 70:

Evaluate $I = \iint_S (xz \, dy \, dz + xy \, dz \, dx + yz \, dx \, dy)$,

where S is the outer surface situated in the first octant and formed by the cylinder $x^2 + y^2 = a^2$ and the planes $x = 0$, $y = 0$, $z = 0$, $z = h$.

Solution:

Let f = xz, g = xy, h = yz, so that

$f_x = z$, $g_y = x$, $h_z = y$. By Gauss's theorem we have

$$I = \iint_E (x + y + z) \, dx \, dy \, dz$$

$$= \iint_D \left[\int_0^h (x + y + z) \, dz \right] dx \, dy,$$

where D : $x^2 + y^2 = a^2$; $x \geq 0$; $y \geq 0$.

$$\therefore \quad I = \iint_D [(x + y)h + \tfrac{1}{2}h^2] \, dx \, dy$$

Changing to polars ($x = r \cos \theta$, $y = r \sin \theta$), we have

$$I = \int_0^{\pi/2} \int_0^a [r(\cos \theta + \sin \theta) \, h \, \tfrac{1}{2} h^2] \, r \, dr \, d\theta$$

$$= \frac{1}{3} a^2 h \, \Big| \sin \theta - \cos \theta \, \Big|_0^{\pi/2} + \tfrac{1}{4} h^2 a^2 . \frac{\pi}{2}$$

$$= \frac{2}{3} a^2 h + \frac{\pi}{8} h^2 a^2 = ha^2 \left(\frac{2a}{3} + \frac{\pi h}{8} \right).$$ **Ans.**

Example 71:

Evaluate $\iint_S x\,dy\,dz + y\,dz\,dx + z\,dx\,dy,$

S being the outer side of the cube [0, a; 0, a; 0, a].

Solution:

Let f = x, g = y, h = z, so that $f_x = g_y = h_z = 1$.

By Gauss's Theorem, we have

$$\iint_S (f\,dy\,dz + g\,dz\,dx + h\,dx\,dy) = \iiint_E (f_x + g_y + h_z)\,dx\,dy\,dz$$

$$= \int_0^a\int_0^a\int_0^a (1 + 1 + 1)\,dx\,dy\,dz = 3a^3. \qquad \textbf{Ans.}$$

Example 72:

Evaluate $\iint_S [(x^3 - yz)\cos\alpha\ 2x^2\ y\cos\beta + 2\cos\gamma]$

dS, taken over the outer surface of the cube bounded by the planes x = 0, x = a; y = 0, y = a ; z = 0, z = a, cos α, cos β, cos γ are the direction cosines of the outward drawn normal.

Solution:

Here we have

$$I = \iint_S [(x^2 - yz)\,dy\,dz - 2x^2\ y\,dz\,dx + 2\,dx\,dy].$$

Let $f = x^3 - yz$, $g = -2x^3y$, $h = 2$. Then

$f_x = 3x^2$, $g_x = -2\,x^2$, $h_z = 0$. (By Gauss's theorem) we have

$$I = \iiint_E (f_x + g_y + h_z)\,dx\,dy\,dz$$

$$= \int_0^a\int_0^a\int_0^a x^2\,dx\,dy\,dz = \int_0^a x^2\,dx \int_0^a dy \int_0^a dz$$

$$= \tfrac{1}{3}a^5. \qquad \textbf{Ans.}$$

Example 73:

Show that $\int_S F\,.\,n\,dS = 0$, $F = (2x, -3y, z)$,

where S is the skin of solid $x^2 + y^2 = 1$, $z = 0$, $z = x + 2$.

Solution:

We have F = (2x, – 3y, z), so that

div F = 2 − 3 + 1 = 0

By Gauss's Divergence Theorem, we obtain

$$\int_S F \cdot n\, dS = \int_V \text{div } F\, dV = 0.$$ **Ans.**

Example 74:

Evaluate $\iint_S F \cdot n\, dS,\ F = (z^2 - x)i - xyj + 3zk,$

where S is the surface of the region bounded by $z = 4 - y^2,\ x = 0,\ x = 3$ *and the xy-plane.*

Solution:

div F = (−1 − x + 3) = 2 − x.

By Gauss's Divergence Theorem, we obtain,

$$\int_S F \cdot n\, dS = \int_V \text{div } F\, dV$$

$$= \iiint_V (2 - x)\, dx\, dy\, dz$$

$$= \int_{-2}^{2} dy \int_0^{4-y^2} dz \int_0^3 (2 - x)\, dx$$

$$= \frac{3}{2}\int_{-2}^{2} (4 - y^2)\, dy = \frac{3}{2}\left|\, 4y - \frac{1}{3}y^2 \right|_{-2}^{2} = 16.$$ **Ans.**

Example 75:

Let p be the length of the perpendicular from the origin to the tangent plane at the point (x, y, z) of the ellipsoid

$x^2/y^2 + y^2/b^2 + z^2/c^2 = 1.$

Prove that (i) $\int_S P\, dS = 4\pi\, abc.$ **(D.U., 1995)**

(ii) $\int_S p(x^4 + y^4 + z^4)\, dS = \frac{4}{5}\pi\, abc\, (a^4 + b^4 + c^4).$

Solution:

We know that

(i) Let F . n = p so that $F.\left(\frac{x}{a^2} i + \frac{y}{b^2} j + \frac{z}{c^2} k\right) = 1,$ ...(1)

Again

$F = (c^2/z)\, k \Rightarrow \text{div } F = -(c^2/z^2).$

Now, $\int_S p\, dS = \int_S F \cdot n\, dS$

$= \int_V \text{div } F\, dV$ (by Gauss's Theorem)

$$= -c^2 \iiint_V \frac{dx\, dy\, dz}{z^2} \qquad ...(2)$$

Taking $x/a = r \sin\theta \cos\phi$, $y/b = r \sin\theta \sin\phi$, $z/c = r\cos\theta$, ...(3)

where $0 \le r \le 1$, $0 \le \theta \le \pi$, $0 \le \phi \le 2\pi$; we obtain

$$|J| = \left|\frac{\partial(x, y, z)}{\partial(r, \theta, \phi)}\right| = abc\, r^2 \sin\theta.$$

From (2), $\int_S p\, dS = -c^2 \iiint \dfrac{abc\, r^2 \sin\theta\, dr\, d\theta\, d\phi}{c^2 r^2 \cos^2\theta}$

$$= -abc \int_0^1 dr \int_0^{2\pi} d\phi \int_0^{\pi} \sec\theta \tan\theta\, d\theta = 4\pi abc.$$

(ii) Let $F \cdot n. = p\,(x^4 + y^4 + z^4)$. We know that

$F \cdot (x/a^2, y/b^2, z/c^2) = x^4 + y^4 + z^4$. ...(4)

The relation (4) will be true if we take

$F = (a^2x^3, b^2y^3, z^2c^3) \Rightarrow \text{div } F = 3(a^2x^2 + b^2y^2 + c^2z^2)$.

Now, $\int_S p(x^4 + y^4 + z^4)\, dS = \int_S F.n\, dS$

$= \int_V \text{div } F\, dV$ (Gauss's Theorem)

$$= 3 \iiint_V (a^2x^2 + b^2y^2 + c^2z^2)\, dx\, dy\, dz$$

$$= 3 \iiint r^2\, [(a^4 \cos^2\phi + b^4 \sin^2\phi) \sin^2\theta) + c^4 \cos^2\theta]\, abc\, r^2 \sin\theta\, dr\, d\theta\, d\phi, \text{ [using (3)]}$$

$$= 3abc \int_0^1 r^4\, dr \int_0^{\pi} \int_0^{2\pi} [(a^4 \cos^2\phi + b^4 \sin^2\phi) \sin^2\theta + c^4 \cos^2\theta] \sin\theta\, d\theta\, d\phi$$

$$= \frac{3}{5}\pi abc \int_0^{\pi} [(a^4 + b^4) \sin^3\theta + 2c^4 \cos^2\theta \sin\theta]\, d\theta$$

$$= \frac{3}{5}\pi abc \left[2 \cdot \frac{2}{3}(a^4 + b^4) + \frac{4}{3} c^4\right]$$

$$= \frac{3}{5}\pi abc\,(a^4 + b^4 + c^4).$$ **Ans.**

Example 75:

Evaluate the surface integral

$$I = \iint_S (x^3\,dy\,dz + y^3\,dz\,dx + z^3\,dx\,dy)$$

over the sphere $S : x^2 + y^2 + z^2 = a^2$. **(D.U., 1985)**

Solution:

By Gauss's theorem, we have

$$\iint_S x^3\,dy\,dz + y^2\,dz\,dx + z^3\,dx\,dy = \iiint_E 3(x^2 + y^2 + z^2)\,dx\,dy\,dz \qquad ...(1)$$

Changing to spherical polar coordinates we obtain

$$x = r\sin\theta\cos\phi,\ y = r\sin\theta\sin\phi,\ z = r\cos\theta;$$

equ. (1) reduces to

$$I = 3\iiint_E r^2 . r^3 \sin\theta\ .\ dr\,d\theta\,d\phi \quad (\because |J| = r^2\sin\theta)$$

$$= 3\int_0^{2\pi} d\phi \int_0^{\pi} \sin\theta\,d\theta \int_0^{a} r^4\,dr$$

$$= 3.2\pi \left| -\cos\theta \right|_0^{\pi} . \frac{a^5}{5} = \frac{12\pi a^5}{5}.$$

Example 76:

Evaluate

$$I = \iint_S (x^3\cos\alpha + y^3\cos\beta + z^3\cos\gamma)\,dS$$

where S is the surface of sphere of radius a with centre at the origin and α, β, γ *are angles of inclination to the coordinate axes of the normal to the surface.*

Solution:

Do your self

[Hint. $I = \iint_S (x^3\,dy\,dz + y^3\,dz\,dx + z^3\,dx\,dy)$.]

Example 77:

Evaluate

$$I = \iint zx\, dx\, dy + xy\, dy\, dz + yz\, dz\, dx$$

taken over the pyramid formed by the planes

$x = 0, y = 0,. z = 0, x + y + z = 1.$ **(D.U., 1988)**

Solution:

Let $f = xy$, $g = yz$, $h = zx$ so that $f_x = y$, $g_y = z$, $h_z = x$.

By Gauss's Theorem, we have

$$I = \iint_D \left[\int_0^{1-x-y} (y + z + x)\, dz \right] dx\, dy,$$

where D is the projection on the xy-plane of the given domain.

$$I = \iint_D \left[\left| yz + \tfrac{1}{2}z^2 + xz \right|_0^{1-x-z} \right] dx\, dy.$$

$$= \iint_D [y(1 - x - y) + \tfrac{1}{2}(1 - x - y)^2 + x(1 - x - y)]\, dx\, dy$$

$$= \int_{x=0}^1 \int_{y=0}^{1-x} [y(1-x) - y^2 + \tfrac{1}{2}(1-x-y)^2 + x(1-x-y)]\, dxdy$$

$$= \int_{x=0}^1 \left| \tfrac{1}{2}y^2(1-x) - \tfrac{1}{3}y^3 - \tfrac{1}{6}(1-x-y)^3 - \tfrac{1}{2}x(1-x-y) \right|_0^{1-x} dx$$

$$= \int_{x=0}^1 \left[\frac{(1-x)^3}{2} - \frac{(1-x)^3}{3} - \{0 - \tfrac{1}{6}(1-x)^3\} - \tfrac{1}{2}\{0 - x(1-x)^2\} \right]$$

$$= \int_0^1 \{\tfrac{1}{3}(1-x)^3 + \tfrac{1}{2}(x + x^3 - 3x^2)\}\, dx$$

$$= \frac{1}{3} \left| -\frac{(1-x)^4}{4} \right|_0^1 + \frac{1}{2}\left[\frac{x^2}{2} + \frac{x^4}{4} - \frac{2x^2}{3} \right]_0^1$$

$$= \frac{1}{12} + \frac{1}{2}\left(\frac{1}{2} + \frac{1}{4} - \frac{2}{3} \right) = \frac{1}{8}.$$ **Ans.**

Example 78:

Evaluate $I = \iint_S (x\, dy\, dz + y\, dz\, dx + z^2\, dx\, dy)$,

where S denotes the closed surface bounded by the cone $x^2 + y^2 = z^2$

and the plane $z = 1$.

Solution:

Let $f = x$, $g = y$, $h = z^2$ so the $f_x = g_y = 1$, $h_z = 2z$.

By Gauss's Theorem, We have

$$I = 2 \iiint_E (1 + z)\, dx\, dy\, dz$$

$$= 2 \int_0^1 (1 + z)\, dz \iint_{x^2+y^2=z^2} dx\, dy$$

$$= 2 \int_0^1 (1 + z) \,.\, \pi z^2\, dz = 2\pi\left(\tfrac{1}{3} + \tfrac{1}{4}\right) = \frac{7\pi}{6}.$$ **Ans.**

Example 79:

Evaluate

$$I = \iint_S (x \cos\alpha + y \cos\beta + z^2 \cos\gamma)\, dS.$$

where S denotes the closed surface bounded by the cone $x^2 + y^2 = z^2$ and the plane $z = 1$, and $\cos\alpha$, $\cos\beta$, $\cos\gamma$ are direction cosines of the outward normal of S.

Solution:

Do your self.

[**Hint.** $I = \iint_S (x\, dy\, dz + y\, dz\, dx + z^2\, dx\, dy)$.]

Example 80:

Evaluate

$$I = \iint_S (x\, dy\, dz + y\, dz\, dx + z\, dx\, dy),$$

where S is the surface of the region bounded by the cylinder $x^2 + y^2 = 9$ and the planes $z = 0$ and $z = 3$.

Solution:

Let $f = x$, $g = y$, $h = z$ so that $f_x = f_y = f_z = 1$.

By Gauss's Theorem, we have

$$I = \iiint_E (1 + 1 + 1)\, dx\, dy\, dz$$

$$= 3 \iint \left[\int_0^3 dz\right] dx\, dy = 9 \iint_{x^2+y^2=9} dx\, dy$$

$= 9\ (\pi\ .\ 3^2) = 81\ \pi.$ **Ans.**

Example 81:

Show that $I = \iint_S (x^2\ dy\ dz + y^2\ dz\ dx + z^2\ dx\ dy) = 0,$
where S denotes the surface of the ellipsoid $x^2/a^2 + y^2/b^2 + z^2c^2 = 1.$

Solution:

Let $f = x^2$, $g = y^2$, $h = z^2$ so that $f_x = 2x.\ g_y = 2y,\ h_z = 2z.$
By Gauss's theorem, we have

$$I = 2\iiint_E (x + y + z)\ dx\ dy\ dz. \qquad ...(1)$$

Let $x/a = r \sin\theta \cos\phi$, $y/b = r \sin\theta \sin\theta$, $z/c = r\cos\theta$. Then

$$J = |\ \partial(x, y, z)/\partial(r, \theta, \phi)| = abc\ r^2 \sin\theta.$$

Now $\iiint_E x\ dx\ dy\ dz = \int_{\theta=0}^{\pi} \int_{\phi=0}^{2\pi} \int_{r=0}^{1} ar \sin\theta \cos\phi$

$$abc\ r^2 \sin\theta\ .\ dr\ d\theta\ d\phi$$

$$= 0 \left(\because \int_0^{2\pi} \cos\phi\ d\theta = 0\right)$$

Similarly, $\iiint_E y\ dx\ dy dz = \iiint_E z\ dx\ dy\ dz = 0,$

Hence, by (1), I = 0. **Ans.**

Example 82:

Show that $\iint_S (x \cos\alpha + y \cos\beta + z\cos\gamma)\ dS = \frac{4}{3}\pi\ abc,$ *S is the surface of the ellipsoid* $x^2/a^2 + y^2/b^2 + z^2/c^2 = 1.$

Solution:

Do your self.

Example 83:

Let p be the length of perpendicular from the origin to the tangent plane of any point of the ellipsoid

$$S : x^4/a^2 + y^2/b^2 + z^2/c^2 = 1.$$

Prove that $\int_S \frac{dS}{P} = \frac{4\pi abc}{3}\left(\frac{1}{a^2} + \frac{1}{b^2} + \frac{1}{c^2}\right)$ **(D.U., 1994)**

Solution:

The tangent plane at any point (x, y, z) to the given ellipsoid S is

$$(xX/a^2) + (yY/b^2) + (zZ/c^2) = 1. \qquad ...(1)$$

The length of perpendicular from the origin (0, 0, 0) to (1) is

$$p = \frac{1}{\sqrt{[(x^2/a^4) + (y^2/b^4) + (z^2/c^4)}}. \qquad ...(2)$$

From (1), the unit normal to the ellipsoid is

$$n = (x/a^2, y/b^2, z/c^2) / \sqrt{[(x^2/a^4) + (y^2/b^4) + (z^2/c^4)]}$$

or $n = p[(x/a^2)\, i + (y/b^2)\, j + (z/c^2)\, k]$. ...(3)

By Gauss's Divergence Theorem, we have

$$\int_S F.n\, dS = \int_V \text{div } F\, dV. \qquad ...(4)$$

If we choose $F = (x/a^2)\, i + (y/b^2)\, j + (z/c^2)\, k$, then ...(5)

$F \,.\, n = p(x^2/a^4) + (y^2/b^4) + (z^2/c^4)] = p/p^2 = 1/p$, by (2) and (3)

$$\text{Thus } \int_S \frac{dS}{p} = \int_S F\,.\,n\, dS = \int_V \text{div } F\, dV, \text{ by (4)}$$

$$= \int_V (1/a^2 + 1/b^4 + 1/c^2)\, dV, \text{ by (5)}$$

$$= (1/a^2 + 1/b^2 + 1/c^2) \times \text{volume of the ellipsoid}$$

$$= \frac{4}{3}\pi\, abc \left(\frac{1}{a^2} + \frac{1}{b^2} + \frac{1}{c^2}\right).$$

Hence the result.

Example 84:

Show that

$$\iint_S (ax\, dy\, dz + by\, dz\, dx + cz\, dx\, dy) = \frac{4}{3}\pi\,(a + b + c),$$

where S is the surface of the sphere $x^2 + y^2 + z^2 = 1$

Solution:

Let f = ax, g = by, h = cz; so that

$f_x = a$, $g_y = b$, $h_z = c$. By Gauss's Theorem,

$$\iint_S (ax\ dy\ dz + by\ dz\ dx + cz\ dx\ dy) = \iiint_E (a + b + c)\ dx\ dy\ dz$$

$$= (a + b + c) \iiint_E r^2 \sin\theta \,.\, dr\ d\theta\ d\phi,$$

(changing to spherical polar coordinates)

$$= (a + b + c) \int_0^{2\pi} d\phi \int_0^{\pi} \sin\theta\ d\theta \int_0^1 r^2\ dr$$

$$= 2\pi\ (a + b + c) \,.\, 2 \,.\, \frac{1}{3} = \frac{4}{3}\pi\ (a + b + c).$$

Hence the result.

Example 85:

Evaluate $I = \iint_S (ax \cos\alpha + by \cos\beta + cz \cos\gamma)\ dS$, *where S is the surface of the sphere* $x^2 + y^2 + z^2 = 1$, *and* $\cos\alpha$, $\cos\beta$, $\cos\gamma$ *are direction cosines of the outward normal of S.*

Solution:

Do your self.

[**Hint.** $I = \iint_S (ax\ dy\ dz + by\ dz\ dx + cz\ dx\ dy)$.]

Example 86:

Show $\iint_S [(2xy + z)\ dy\ dz + y^2\ dz\ dx - (x + 3y)\ dx\ dy] = 27$

taken over thee region bounded by

$2x + 2y + z = 6,\ x = 0,\ y = 0,\ z = 0.$

Solution:

Do your self.

Example 87:

Evaluate the surface integral

$$I = \iint_S (y^2z\ dx\ dy + xz\ dy\ dz + x^2y\ dz\ dx),$$

where S is the outer side of the surface situated in the first octant and formed by the paraboloid of revolution $z = x^2 + y^2$, *cylinder* $x^2 + y^2 = 1$ *and the co-ordinate planes.*

Solution:

Let $f = xz$, $g = x^2y$, $h = y^2z$ so that

$f_x = z$, $g_y = x^2$, $h_z = y^2$.

By Gauss's Theorem, we have

$$I = \iint_D \left[\int_0^{x^2+y^2} (x^2 + y^2 + z)\, dz \right] dx\, dy,$$

Where D is the projection on the xy-plane of the domain enclosed by S.

$$I = \iint_D \left[\left| x^2 z + y^2 z + \tfrac{1}{2} z^2 \right|_0^{x^2+y^2} \right] dx\, dy,$$

$$= \iint_D [x^2(x^2 + y^2) + y^2 (x^2 + y^2) + \tfrac{1}{2}(x^2 + y^2)^2]\, dx\, dy$$

$$= \iint_D [(x^2 + y^2)^2 + \frac{1}{2} (x^2 + y^2)^2]\, dx\, dy$$

$$= \tfrac{3}{2} \iint_D (x^2 + y^2)^2\, dx\, dy.$$

Putting $x = r \cos \theta$, $y = r \sin \theta$, we get

$$I = \tfrac{3}{2} \iint_D r^4 \, . \, r\, dr\, d\theta$$

$$= \tfrac{3}{2} \int_0^{\pi/2} d\theta \int_0^1 r^2\, dr = \frac{3}{2} \cdot \frac{\pi}{2} \cdot \frac{1}{6} = \frac{\pi}{8}.$$ **Ans.**

EXERCISES

1. Show that the surface integral:

   $$\iint_S (z\, dy\, dz + x\, dz\, dx + y\, dx\, dy) = 0,$$

   over the closed surface S.

   [**Hint.** $F = (z, x, y) \Rightarrow \text{div } F = 0$.]

2. Use Gauss's theorem to evaluate:

   $$\iint_S (ax \cos \alpha + by \cos \beta + cz \cos \gamma)\, dS, \text{ where}$$

   (i) S is the surface of the cube [0, a; 0, a; 0, a].

   (ii) S is the surface of the ellipsoid $x^2/a^2 + y^2/b^2 + z^2/c^2 = 1$.

   $$\left[\textbf{Ans.}\ \text{(i)}\ (a + b + c)a^3,\ \text{(ii)}\ \frac{4}{3} \pi abc\, (a + b + c) \right]$$

Hint. F = (ax, by, cz) $\Rightarrow$ div F = (a + b + c).

By Gauss's theorem,

(i) $\iint_S \text{F.n dS} = \iiint_V \text{div F dV} = (a + b + c)$ [Volume of the cone.]

4.19 Integral of a Bounded Function Over a Rectangle R = [a, b; c, d; e, f]

Let us consider three partition of the intervals [a, b], [c, d] and [e, f]:

$P_1 = \{a = x_0, x_1 \text{ ....}, x_{t-1}, x_t, \text{ .....}, x_m = b\}$,

$P_2 = \{c = y_0, y_1 \text{ ....}, y_{t-1}, y_t, \text{ .....}, y_n = d\}$,

$P_3 = \{e = z_0, z_1 \text{ ....}, z_{t-1}, x_t, \text{ .....}, x_p = f\}$.

These three partitions give rise to a partition say, P of the rectangle R into mnp sub-rectangles

$[x_{r-1}, x_r; y_{t-1}, y_t; z_{t-1}, z_t]$ ..(1)

r, s, t varying from 1 to m, 1 to n, and 1 to p respectively,

Let $w_{rst} = (x_r = x_{r-1})(y_s = y_{s-1})(z_t = z_{t-1})$,

which is the volume of the rectangle (1).

Let M_{rst}, m_{rst} denote the supremum and the influence of f in the rectangle (1).

We write

$$U(P, f) = \sum_{t=1}^{p} \sum_{s=1}^{n} \sum_{r=1}^{m} M_{rst}\, w_{rst}$$

$$L(P, f) = \sum_{t=1}^{p} \sum_{s=1}^{n} \sum_{r=1}^{m} m_{rst}\, w_{rst}$$

Then the infimum of the set of sums U(P, f) and the supremum of the set of sums L(P, f) are respectively known as the upper and lower integrals of f over R and denoted by

$$\iiint_R f(x,y,z)\, dx\, dy\, dz,\ \iiint_R f(x,y,z)\, dx\, dy\, dz.$$

If these two integrals are equal, we say that the function is integrable over R and the common value, called the integral of f over R is denoted by

$$\iiint_R f(x,y,z)\, dx\, dy\, dz.$$

Remark:

The volume V of the solid of any shape in R^3 is given by

$$V = \iiint dx\,dy\,dz.$$

4.20 Calculation of Triple Integrals

If (i) $\iiint_R f(x,y,z)\,dx\,dy\,dz.$

and (ii) $\iiint_S f(x,y,z)\,dx\,dy$

where $e \le z \le f$ and $S = [a, b; c, d]$ both exist, then

$$\int_e^f \left[\iint_S f(x, y, z)\,dx\,dy \right] dz$$

also exists and is equal to the triple integral.

$S = [a, b; c, d]$ is the projection of the rectangle R on the xy-plane.

Further, if a function is continuous over R, then the triple integral can be evaluated by expressing it as an literated (repeated) integral of xyz in any order i.e.,

$$\iiint_R f(x, y, z)\,dx\,dy\,dz = \int_a^b dx \int_c^d dy \int_e^f f(x, y, z)\,dz$$

$$= \int_e^f dz \int_c^d dy \int_e^f f(x, y, z)\,dx \text{ etc.}$$

However, in the general case of an arbitrary integrable function f, the orders are not always interchangeable.

4.21 Calculation of a Triple Integral on a Bounded Domain E

If f is continuous in a region E bounded by surfaces

$z = \phi(x, y)$, $z = \psi(x, y)$; $y = g(x)$, $y = h(x)$; $x = a$, $x = b$; then

$$\left[\iiint_E f(x, y, z)\,dx\,dy\,dz == \int_a^b \int_{g(x)}^{h(x)} \left\{ \int_{\psi(x,y)}^{\phi(x,y)} f(x, y, z)\,dz \right\} dy \right] dx$$

4.22 Change of Variables in a Triple Integral

Let f be continuous in a bomain E in xyz-space.

Let the functions

$x = x(u, v, w)$, $y = y(u, v, w)$, $z = z(u, v, w)$

map the domain E in a one-to-one manner on to a domain E_1 in uvw-space. We further suppose that these functions possess continuous first order partial derivatives at each point of E_1.

Let f(x, y, z) = f(x(u, v, w), y(u, v, w), z(u, v, w)) = F(u, v, w).

Then

$$\iiint_E f(x, y, z)\, dx\, dy\, dz = \iiint_{E_1} F(u, v, w)\, |J|\, du\, dv\, dw,$$

where $$J = \frac{\partial(x, y, z)}{\partial(u, v, w)} = \begin{vmatrix} \frac{\partial x}{\partial u} & \frac{\partial x}{\partial v} & \frac{\partial x}{\partial w} \\ \frac{\partial y}{\partial u} & \frac{\partial y}{\partial v} & \frac{\partial y}{\partial w} \\ \frac{\partial z}{\partial u} & \frac{\partial z}{\partial v} & \frac{\partial z}{\partial w} \end{vmatrix}$$

is the Jacobian.

Remark 1:

Cylindrical Polar Coordinates

These are given by $x = r\cos\theta$, $y = r\sin\theta$, $z = z$.

It is easy to verify that $J = \dfrac{\partial(x, y, z)}{\partial(r, \theta, z)} = r$.

We use cylindrical polar co-ordinates when the domain of integration is a right circular cylinder.

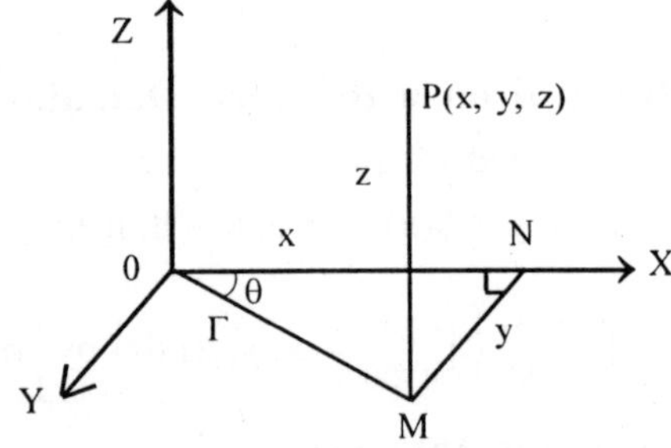

Remark 2:

Spherical Polar Co-ordinates

These are given by

$x = r\sin\theta\cos\phi$,

$y = r\sin\theta\sin\phi$,

$z = r\cos\theta$.

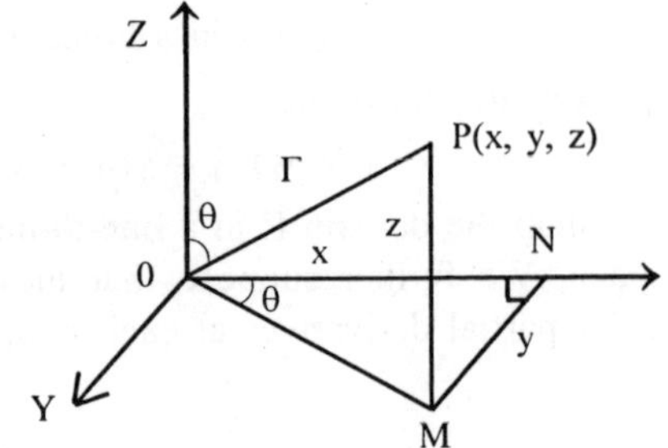

It is easy to verify that

$$J = \frac{\partial(x, y, z)}{\partial(r, \theta, \phi)} = r^2 \sin\theta.$$

We use spherical polar co-ordinates when the domain of integration is part of a sphere.

Example 88:

Show that $\iiint \frac{dx\,dy\,dz}{\sqrt{1 - x^2 - dy - z^2}} = \frac{\pi^2}{8}$, *integral being extended to all the positive values of the variables for which the expression is real.*

Solution:

The domain of integration is the positive octant of the sphere $x^2 + y^2 + z^3 \le 1$.

Changing to spherical polar co-ordinates; the region

$$E = \{(x, y, z) : x \ge 0, y \ge 0, z \ge 0, x^2 + y^2 + z^2 \le 1\}$$

transforms into

$$E_1 = \{(r, \theta, \phi) : 0 \le r \le 1, 0 \le \theta \le \tfrac{1}{2}\pi, 0 \le \phi \le \tfrac{1}{2}\pi\}.$$

Thus the given integral become

$$= \int_0^{\pi/2} \int_0^{\pi/2} \int_0^1 \frac{r^2 \sin\theta\, dr\, d\theta\, d\phi}{\sqrt{1 - r^2}}$$

$$= \int_0^{\pi/2} d\phi \int_0^{\pi/2} \sin\theta\, d\theta \int_0^1 \frac{r^2}{\sqrt{1 - r^2}}\, dr$$

$$= \frac{\pi}{2} \left| -\cos\theta \right|_0^{\pi/2} \int_0^{\pi/2} \sin^2 t\, dt, \; r = \sin t$$

$$= \frac{\pi}{2} \cdot \frac{1}{2} \cdot \frac{\pi}{2} = \frac{\pi^2}{8}.$$

Ans.

Example 89:

Evaluate

$$\iiint \frac{dx\,dy\,dz}{\sqrt{a^2 - x^2 - y^2 - z^2}} \text{ over } x \ge 0, y \ge 0, z \ge 0$$

for which this expression is valid. **(D.U., 1994)**

Solution:

Do your self. **(Ans.** $\pi^2a^2/8$)

Example 90:

Evaluate $I = \iiint_R \sqrt{1 - x^2\ y^2\ z^2}\ dx\ dy\ dz,$

where R is the region interior to the sphere $x^2 + y^2 + z^2 = 1$.

Solution:

Do your self. **(Ans.** $\pi^2/32$)

$$\left[\textbf{Hint. } I = \int_0^{\pi/2} d\phi \int_0^{\pi/2} \sin\theta\ d\theta \int_0^1 r^2 \sqrt{1 - r^2}\ dr\right.$$

$$\left. = \frac{\pi}{2}\int_0^{\pi/2} \sin^2\theta \cos^2\theta\ d\theta = \frac{\pi}{2}\cdot\frac{1}{4.2}\cdot\frac{\pi}{2}(r = \sin\theta)\right]$$

Example 91:

Show that

$$\iiint_E \frac{dx\ dy\ dz}{\sqrt{x^2 + y^2 + (z - b)^2}} = \frac{4\pi a^3}{3b},\ 0 < a < b$$

where E is the volume of $x^2 + y^2 + z^2 \le a^2$. **(D.U., 1995)**

Solution:

Let $x = r \sin\theta \cos\phi$, $y = r \sin\theta \sin\phi$, $z = r \cos\theta$.

$\therefore$ $I = r^2 \sin\theta$. The given integral is obtain as

$$= \int_0^{2\pi}\int_0^{\pi}\int_0^{a} \frac{r^2 \sin\theta\ dr\ d\theta\ d\phi}{\sqrt{[r^2 \sin^2\theta + (r\cos\theta - b)^2}}$$

$$= 2\pi \int_0^a r\ dr \int_0^{\pi} \frac{r \sin\theta\ d\theta}{\sqrt{(r^2 - 2br\cos\theta + b^2)}}$$

$$= \frac{2\pi}{b}\int_0^a r \left|\sqrt{[r^2 - 2br\cos\theta + b^2]}\right|_0^{\pi} dr$$

$$= \frac{2\pi}{b}\int_0^a r\,[(r + b) - (b - r)]\ dr \qquad (\because r < b)$$

$$= \frac{4\pi}{b}\int_0^a r^2\ dr = \frac{4\pi a^2}{3b}$$ **Ans.**

Example 92:

Show that $\iiint (lx^2 + my^2 + nz^2)$ *dx dy dz taken throughout the sphere*

$x^2 + y^2 + z^2 = a^2$, *is* $\frac{4\pi}{15}$ $(l + m + n)\, a^5$.

Solution:

Do your self.

Example 93:

Show that

$$\iiint_E (x^3 + y^3 + z^3)\, dx\, dy\, dz = \frac{32}{5} \pi a^6,$$

where E is the interior of the sphere

$$x^2 + y^2 + z^2 - 2a (x + y + z) + 2a^2 = 0.$$

Solution:

The given sphere can be written as

$$(x - a)^2 + (y - a)^2 + (z - a)^2 = a^3$$

or $X^2 + Y^2 + Z^2 = a^2$; $X = x - a$, $Y = y - a$, $Z = z - a$.

Now, $J = \frac{\partial(x, y, z)}{\partial(X, Y, Z)} = 1$. The given integral is

$$I = \iiint_D \left[(X + a)^3 + (Y + a)^2 + (Z + a)^3\right] . 1 . dX\, dY\, dZ.$$

where D is the interior of the sphere $X^2 + Y^2 + Z^2 = a^2$.

or $I = \iiint_D (X^2 + Y^2 + Z^2)\, dX\, dY\, dZ$

$$+ 3a = \iiint_D (X^2 + Y^2 + Z^2)\, dX\, dY\, dZ + 3a^2 \iiint_D (X + Y + Z)$$

$$dX\, dY\, dZ + 3a^2 \iiint_D dX\, dY\, dZ. \quad ...(1)$$

Put, $X = r \sin\theta \cos\phi$, $Y = r \sin\theta \sin\phi$, $Z = r\cos\theta$ so that $J = r^2 \sin\theta$.

Now, $\iiint_D (X + Y + Z)\, dX\, dY\, dZ$

$$= \iiint [r \sin\theta (\cos\phi + \sin\phi) + r\cos\theta]\, r^2 \sin\theta\, dr\, d\theta\, d\phi\delta$$

$$= \int_0^a r^3\,dr \int_0^{2\pi} d\phi \int_0^{\pi} \left[\sin^2\theta\,(\cos\phi + \sin\phi) + \sin\theta\cos\theta\right] d\theta$$

$$= \frac{1}{4}a^4 \int_0^{2\pi} \left[2 \cdot \frac{1}{2} \cdot \frac{\pi}{2}\,(\cos\phi + \sin\phi) + 0\right] d\phi = 0.$$

Similarly we can verify that

$$\iiint_D (X^3 + Y^3 + Z^3)\, dX\, dY\, dZ = 0$$

Now $\iiint_D (X^2 + Z^2 + Z^2)\, dX\, dY\, dZ = \dfrac{4\pi a^5}{5}$

Also $\iiint_D dX\, dY\, dZ = \dfrac{4}{5}\pi a^3$

being the volume of the sphere $X^2 + Y^2 + Z^2 = a^2$.

Hence, by (1),

$$I = 0 + 3a\left(\frac{4\pi a^5}{5}\right) + 0 + 3a^3\left(\frac{4\pi a^3}{3}\right) = \frac{32\pi a^5}{5}.$$

Example 94:

Express $I = \iiint f(x, y, z)\, dx\, dy\, dz$ *over the sphere* $x^2 + y^2 + z^2 = a^2$ *as a repeated integral.*

Solution:

We have

$$I = \int_{-a}^{a} dx \int_{-\sqrt{a^2-x^2}}^{\sqrt{a^2-x^2}} dy \int_{-\sqrt{a^2-x^2-y^2}}^{\sqrt{a^2-x^2-y^2}} f(x, y, z)\, dz.$$

Example 95:

Evaluate $I = \iiint (x + y + z + 1)^2\, dx\, dy\, dz$ *over the region* $x \geq 0$, $y \geq 0$, $z \geq 0$, $x + y + z \leq 1$.

Solution:

The given region is bounded below and above by $z = 0$ and $z = 1 - x - y$ respectively. Its projection D on xy-plane is a triangle bounded by $x = 0$, $y = 0$, $y = 1 - x$. Thus

$$I = \int_0^1 dx \int_0^{1-x} dy \int_0^{1-x-y} (x + y + z + 1)^2\, dz$$

$$= \frac{1}{3} \int_0^1 dx \int_0^{1-x} \left|(x + y + z + 1)^3\right|_0^{1-x-y} dy$$

$$= \frac{1}{3} \int_0^1 dx \int_0^{1-x} [8 - (x + y + 1)^3]\, dy$$

$$= \frac{1}{3} \int_0^1 \left[8y - \frac{1}{4}(x + y + 1)^4\right]_0^{1-x} dx$$

$$= \frac{1}{3} \int_0^1 \left[8(1-x) - \frac{1}{4}\{16-(x + 1)^4\}\right] dx$$

$$= \frac{1}{3}\left[4x - 4x^2 + \frac{1}{20}(x+1)^5\right]_0^1 = \frac{1}{60}(32-1) = \frac{31}{60}.$$ **Ans.**

Example 96:

Show that

$$\iiint (ax^2 + by^2 + cz^2)\, dx\, dy\, dz = \frac{4}{15}\pi\, (a + b + c)\, R^5$$

taken throughout $\{(x, y, z) + x^2 + y^2 + z^2 \le R^5\}$.

Solution:

Changing to spherical polar coordinates we have

$x = r \sin\theta \cos\phi,\ y = r \sin\theta \sin\phi,\ z = r\cos\theta.$

$$\iiint (ax^2 + by^2 + cz^2)\, dx\, dy\, dz$$

$$= \int_0^{2\pi} \int_0^{\pi} \int_0^{R} (a \sin^2\theta \cos^2\phi + b \sin^2\theta \sin^2\phi + c \cos^2\theta)\, r^2 \times (r^2 \sin\theta)\, dr\, d\theta\, d\phi$$

$$= \frac{R^5}{5}\left[4 \times 2 \times a \int_0^{\pi/2} \sin^2\theta\, d\theta \int_0^{\pi/2} \cos^2\phi\, d\phi + 4 \times 2 \times b \int_0^{\pi/2} \sin^2\theta\, d\theta \int_0^{\pi/2} \sin^2\phi\, d\phi - \frac{1}{3} c \left|\cos^3\theta\right|_0^{\pi} .2\pi\right]$$

$$= \frac{R^5}{5}\left[8a \times \frac{2}{3} \times \frac{1}{2} \times \frac{\pi}{2} + 8b \times \frac{2}{3} \times \frac{1}{2} \times \frac{\pi}{2} + \frac{4}{3}\pi c\right]$$

$$= \frac{4\pi}{15}(a + b + c)R^5.$$ **Ans.**

Example 97:

Evaluate $I = \iiint_V x^2 \, dx\, dy\, dz$, *where V the sphere* $x^2 + y^2 + z^2 \le 1$.

Solution:

Changing to spherical polar co-ordenates; we have,

$$I = \int_0^{2\pi} \int_0^{\pi} \int_0^{1} r^2 \sin^2 \theta \cos^3 \phi \, (r^2 \sin \theta)\, dr\, d\theta\, d\phi$$

$$I = \int_0^{2\pi} \cos^2 \phi \, d\phi \, . \int_0^{\pi} \sin^2 \theta \, d\theta \, . \left|\frac{1}{5} r^5\right|_0^1$$

$$= \frac{1}{5} \times 4 \times 2 \int_0^{2\pi} \cos^2 \phi \, d\phi \, . \int_0^{\pi} \sin^2 \theta \, d\theta$$

$$= \frac{8}{5} . \frac{1}{2} . \frac{\pi}{2} . \frac{2}{3} = \frac{4\pi}{15}.$$

Ans.

Example 98:

Evaluate $I = \iiint_V z^2 \, dx\, dy\, dz$, *where V is the sphere* $x^2 + y^2 + z^2 \le 1$.

Solution:

Do your self. **(Ans.** $4\pi/15$)

$$\left[\textbf{Hint. } I = \int_0^{2\pi} \int_0^{\pi} \int_0^{1} r^2 \cos^2 \theta \, (r^2 \sin \theta)\, dr\, d\theta\, d\phi.\right]$$

Example 99:

Show that for $m > 0$, $\iiint_V (x^2 + y^2 + z^2)^m \, dx\, dy\, dz = \dfrac{4\pi}{2m + 3}$,
where V is the sphere $x^2 + y^2 + z^2 \le 1$.

Solution:

Do your self.

Example 100:

Evaluate $\iiint_V \sqrt{x^2 + y^2 + z^2} \, dx\, dy\, dz$, *where V is the sphere* $x^2 + y^2 + z^2 \le 1$.

Solution:

Do your self. **(Ans. $\pi/8$)**

Example 101:

Show that $\iiint e^{\sqrt{x^2/a^2 + y^2/b^2 + z^2/c^2}}\, dx\, dy\, dz = 4\pi abc\,(e - 2)$, integral being taken over the region $x^2/a^2 + y^2/b^2 + z^2/c^2 \leq 1$.

Solution:

Changing to spheral polar co-ordinates, where $x/a = r \sin\theta \cos\phi$, $y/b = r \sin\theta \sin\phi$, $z/c = r \cos\theta$; then the given integral

$$= \int_0^{2\pi}\int_0^{\pi}\int_0^1 e^r\,(abc\, r^2 \sin\theta)\, dr\, d\theta\, d\phi \qquad (J = abc\, r^2 \sin\theta)$$

$$= abc \int_0^{2\pi} d\phi \int_0^{\pi} \sin\theta\, d\theta \int_0^1 r^2 e^r\, dr$$

$$= 2\pi\, abc \left|-\cos\theta\right|_0^{\pi} \left[\left|r^2 e^r\right|_0^1 - 2\int_0^1 r\, e^r\, dr\right]$$

$$= 4\pi\, abc \left[e - 2\left\{\left|r\, e^r\right|_0^1\ 1 \,.\, e^r\, dr\right\}\right]$$

$$= 4\pi\, abc \left[e - 2\left\{e - \left|e^r\right|_0^1\right\}\right]$$

$$= 4\pi\, abc\,[e - 2\,\{e\,(e - 1)\}] = 4\pi\, abc\,(e - 2).$$ **Ans.**

Example 102:

Show that

$$\iiint (lx + my + nz)^2\, dx\, dy\, dz = \frac{4\pi}{15}(l^2 + m^2 + n^2),$$

taken throughout the region $\{(x, y, z);\ x^2 + y^2 + z^2 \leq 1\}$.

Solution:

Changing x, y, z to spherical polar co-ordinates $x = r \sin\theta \cos\phi$, $y = r\ \sin\theta, \sin\phi$, $z = r \cos\theta$; we have the following

$$\iiint (lx + my + nz)^2\, dx\, dy\, dz$$

$$= \int_0^{2\pi}\int_0^{\pi}\int_0^1 r^2\,(l \sin\theta \cos\phi + m \sin\theta \sin\phi + n \cos\theta)^2 \times r^2 \sin\theta\, dr\, d\theta\, d\phi$$

$$= \frac{1}{5} \int_0^{2\pi} \int_0^{\pi} (l^2 \sin^3\theta \cos^2\phi + m^2 \sin^2\theta \sin^3\phi + n^2 \cos^2\theta \sin\theta$$
$$+ lm \sin^2\theta \sin\phi \cos\phi + 2mn \sin^2\theta \cos\theta \sin\phi$$
$$+ 2\, ln \sin^2\theta \cos\theta \cos\phi)\, d\theta\, d\phi$$

$$= \frac{1}{5} \int_0^{2\pi} \int_0^{\pi} (l^2 \cos^2\phi + m^2 \sin^2\phi) \sin^3\theta\, d\theta\, d\phi$$
$$- \frac{n^2}{3} \left|\cos^2\theta\right|_0^{\pi} 2x + 2lm \int_0^{\pi} \left|\frac{1}{2} \sin^3\phi\right|_0^{2\pi} \sin^2\theta\, d\theta$$
$$+ 2mn \left|\frac{1}{3} \sin^3\theta\right|_0^{\pi} \left|-\cos\theta\right|_0^{2\pi} + 2ln \left|\frac{1}{3} \sin^3\theta\right|_0^{\pi} \left|\sin\phi\right|_0^{2\pi}$$

$$= \frac{1}{5}\Bigg[4 \times 2 \int_{\phi=0}^{\pi/2} (l^2 \cos^2\phi - m^2 \sin^2\phi)\, d\phi \times \int_0^{\pi/2} \sin^3\theta\, d\theta$$
$$+ \frac{4}{3}\pi n^2 + 0 + 0 + 0\Bigg]$$

$$= \frac{1}{5}\left[8\left(l^2 \cdot \frac{1}{2} \cdot \frac{\pi}{2} + m^2 \cdot \frac{1}{2} \cdot \frac{\pi}{2}\right) \cdot \frac{2}{3} + \frac{4}{3}\pi n^2\right]$$

$$= \frac{4\pi}{15}(l^2 + m^2 + n^2).$$ **Ans.**

Example 103:

Show that

$$\iiint_E (ax + by + cz)^2\, dx\, dy\, dz = \frac{4}{15}\pi\,(a^2 + b^2 + c^2).$$

where domain E is the sphere $x^2 + y^2 + z^2 \le 1$.

Solution:

Do your self.

Example 104:

Show that

$$\iiint \frac{dx\, dy\, dz}{(x + y + z + 1)^2} = \frac{1}{16} \log \frac{256}{e^5},$$

taken throughout the integral of the tetrahedron bounded by the planes $x = 0,\ y = 0,\ z = 0,\ x + y + z = 1$. **(D.U., 1996)**

Solution:

The given integral is written as

$$= \int_0^1 dx \int_0^{1-x} dy \int_0^{1-x-y} \frac{dz}{(x+y+z+1)^3}$$

$$= -\frac{1}{2}\int_0^1 dx \int_0^{1-x} \left|\frac{1}{(x+y+z+1)^2}\right|_0^{1-x-y} dy$$

$$= -\frac{1}{2}\int_0^1 dx \int_0^{1-x} \left[\frac{1}{4} - \frac{1}{(x+y+1)^2}\right] dy$$

$$= -\frac{1}{2}\int_0^1 \left[\frac{1}{4}y + \frac{1}{(x+y+1)}\right]^{1-x} dx$$

$$= -\frac{1}{2}\int_0^1 \left[\frac{1}{4}(1-x) + \frac{1}{2} - \frac{1}{x+1)}\right] dx$$

$$= -\frac{1}{2}\left[\frac{3}{4}x - \frac{1}{8}x^2 - \log(x+1)\right]_0^1$$

$$= -\frac{1}{2}\left[\frac{3}{4} - \frac{1}{8} - \log 2\right]$$

$$= \frac{1}{16}(8\log 2 - 5) = \frac{1}{16}(\log 2^8 - 5\log e) = \frac{1}{16}\log\frac{256}{e^5}. \quad \textbf{Ans.}$$

Example 105:

Evaluate $I = \iiint (x^2 + y^2 + z^2)$ *dx dy dz over the sphere* $x^2 + y^2 + z^2 \leq a^2$.

Solution:

Changing to spherical polar co-ordinates $x = r\sin\theta\cos\phi$, $y = r\sin\theta\sin\phi$, $r = z\cos\theta$; we have

$$J = r^2 \sin\theta$$

To describe the sphere $x^2 + y^2 + z^2 \leq a^2$, r varies from 0 to a, θ varies from 0 to π; ϕ from 0 to 2π. Thus

$$I = \int_0^{2\pi} \int_0^{\pi} \int_0^{a} r^2\, J\, dr\, d\theta\, d\phi, \; |J| = r^2 \sin\theta$$

$$= \int_0^{2\pi} d\phi \int_0^{\pi} \sin\theta \, d\theta \left| \frac{1}{5} r^5 \right|_0^a$$

$$= \frac{a^6}{5} \cdot 2\pi \left[-\cos\theta \right]_0^{\pi} = \frac{4\pi a^5}{5}.$$ **Ans.**

Example 106:

Evaluate

$$\iiint (1-x-y-z)^{l-1} x^{m-1} y^{n-1} z^{p-1} \, dx \, dy \, dz,$$

over the interior of the tetrahedron bounded by the planes

$x = 0, y = 0, z = 0, x + y + z = 1$.

Solution:

We substitute $x + y + z = u$, $x + y = uv$, $x = uvw$

i.e., $x = uvw$, $y = uv \quad (1 - w)$ $z = u(1 - v)$.

It may be seen that when x, y, z are all positive and $x + y + \leq 1$, then u, v, w all lie between 0 and 1 and conversely. Thus the given region corresponds to the region bounded by the planes

$u = 0, u = 1; v = 0, v = 1; w = 0, w = 1.$

Also $J = \dfrac{\partial(x, y, z)}{\partial(u, v, w)} = -u^2 v.$ Thus $|J| = u^2 v$.

The given integral

$$= \left[\int_0^1 (1-u)^{l-1} u^{m+n+p-1} du\right]\left[\int_0^1 v^{m+n-1} (1-v)^{p-1} dv\right]$$

$$\times \left[\int_0^1 w^{m-1} (1-w)^{n-1} dw\right]$$

$$= \beta(m + n + p, 1)\, \beta(m + n, p)\, \beta(m, n)$$

$$\frac{\Gamma(m+n+p)\Gamma(l)}{\Gamma(m+n+p+l)} \cdot \frac{\Gamma(m+n)\Gamma(p)}{\Gamma(m+n+p)} \cdot \frac{\Gamma(m)\Gamma(n)}{\Gamma(m+n)}$$

$$= \frac{\Gamma(m)\Gamma(n)\Gamma(p)}{(l+m+n+p)}.$$ **Ans.**

Example 107:

Evaluate $I = \iiint xyz \, dx \, dy \, dz$ *over the domain bounded by $x = 0$, $y = 0$, $z = 0x + y + z = 1$.*

Solution:

We know that

$$I = \int_0^1 x\,dx \int_0^{1-x} y\,dy \int_0^{1-x-y} z\,dz$$

$$= \frac{1}{2}\int_0^1 x\,dx \int_0^{1-x} y(1-x-y)^2\,dy$$

$$= \frac{1}{2}\int_0^1 x\,dx \left[\left|\frac{-1}{3} y(1-x-y)^3\right|_0^{1-x} + \frac{1}{3}\int_0^{1-x} y(1-x-y)^3 . 1\,dy\right]$$

$$= \frac{1}{2}\int_0^1 x\left[0 + \frac{1}{12}\left|(1-x-y)^4\right|_0^{1-x}\right] dx$$

$$= \frac{1}{24}\int_0^1 (1-x)^4\,dx$$

$$= \frac{1}{24}\left[-\frac{1}{5}\left|x(1-x)^5\right|_0^1 + \frac{1}{5}\int_0^1 (1-5)^5\,dx\right]$$

$$= \frac{1}{24}\left[0 - \frac{1}{30}\left|(1-x)^6\right|_0^1\right] = \frac{1}{720}.$$ **Ans.**

Example 108:

Show that $\iiint_E \frac{dx\,dy\,dz}{x^2 + y^2 + (z-2)^2} = \frac{2\pi}{3}$, *where E is the spherical volume of* $x^2 + y^2 + z^2 \le 1$.

Solution:

Do your self. $\left[\textbf{Ans. } \frac{2\pi}{3}\right]$

Example 109:

Compute $I = \iiint\sqrt{I = \frac{x^2}{a^2} - \frac{y^2}{b^2} - \frac{z^2}{c^2}}$ *dx dy dz taken over the region* $x^2/a^2 + y^2/b^2 + z^2/c^2 \le 1$.

Solution:

Changing to spherical polar co-ordinates, where $x/a = r \sin\theta \cos\phi$, $y/b = r \sin\theta, \sin\phi$, $z/c = r\cos\theta$; we obtain as

$$= \frac{\partial(x, y, z)}{\partial(r, \theta, \phi)} = abc\ r^2 \sin\theta.$$

Thus, $I = \int_0^{2\pi} \int_0^{\pi} \int_0^{1} \sqrt{1-r^2}\ (abc\ r^2 \sin\theta)\ dr\ d\theta\ d\phi$

$$= abc \int_0^{2\pi} d\phi \int_0^{\pi} \sin\theta\ d\theta \int_0^{1} r^2 \sqrt{1-r^2}\ dr$$

$$= 2\pi\ abc \left| -\cos\theta \right|_0^{\pi} \int_0^{\pi/2} \sin^2 t \cos^2 t\ dt,\ r = \sin t$$

$$= 4\pi\ abc\ \frac{1.1}{4.2} \cdot \frac{\pi}{2} = \frac{\pi^2}{4}\ abc.$$ **Ans.**

Example 110:

Evaluate

$$I = \iiint \sqrt{(a^2b^2c^2 - b^2c^2x^2 - c^2a^2y^2 - a^2b^2z^2}\ dx\ dy\ dz$$

taken throughout the domain $\{(x, y, z);\ x^2/a^2 + y^2/b^2 + z^2/c^2 \leq 1\}$.

Solution:

We have

$$I = abc \iiint \sqrt{I - \frac{x^2}{a^2} - \frac{y^2}{b^2} - \frac{z^2}{c^2}}\ dx\ dy\ dz$$

$$= abc\left(\frac{1}{4}\pi^2\ abc\right) = \frac{1}{4}\pi^3\ a^2b^2c^2.$$ **Ans.**

Example 111:

Evaluate

$$I = \iiint_E (y^2z^2 + z^2x^2 + x^2y^2)\ dx\ dy\ dz$$

taken over the domain bounded by the cylinder $x^2 + y^2 = 2ax$, *and the cone* $z^2 = k^2(x^2 + y^2)$.

Solution:

The domain E is bounced above and below by the surfaces

$$z = k\sqrt{x^2 + y^2} \text{ and } z = -k\sqrt{x^2 + y^2}.$$

Further the projection of E on the xy-plane is the circular domain

$D : x^2 + y^2 = 2ax.$

Since the integrand is an even function of z and E is symmetric about the plane z = 0, so

$$I = 2 \iint dx\, dy \int_0^{k\sqrt{(x^2+y^2)}} [x^2y^2 + (x^2+y^2)z^2]\, dz$$

$$= 2 \iint_D \left[k\sqrt{x^2+y^2}\, .\, x^2y^2 + \frac{k^3}{3}(x^2+y^2)^{5/2} \right] dx\, dy.$$

The domain D is symmetrical about the line y = 0. Hence changing to polar coordinates, we obtain

$$I = 4 \int_0^{\pi/2} \int_0^{2a\cos\theta} \left(kr.r^4 \cos^2\theta \sin^2\theta + \frac{k^3}{3} r^5 \right) r\, dr\, d\theta$$

$$= \frac{4k}{3} \int_0^{\pi/2} (3\cos^2\theta \sin^2\theta + k^2)\, d\theta \int_0^{2a\cos\theta} r^6\, dr$$

$$= \frac{4k}{21}(2a)^7 \int_0^{\pi/2} (2\cos^9\theta \sin^2\theta + k^2 \cos^7\theta)\, d\theta$$

$$= \frac{512\, ka^7}{21}\left[3\, .\, \frac{8.6.4.2.1}{11.9.7.5.3.1} + k^2 \frac{6.4.2}{7.5.3.1} \right]$$

$$= \frac{8192\, ka^7}{735}\left[k^2 + \frac{8}{33} \right].$$ **Ans.**

Example 112:

Show that $\iiint z^2\, dx\, dy\, dz = \frac{30\pi - 32}{225} a^3$, *the field of integration being the region bounded by the surfaces*

$x^2 + y^2 + z^2 = a^2,\ x^2 + y^2 = ax.$ **(D.U., 1986)**

Solution:

The given region is bounded by the surfaces

$$z = -\sqrt{a^2 - x^2 - y^2} \text{ and } z = \sqrt{a^2 - x^2 - y^2}$$

and its projection on the xy-plane is the circular domain

$D : x^2 + y^2 = ax.$ Thus, we have

$$I = \iiint z^2\, dx\, dy\, dz = \iint_D dx\, dy \int_{-\sqrt{a^2-x^2-y^2}}^{\sqrt{a^2-x^2-y^2}} z^2\, dz$$

$$I = \frac{2}{3} \iint_D (a^2 - x^4 - y^2)^{3/2}\, dx\, dy.$$

Changing to polars, the region D becomes

$$r = a \cos\theta,\ 0 \le \theta \le \pi;\ 0 \le r \le a\cos\theta.$$

$$\therefore \qquad I = \frac{2}{3} \int_0^{\pi} d\theta \int_0^{a\cos\theta} (a^2 - r^2)^{3/2}\, r\, dr$$

$$= \frac{2}{3} \int_0^{\pi} \left| -\frac{1}{5}(a^2 - r^2)^{5/2} \right|_0^{a\cos\theta} d\theta = \frac{2a^5}{15} \int_0^{\pi} (1 - \sin^5\theta)\, d\theta$$

$$= \frac{2a^5}{15}\left[\pi - 2\int_0^{\pi/2} \sin^5\theta\, d\theta\right] = \frac{2a^5}{15}\left[\pi - 2.\frac{4}{5}.\frac{2}{3}\right]$$

Hence $I = \dfrac{30\pi - 32}{225} a^5.$ **Ans.**

Example 113:

Prove that

$$\iiint_E \sqrt{\frac{1 - x^2 - y^2 - z^2}{1 + x^2 + y^2 + z^2}}\, dx\, dy\, dz = \frac{\pi}{8}\left[\beta\left(\frac{3}{4}, \frac{1}{2}\right) - \beta\left(\frac{5}{4}, \frac{1}{2}\right)\right],$$

where E is the domain for which x, y, z are all positive and $x^2 + y^2 + z^2 \le 1$.

Solution:

Changing to spherical polar coordinates, thee given region E transforms into

$$E_1 = \{(r, \theta, \phi) : 0 \le r \le 1,\ 0 \le \theta \le \tfrac{1}{2}\pi,\ 0 \le \pi \le \tfrac{1}{2}\pi\}$$

Thus thee given integral

$$= \int_0^{\pi/2} \int_0^{\pi/2} \int_0^{1} \sqrt{\frac{1 - r^2}{1 + r^2}}\, (r^2 \sin\theta)\, dr\, d\theta\, d\phi$$

$$= \int_0^{\pi/2} d\phi \int_0^{\pi/2} \sin\theta\, d\theta \int_0^{1} \sqrt{\frac{1 - r^2}{1 + r^2}}\, r^2\, dr$$

$$= \frac{\pi}{2} \times 1 \times \int_{\pi/2}^{0} \sqrt{\frac{1-\cos t}{1+\cos t}} \frac{\cos t\,(-\sin t)\,dt}{2\sqrt{\cos t}}, \; r^2 = \cos t$$

$$= \frac{\pi}{4} \int_{0}^{\pi/2} \frac{\sin t/2}{\cos t/2} \sqrt{\cos t}\,.\,2 \sin t/2 \cos t/2 \; dt$$

$$= \frac{\pi}{4} \int_{0}^{\pi/2} \sqrt{\cos t}\,(2 \sin^2 t/2)\; dt = \frac{\pi}{4} \int_{0}^{\pi/2} \sqrt{\cos t}\,(1-\cos t)\; dt$$

$$= \frac{\pi}{4} \left[\int_{0}^{\pi/2} \sin^0 t \cos^{1/2} t\, dt - \int_{0}^{\pi/2} \sin^0 t \cos^{3/2} t\, dt \right.$$

$$= \frac{\pi}{8} \left[\beta\left(\frac{0+1}{0}, \frac{\frac{1}{2}+1}{2}\right) - \beta\left(\frac{0+1}{2}, \frac{(3/2)+1}{2}\right) \right]$$

$$= \frac{\pi}{8} \left[\beta\left(\frac{1}{2}, \frac{3}{4}\right) - \beta\left(\frac{1}{2}, \frac{5}{4}\right) \right]$$

$$\left[\because \int_{0}^{\pi/2} \sin^p \theta \cos^q \theta \, d\theta = \frac{1}{2}\left(\frac{p+1}{2}, \frac{q+1}{2}\right), p > -1, q > -1. \right]$$

Example 114:

Show that $\iiint (x+y+z)\, x^2 y^2 z^2 \; dx\, dy\, dz = \frac{1}{50400}$

throughout the region $x + y + z \le 1,\ x \ge 0,\ y \ge 0,\ z \ge 0.$

Solution:

We know that

$$x + y + z = u,\ x + y = uv,\ x = uvw$$

i.e. $x = uvw,\ y = uv\,(1-w),\ z = u\,(1-v),\ |J| = u^2 v.$

Given integral

$$= \int_0^1 \int_0^1 \int_0^1 u(u^2 v^2 w^2)[u^2 v^2 (1-w)^2][u^2 (1-v)^2][u^2 v]\,du\,dv\,dw$$

$$= \int_0^1 u^2\, du \int_0^1 v^5\,(1-v)^2\, dv \int_0^1 w^2 (1-w)^2\, dw$$

$$= \frac{1}{10}\, \beta(6, 3)\, \beta(3, 3) = \frac{1}{10} \frac{\Gamma(6)\,\Gamma(3)}{\Gamma(9)} \frac{\Gamma(3)\,\Gamma(3)}{\Gamma(6)}$$

$$= \frac{1}{10} \frac{[\Gamma(3)]^3}{\Gamma(9)} = \frac{1}{10} \cdot \frac{(2!)^2}{8!}$$

[$\because$ $\Gamma(n) = (n - 1)$ 1, n being a positive integer]

$$= \frac{1}{10} \cdot \frac{2.2.2}{8.7.6.5.4.3.2} = \frac{1}{50400}.$$ **Ans.**

Example 115:

Show that $\iiint (x + y + z)^n \; xyz \; dx \; dy \; dz = \dfrac{1}{120(n+6)}$

when $x \geq 0, \; y \geq 0, \; z \geq 0,$ *and* $x + y + z \leq 1.$

Solution:

Do your self.

Example 116:

Evaluate $\iiint x^{l-1} \; y^{m-1} \; z^{n-1} \, dx \; dy \; dz$

when $x \geq 0, \; y \geq 0, \; z \geq 0,$ *and* $x + y + z \leq 1.$

Solution:

Do your self. $\left[\textbf{Ans. } \dfrac{\Gamma(l)\Gamma(m)\Gamma(n)}{\Gamma(l+m+n+1)} \right]$

Example 117:

Evaluate $\iiint e^{x+y+z} \, dx \; dy \; dz,$ *taken over positive octant such that* $x + y + z \leq 1.$ **(D.U., 1993)**

Solution:

We substitute $x + y + z = u$, $x + y = uv$, $x = uvw$ then we have i.e. $x = uvw$; $y = uv, (1 - w)$, $z = u(1 - v)$, $| J | = u^2v$.

$$\text{Given integral} = \int_0^1 \int_0^1 \int_0^1 e^u \; (u^2 v) \; du \; dv \; dw$$

$$= \int_0^1 u^2 \, e^u \, du \times \left| \frac{1}{2} v^2 \right|_0^1 \times |w|_0^1 = \frac{1}{2} \int_0^1 u^2 \; e^u \; du$$

$$= \frac{1}{2}\left[\left|u^2 e^u\right|_0^1 - 2\int_0^1 u\, e^u\, du\right]$$

$$= \frac{1}{2}\left[e - 2\left\{\left|u\, e^u\right|_0^1 - \int_0^1 e^u\, du\right\}\right]$$

$$= \frac{1}{2}[e - 2)\{e - (e - 1)\}] = ½\,(e - 2).$$ **Ans.**

Example 118:

Evaluate

$$I = \iiint x^l\, y^m\, z^n\,(1 - ax - by - cz)^p\, dx\, dy\, dz,$$

over the interior of the tetrahedron bounded by the planes
$x = 0,\ y = 0,\ z = 0,\ ax + by + cz = 1.$ **(D.U., 1988)**

Solution:

Let ax = X, by = Y, cz = Z so that J = 1/abc.

$$\therefore \quad I = \frac{1}{a^{l+1}\, b^{m+1}\, c^{n+1}} \iiint_D X^l\, Y^m\, z^n\,(l - X - Y - Z)^p\, dX\, dY\, dZ$$

where D is the interior of X = 0, Y = 0, Z = 0 and X + Y + Z = 1.

$$I = \frac{1}{a^{l+1}\, b^{m+1}\, c^{n+1}} \frac{\Gamma(l+1)\,\Gamma(m+1)\,\Gamma(n+1)\,\Gamma(p+1)}{\Gamma(l + m + n + p + 4)}.$$

Example 119:

Show that

$$\iiint x^{l-1}\, y^{m-1}\, z^{n-1}\,(1 - ax - by - cz)^{p-1}\, dx\, dy\, dz$$

$$= \frac{\Gamma(l)\,\Gamma(m)\,\Gamma(n)\,\Gamma(p)}{a^l\, b^m\, c^n\, \Gamma(l + m + n + p)}.$$

Solution:

Do your self.

Example 120:

Prove that

$$I = \iiint_E z^{-3}\left(\frac{x^2}{a^2} + \frac{y^2}{b^2} + \frac{z^2}{c^2}\right)^{1/2} dx\, dy\, dz = \frac{\pi}{pq},$$

where E is the smaller region bounded by the standard ellipsoid $x^2/a^2 + y^2/b^2 + z^2c^2 = 1$ *and the half cone* $z^2 = p^2x^2 + q^2y^2$, $z > 0$.

(D.U., 1989)

Solution:

Let us put x = aX, y = bY, z = cZ so that

$$J = \frac{\partial(x, y, z)}{\partial(X, Y, Z)} = abc.$$

$$\therefore \quad I = \frac{ab}{c^2} \iiint_D z^{-3} (X^2 + Y^2 + Z^2)^{1/2} \, dX \, dY \, dZ, \qquad ...(1)$$

where D is the smaller region bounded by the sphere $X^2 + Y^2 + Z^2 = 1$ and the half cone $c^2Z^2 = (ap)^2X^2 + (bq)^2Y^2$. Now we put

$$X = r \sin\theta \cos\phi, \; Y = r \sin\theta \sin\phi, \; Z = r \cos\theta$$

so that $J = r^2 \sin\theta$. From (1), we have

$$I = \frac{ab}{c^2} \iiint \frac{r . r^2 \sin\theta}{r^3 \cos^3\theta} . \, dr \, d\theta \, d\phi$$

$$= \frac{ab}{c^2} \int_0^1 dr \int_0^{2\pi} d\phi \int_0^{\alpha} \tan\theta \sec^2\theta \, d\theta, \qquad ...(2)$$

where α is given by the intersection of the surfaces of the sphere r = 1 and the cone then,

$$c^2r^2 \cos^2\theta = r^2 \sin^2 (a^2 p^2 \cos^2\phi + b^2 \theta^2 \sin^2\phi)$$

or $\qquad \tan^2\alpha = c^2/(a^2 p^2 \cos^2\phi + b^2 q^2 \sin^2\phi) \; (\theta = \alpha)$

From (2) we have, $I = \dfrac{ab}{2c^2} \displaystyle\int_0^{2\pi} \tan^2\alpha \, d\phi$

$$= \frac{ab}{2} \int_0^{2\pi} \frac{d\phi}{a^2p^2 \cos^2\phi + b^2q^2 \sin^2\phi}$$

$$= 4\left(\frac{ab}{2}\right) \int_0^{\pi/2} \frac{\sec^2\phi \, d\phi}{a^2p^2 + b^2q^2 \tan^2\phi} = \frac{2}{pq} \int_0^{\infty} \frac{dt}{1+t^2}$$

where bq an ϕ = apt

$$= \frac{2}{pq} . \left| \tan^{-1} t \right|_0^{\infty} = \frac{2}{pq} . \frac{\pi}{2} = \frac{\pi}{pq}.$$

Ans.

Example 121:

Find the volume of thee ellipsoid

$x^2/a^2 + y^2/b^2 + z^2/c^2 = 1$.

Solution:

The projection of the ellipsoid on the xy-plane is the ellipse D : $x^2/a^2 + y^2/b^2 = 1$.

$$\text{Volume} = \iiint_{-c\sqrt{(1-x^2/a^2-y^2/b^2)}}^{c\sqrt{(1-x^2/a^2-y^2/b^2)}} dx\, dy\, dz$$

We know that

$$= 2c \iint_D \sqrt{1 - \frac{x^2}{a^2} - \frac{y^2}{b^2}}\, dx\, dy.$$

Putting $\frac{x}{a} = r \cos q$, $\frac{y}{b} = r \sin \theta$; we obtain J = abv.

$$\therefore \quad \text{Volume} = 2c \int_0^{2\pi} d\theta \int_0^1 \sqrt{1 - r^2}\,(abr)\, dr\, dx$$

$$= 2abc, 2\pi \left[-\frac{1}{3}(1 - x^2)^{3/2} \right]_0^1 = \frac{4}{3}\pi abc. \qquad \textbf{Ans.}$$

Example 122:

Prove that

$$I = \iiint_E z\, dx\, dy\, dz = \frac{\pi}{4} h^4 \cot \alpha \cot \beta$$

where E is the domain bounded by the cone $z^2 = x^2 \tan^2 \alpha + y^2 \tan^2 \beta$ and the plane $z = 0$, $z = h$. **(D.U., 1987)**

Solution:

Let us substitute $x \tan \alpha = X$, $y \tan \beta = Y$, $z = Z$ so that

$$J = \frac{\partial(x, y, z)}{\partial(X, Y, Z)} = \cot \alpha \cot \beta.$$

$$\therefore \quad I = \iiint Z.\cot \alpha \cot \beta.\, dX\, dY\, dZ$$

over the cone $Z^2 = X^2 + Y^2$ and $Z = 0$, $Z = h$

or $I = \cot\alpha \cot\beta \int Z\,dZ \iint dX\,dY$. ...(1)

It is clear that area of a circular disc cut parallel to the plane Z = 0 is $\iint dX\,dY = pZ^2$ ($\because X^2 + Y^2 = Z^2$ is a circle of radius Z).

From (1), we obtain

$$I = \cot\alpha \cot\beta \int_0^h Z\,dZ.\,\pi Z^2$$

$$= \frac{\pi}{4} h^4 \cot\alpha \cot\beta.$$ **Ans.**

Example 123:

Show that $I = \iiint_E z\,dx\,dy\,dz = \frac{\pi}{8}$, *where E is* $x^2 + y^2 + z^2 \leq 1$, $x^2 + y^2 \leq z$, $z \geq 0$.

Solution:

Put $x = r\sin\theta\cos\phi$, $y = r\sin\theta\sin\phi$, $z = r\cos\theta$.

$$\therefore \quad I = \iiint_D r\cos\theta,\, r^2\sin\theta\,dr\,d\theta\,d\phi,$$

where D is the region bounded by the sphere r = 1 and the cone

$$r^2\cos^2\theta = r^2\sin^2\theta(\cos^2\phi + \sin^2\phi) \text{ or } \theta = \frac{\pi}{4}.$$

$$\therefore \quad I = \int_0^1 r^2\,dr \int_0^{2\pi} d\phi \int_0^{\pi/4} \cos\theta\sin\theta\,d\theta$$

$$= \frac{1}{4}.2\pi.\left|\frac{\sin^2\theta}{2}\right|_0^{\pi/4} = \frac{\pi}{8}.$$ **Ans.**

EXERCISES

1. Compute $I = \iiint_E xy\,dy\,dz$, when E is the domain bounded by $z = xy$, $x + y = 1$, $t = 0$ $(z \geq 0)$. **(Ans.** 1/180)

   Hint. $I = \int_0^1 \int_0^{1-x} \int_0^{xy} xy\,dx\,dy\,dz = \int_0^1 \int_0^{1-x} x^2y^2\,dx\,dy$

   $= \frac{1}{3}\int_0^1 x^2(1-x)^3\,dx$. Now put $x = \sin^2\theta$.

2. Compute $I = \iiint_E (x^2 + y^2)\, dx\, dy\, dz$, where E is specified by $z \geq 0,\ a^2 \leq x^2 + y^2 + z^2 \leq b^2$. $\left[\textbf{Ans. } \frac{4\pi}{15}(b^5 - a^5)\right]$

3. Evaluate $\int_0^a dx \int_0^x dy \int_0^y xyz\, dz$. $\left[\textbf{Ans. } \frac{a^6}{48}\right]$

4. Evaluate $\int_1^3 \int_{1/x}^1 \int_0^{\sqrt{xy}} xyz\, dz\, dy\, dx$. $\left[\textbf{Ans. } \frac{13}{9} - \frac{1}{6}\log 3\right]$

5. Compute $\iiint_E xy\, dx\, dy\, dz$, where E is the domain bounded by $x = 0,\ y = 0,\ z = 0,\ x + y + z = 1$. **[Ans.** 1/120]

6. Evaluate $\iiint_E y\cos(z + x)\, xy\, dx\, dy\, dz$, E being the domain bounded by the cylinder $y = \sqrt{x}$ and the planes $y = 0,\ z = 0,\ x + z = \frac{1}{2}\pi$.
[Ans. $(\pi^2 - 8)/16$]

$\left[\textbf{Hint. } I = \int_{x=0}^{\pi/2} \int_{y=0}^{\sqrt{x}} dy \int_{z=0}^{\frac{1}{2}\pi - x} \cos(z + x)\, dz.\right]$

4.23 Volume of a Solid by a Triple Integral

The volume V os a solid in a given by

$$V = \iiint dx\, dy\, dz \qquad \text{...(A)}$$

It can be shown that on a closed (regular) domain E bounded below and above by the surfaces

$$z = \phi(x, y), \quad z = f\psi(x, y); \quad \psi(x, y) \geq \phi(x, y)$$

$$V = \iiint dx\, dy\, dz = \iint_D \left[\int_{\phi(x, y)}^{\psi(x, y)} dz\right] dx\, dy \qquad \text{...(B)}$$

where D is the projection of E on xy-plane.

In order to find the volume of cylindrical solids, we proceed as follows:

Suppose a cylindrical solid is bounded above by a surface $z = \psi(x, y)$, below by a plane region D (in the xy-plane) and on the sides by lines parallel to z-axis. Its volume V is given by

$$V = \iint_D z\, dx \qquad ...(C)$$

If a solid is bounded above by the surface $z = z_2(x, y) \geq 0$ and below by the surface $z = z_1(x, y) \geq 0$, and if D in the domain of projection of both the surfaces on the xy-plane, then the volume V of the solid is equal to the difference between the volumes of the two cylindrical bodies i.e.,

$$V = \iint_D (z_2 - z_1)\, dx\, dy \qquad ...(D)$$

Example 123:

Find the volume of the solid bounded by the surface $z = 1 - 4x^2 - y^2$ and the plane $z = 0$.

Solution:

The paraboloid $z = 1 - 4x^2 - y^2$ cuts the xy-plane ($z = 0$) along the ellipse $D : 4x^2 + y^2 = 1$. Moreover thee solid being symmetrical, its volume V is four times the volume lying in the first octant. Then we have

$$\therefore \quad V = 4 \iint_D z\, dx\, dy$$

$$= 4 \int_0^{1/2} dx \int_0^{\sqrt{(1-4x^2)}} (1-4x^2-y^2)\, dy$$

$$= 4 \int_0^{1/2} \left| (1-4x^2)y - \frac{1}{3}y^3 \right|_0^{\sqrt{1-4x^2}}$$

$$= \frac{8}{3} \int_0^{1/2} (1-4x^2)^{3/2}\, dx$$

$$= \frac{4}{3} \int_0^{\pi/2} \cos^4 t, \text{ where } x = \frac{1}{2} \sin t$$

$$= \frac{4}{3} \cdot \frac{3}{4} \cdot \frac{1}{2} \cdot \frac{\pi}{2} = \frac{\pi}{4}. \qquad \textbf{Ans.}$$

Example 124:

Find the volume cut off from the sphere $x^2 + y^2 + z^2 = a^2$ by the cylinder $x^2 + y^2 = ax$.

Solution:

We have

$$V = \iint_D dx\,dy \int_{-\sqrt{(a^2-x^2-y^2)}}^{\sqrt{(a^2-x^2-y^2)}} dz = 2\iint_D \sqrt{a^2-x^2-y^2}\,dx\,dy$$

or $$V = 2\int_0^{\pi} d\theta \int_0^{a\cos\theta} \sqrt{a^2-r^2}\,.\,r\,dr = -\frac{2}{3}\int_0^{\pi} \left|(a^2-r^2)^{3/2}\right|_0^{a\cos\theta}$$

$$= \frac{2}{3}a^3 \int_0^{\pi} (1-\sin^3\theta) = \frac{2}{3}a^3\left[\pi - 2\,.\,\frac{2}{3}\right]$$

$$= \frac{2}{3}a^3\left(4-\frac{4}{3}\right).$$ **Ans.**

Example 125:

Find the volume of thee sphere $x^2 + y^2 + z^2 = a^2$.

Solution:

The volume V of the sphere $x^2 + y^2 + z^2 = a^2$ is

$$V = \int_{a}^{a} \int_{-\sqrt{a^2-x^2}}^{\sqrt{a^2-x^2}} \int_{-\sqrt{a^2-x^2-y^2}}^{\sqrt{a^2-x^2-y^2}} dx\,dy\,dz$$

$$= 2\int_{a}^{a} dx \int_{-\sqrt{a^2-x^2}}^{\sqrt{a^2-x^2}} \sqrt{a^2-x^2-y^2}\,dy$$

$$= 2\int_{-a}^{a}\left[\frac{1}{2}(a^2-x^2)\sqrt{a^2-x^2-y^2} + \frac{1}{2}(a^2-x^2)\sin^{-1}\frac{y}{\sqrt{a^2-x^2}}\right]_{-\sqrt{a^2-x^2}}^{\sqrt{a^2-x^2}} dx$$

$$= 2\int_{a}^{a}\left\{0 + \frac{1}{2}(a^2-x^2)\left(\frac{\pi}{2}+\frac{\pi}{2}\right)\right\} dx$$

$$= \pi\left|a^2x - \frac{1}{3}x^3\right|_{-a}^{a}$$

$$= a\left[a^2.2a - \frac{2}{3}a^3\right] = \frac{3}{4}\pi a^3.$$ **Ans.**

Example 126:

Find the volume enclosed by the solid common to

$x^2 + y^2 = a^2$, $x^2 + z^2 = a^2$. **(D.U., 1995)**

Solution:

The required volume is

$$V = \int_{a}^{a} \int_{-\sqrt{(a^2 - x^2)}}^{\sqrt{(a^2 - x^2)}} \int_{-\sqrt{(a^2 - x^2)}}^{\sqrt{(a^2 - x^2)}} dx\, dy\, dz$$

$$= 2 \int_{-a}^{a} dx \int_{-\sqrt{(a^2 - x^2)}}^{\sqrt{(a^2 - x^2)}} \sqrt{a^2 - x^2}\, dy$$

$$= 4 \int_{-a}^{a} (a^2 - x^{24})\, dx = 8 \int_{0}^{a} (a^2 - x^2)\, dx = \frac{16a^2}{3}.$$ **Ans.**

Example 127:

Find the volume common to the ellipsoid of revolution $x^2/a^2 + y^2/a^2$ $z^2/b^2 = 1$ and the cylinder $x^2 + y^2 - ay = 0$.

Solution:

The required volume is double the volume that lies above the xy-plane. The given solid is bounded above by $z = \frac{b}{a}\sqrt{a^2 - x^2 - y^2}$ and below by the circular base D : $x^2 + y^2 - ay = 0$ on the xy-plane. Thus the required volume is given by

$$V = 2 \iint_D z\, dx\, dy = \iint_D \frac{b}{a} \sqrt{a^2 - x^2 - y^2}\, dx\, dy.$$

Changing to polars, we obtain

$$V = \frac{2b}{a} \int_{\theta=0}^{\pi} \int_{r=0}^{a \sin\theta} \sqrt{a^2 - r^2}\, (r\, dr\, d\theta)$$

$(x^2 + y^2 = ay \Rightarrow r^2 = ar \sin\theta \Rightarrow r = a \sin\theta)$

$$= \frac{2b}{a} \int_{0}^{\pi} \left| -\frac{1}{3}(a^2 - r^2)^{3/2} \right|_{0}^{a \sin\theta} d\theta$$

$$= \frac{4a^2 b}{3} \int_{0}^{\pi/2} (1 - \cos^2\theta)\, d\theta = \frac{4a^3 b}{3}\left[\frac{\pi}{2} - \frac{2}{3}\right]$$

$$= \frac{2a^2 b}{9}(3\pi - 4).$$ **Ans.**

Example 128:

Find the volume of the region above the xy-plane bounded by the paraboloid $z = x^2 + y^2$ and the cylinder $x^2 + y^2 = a^2$.

Solution:

The given solid is bounded above by $z = x^2 + y2$ and below by $x^2 + y^2 = a^2$. Moreover the solid being symmetrical, its volume V is four times the volume lying in the first octant.

$$\therefore \quad V = 4 \iint_D z \, dx \, dy = 4 \iint_D (x^2 + y^2) \, dx \, dy$$

Changing to polars, we get

$$V = 4 \int_{a=0}^{\pi/2} \int_{r=0}^{a} r^2 \,(r \, dr \, d\theta) = 4 . \frac{\pi}{2} . \frac{a^4}{4} = \frac{\pi a^4}{2}. \qquad \textbf{Ans.}$$

Example 129:

Show that the volume of the solid bounded above by $z = 2 - r$ and below by the plane region, $0 \le r \le 2 \cos \theta$, $-\frac{1}{2}\pi \le \theta \le \frac{1}{2}\pi$ is $\frac{2}{9}(9\pi - 16)$.

Solution:

The required volume is given by

$$V = \iint_D z \, dx \, dy \int_{-\pi/2}^{\pi/2} \int_0^{2 \cos \theta} (2 - r)\,(r \, dr \, d\theta)$$

$$= \int_{-\pi/2}^{\pi/2} \left| r^2 - \frac{1}{2} r^2 \right|_0^{3 \cos \theta} d\theta$$

$$= \int_{-\pi/2}^{\pi/2} \left(4 \cos^2 \theta - \frac{8}{3} \cos^3 \theta \right) d\theta$$

$$= \int_{-\pi/2}^{\pi/2} \left[2(1 + \cos 2\theta) - \frac{2}{3} (\cos 3\theta + 3 \cos \theta) \right] d\theta$$

$$= 2 (\pi + 0) + \frac{4}{9} - 4 = \frac{2}{9} (9\pi - 16). \qquad \textbf{Ans.}$$

Example 130:

Find the volume of the region enclosed by the paraboloid $(x - 1)^2 + y^2 = z$ and the point $2x + z = 2$. **(D.U., 1988)**

Solution:

The required volume is given by

$$V = \iint_E (z_2 - z_1)\, dx\, dy$$
$$= \iint_E [(2 - 2x) - \{(x - 1)^2 + y^2\}]\, dx\, dy,$$

where E is the region given by eliminating z between two given surfaces.

We have $(x - 1)^2 + y^2 = 2 - 2x \Rightarrow x^2 + y^2 = 1$. Thus

$$V = \iint_E [1 - (x^2 + y^2)]\, dx\, dy.$$

Changing to polars, we obtain

$$V = \int_0^{2\pi} \int_0^1 (1 - r^2)\, r\, d\theta\, dr = 2\pi \left[\frac{1}{2} - \frac{1}{4}\right] = \frac{\pi}{2}.$$

Example 131:

Find the volume of the region enclosed by the cone $x^2 + y^2 = z^2$ and the plane $z = 1$.

Solution:

Do your self. **(Ans.** $\pi/3$)

Example 132:

Find the volume of the solid obtained by the cylinders $z = 4 - y^2$ and $z = y^2 + 2$, and the planes $x = -1$, $x = 2$.

Solution:

Do your self. **(Ans.** 8)

Example 133:

Evaluate the volume of the wedge intercepted between the cylinder $x^2 + y^2 = 2ax$ and the planes $z = x \tan \alpha$ and $z = x \tan \beta$.

Solution:

The required volume is given by

$$V = \iint_E (z_2 - z_1)\, dx\, dy = (\tan \beta - \tan \alpha) \iint_E x\, dx\, dy$$

where E is the region : $x^2 + y^2 = 2ax$. Putting $x = r \cos \theta$, $y = r \sin \theta$;

$$V = (\tan\beta - \tan\alpha) \int_{\theta=0}^{\pi} \int_{r=0}^{2a\cos\theta} r\cos\theta \; r \, dr \, d\theta$$

$$= \frac{1}{3}(\tan\beta - \tan\alpha) \int_{0}^{\pi} 8a^3 \cos^3\theta \, . \cos\theta \, d\theta$$

$$= \frac{16}{3} a^3 (\tan\beta - \tan\alpha) \int_{0}^{\pi/2} \cos^4\theta \, d\theta$$

$$= \frac{16a^3}{3}(\tan\beta - \tan\alpha) \, . \frac{3}{4} . \frac{1}{2} . \frac{\pi}{2} = a^3 \, \pi \, (\tan\beta - \tan\alpha).$$ **Ans.**

Example 134:

Find the volume of the portion cut off from the cylinder $x^2 + y^2 = 2ax$ and the planes $z = mx$ and $z = nx$.

Solution:

Do your self. (**Ans.** $a^3\pi \, (m - n)$)

Example 135:

Show that the part of the volume of a sphere of radius a that lies in side a right circular cone of semi-vertical angle a whose vertex is on the sphere and whose axis is a diameter of the sphere is $\frac{4}{3}\pi a^2 \, (1 - \cos^2\alpha)$.

Solution:

Let the equation of the sphere be

$$x^2 + y^2 + (z - a)^2 - a^2 \qquad ...(1)$$

The equation of the right circular cone of semi-vertical angle α with vertex at the origin and z-axis as its axis is

$$x^2 + y^2 = z^2 \tan^2\alpha \qquad ...(2)$$

Let $x = r \sin\theta \cos\phi$, $y = r \sin\theta \sin\phi$ and $z = r\cos\theta$. The $J = r^2 \sin\theta$.

The equations (1) and (2) become

$r = 2a \cos\theta$ and $\theta = \alpha$ respectively.

The required volume is

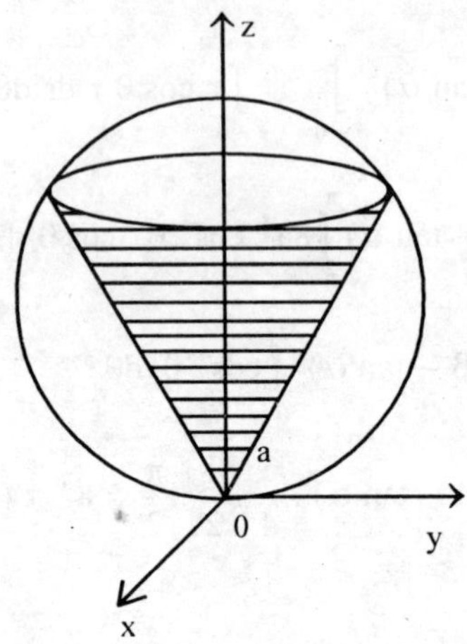

$$V = \iiint dx\, dy\, dz = \int_0^{\alpha} \sin\theta\, d\theta \int_0^{2\pi} d\theta \int_0^{2a\cos\theta} r^2\, dr$$

$$= \frac{2\pi}{3}\int_0^{\alpha} 8a^3 \cos^3 \sin\theta\, d\theta = -\frac{16\pi\, a^3}{3}\left|\frac{\cos^4\theta}{4}\right|_0^{\alpha} = \frac{4\pi a^3}{3}(1-\cos^4\alpha).\quad \textbf{Ans.}$$

Example 136:

Show that the volume common to the surfaces $y^2 + z^2 = 4ax$ *and* $x^2 + y^2 = 2ax$ *is* $\frac{2}{3}(3\pi + 8)a^n$.

Solution:

The paraboloid $y^2 + z^2 = 4ax$ is symmetrical about th plane $z = 0$. Thus the required volumee is given by

$$V = 2\iiint_{z=0}^{\sqrt{4ax-y^2}} dx\, dy\, dz$$

$$= 4\int_0^{2a} dx \int_0^{\sqrt{(2ax-x^2)}} \sqrt{4ax - y^2}\, dy$$

$$= 4\int_0^{2a} \left[\frac{1}{2}\, y\sqrt{4ax - y^2} + 2ax \sin^{-1}\frac{y}{\sqrt{4ax}}\right]_0^{\sqrt{2ax-x^2}} dx$$

$$= 2\int_0^{2a} \left[(2ax - x^2)(2ax + x^2)^{1/2} + 4ax \sin^{-1}\sqrt{\frac{2ax-x}{4ax}}\right] dx$$

$$= 2 \int_0^{2a} \left[\sqrt{4a^2 - x^2} \;. x + 4ax \sin^{-1} \sqrt{\frac{2a - x}{4ax}} \right] dx$$

Put x = 2a cos 2θ so that dx = –4a sin 2θ dθ.

$$\therefore \quad V = 2 \int_0^{\pi/4} (4a^2 \sin 2\theta \cos 2\theta + 8a^2 \theta \cos 2\theta)\, 4a \sin 2\theta \, d\theta$$

$$= 16a^3 \int_0^{2a} (\sin^2 \phi \cos \phi + \phi \cos \phi \sin \phi)\, d\phi, \; \phi = 2\theta$$

$$= 16a^3 \left[\left| \frac{\sin^3 \phi}{3} \right|_0^{\pi/2} + \left| \phi_n \frac{\sin^2 \phi}{2} \right|_0^{\pi/2} - \frac{1}{2} \int_0^{\pi/2} \sin^2 \phi \, d\phi \right]$$

$$= 16a^3 \left[\frac{1}{3} + \frac{\pi}{4} - \frac{1}{2} . \frac{1}{2} . \frac{\pi}{2} \right] = 16a^3 \left[\frac{1}{3} + \frac{\pi}{8} \right]$$

$$= \frac{2}{3} a^3 (3\pi + 8).$$ **Ans.**

Example 137:

Prove that the volume common to the sphere

$x^2 + y^2 + z^2 = a^2$ and the cylindeer $x^2 + y^2 = ay$ is

$a^3 (3\pi - 4)$.

Solution:

Do your self.

Example 138:

Show that entire volume of the solid

$(x/a)^{2/3} + (y/b)^{2/3} + (z/c)^{2/3}$ 1 is $4\pi abc/35$. **(D.U., 1990)**

Solution:

The required volume is given by

$$V = 2 \iiint dx\, dy\, dz$$

where z varies from 0 to $c[1 - (x/a)^{2/3} - (y/b)^{2/3}]^{3/2}$.

$$\therefore \quad V = 2c \iint_D \left[1 - (x/a)^{2/3} - (y/b)^{2/3} \right]^{3/2} dx\, dy,$$

where D is the projection of the given solid on the xy-plane

Putting $x/a = r^3 \cos^3\theta$, $y/b = r^3 \sin^3\theta$, we obtain

$$J = \frac{\partial(x, y)}{\partial(r, \theta)} = 9abr^5 \sin^2\theta \cos^2\theta$$

$$\therefore \quad V = 18abc \int_0^{2\pi} \sin^3\theta \cos^2\theta \, d\theta \int_0^1 (1-r^2)^{3/2} r^5 \, dr$$

$$= 18 \text{ abc} \times 4\left(\frac{1}{4.2}.\frac{\pi}{2}\right) \int_0^{\pi/2} \sin^5 t \cos^4 t \, dt, \; r = \sin t$$

$$= \frac{9}{2} \pi abc . \frac{4.2.3.1}{9.7.5.3} = \frac{4\pi abc}{35}. \qquad \textbf{Ans.}$$

Example 139:

Show that the entir volume of the solid

$(x/a)^{1/2} + (y/b)^{1/2} + (z/c)^{1/2} = 1$ is abc/90.

Solution:

Do your self.

Example 140:

Find the volume of the solid bounded by the paraboloids $z = x^2 + y^2$ and $z = x^2 + 2y^2$, and the planes $y = x$, $y = 2x$, $x = 1$.

Solution:

The required volume is

$$V = \iint_E (z_2 - z_1) \, dx \, dy$$

$$= \int_0^1 \int_x^{2x} \left[(x^2 + 2y^2) - (x^2 + y^2)\right] dx \, dy$$

$$= \frac{1}{3} \int_0^1 [(2x)^3 - x^3] \, dx = \frac{1}{3} \int_0^1 7x^3 = \frac{7}{12}. \qquad \textbf{Ans.}$$

Example 141:

Compute the volume of the solid bounded by the sphere $x^2 + y^2 + z^2 = 4$ and the surface of the paraboloid $x^2 + y^2 = 3z$.

Solution:

The required volue is

$$V = \iint_E (z_2 - z_1)\, dx\, dy$$

$$= \iint_E \left[\sqrt{4 - x^2 - y^2} - \frac{1}{3}(x^2 + y^2)\right] dx\, dy,$$

where E is the region given by eliminating z between two given surfaces i.e.,

$$z^2 + 3z - 4 = 0 \quad \text{or} \quad (z + 4)(z - 1) = 0$$

Thus z = 1 gives E : $x^2 + y^2 = 3$.

Changing to polars, we obtain

$$V = \int_0^{2\pi} \int_0^{\sqrt{3}} \left(\sqrt{4 - r^2} - \frac{1}{3} r^2\right) r\, dr\, d\theta$$

$$= 2\pi \left| -\frac{1}{3}(4 - r^2)^{3/2} - \frac{1}{12} r^4 \right|_0^{\sqrt{3}}$$

$$= 2\pi \left[\frac{1}{3}(8 - 1) - \frac{9}{12}\right] = \frac{19\pi}{6}.$$ **Ans.**

Example 142:

Find the volume of the solid bounded by the sphere $x^2 + y^2 + z^2 = a^2$ and the cone $z^2 = (x^2 + y^2) \cot^2 \alpha$, where α is a constant such that $0 \le \alpha \le \pi$.

Solution:

The required volume is given by

$$V = \iiint (z_2 - z_1)\, dx\, dy$$

$$z_1 = \sqrt{x^2 + y^2} \cot \alpha,\ z_2 = \sqrt{a^2 - x^2 - y^2}.$$

$$\therefore \quad V = \iint_E \left(\sqrt{a^2 - x^2 - y^2} - \sqrt{x^2 + y^2} \cot\right) dx\, dy$$

where E is the region given by eliminating z between two given surfaces

We have $a^2 - x^2 - y^2 = (x^2 + y^2) \cot^2 \alpha$

$\Rightarrow \quad a^2 = (x^2 + y^2)(1 + \cot^2 \alpha) = (x^2 + y^2) \operatorname{cosec}^2 \alpha$

$\therefore \quad E : x^2 + y^2 = a^2 \sin^2 \alpha.$

Changing to polars, we obtain

$$V = \int_0^{2\pi} \int_0^{a \sin \alpha} \left(\sqrt{a^2 - r^2} - r \cos \alpha\right) r \, d\theta \, dr$$

$$= 2\pi \left[-\frac{1}{3}\left(a^2 - r^2\right)^{3/2} - \frac{1}{3} r^2 \cot \alpha \right]_0^{a \sin \alpha}$$

$$= \frac{2\pi a^3}{3}\left[(1 - \cos^3 \alpha) - \sin^2 \alpha \cos x\right] = \frac{2\pi a^3}{3}(1 - \cos \alpha). \quad \textbf{Ans.}$$

EXERCISES

1. Use Gauss's theorem to show that

   $$I = \iint_S xz^2 \, dy \, dz + (x^2 y - z^3) \, dz \, dx + (2xy + y^2 z) \, dx \, dy = \frac{2}{5}\pi a^5$$

   where S is the surface of the hemispherical region bounded by

   $z = \sqrt{a^2 - x^2 - y^2}$ and $z = 0$.

2. Show that $\iint_S F.n \, dS = \frac{15\pi}{4}$, $F = (x^2, y^2, z^2)$;

   S is part of the cone $z^2 = x^2 + y^2$, $1 \leq z \leq 2$, $n . k > 0$, $n . i > 0$.

3. Evaluate $\iint_S (x \, dy \, dz + dz \, dx + xz^2 \, dx \, dy)$,

   where S is the outer side of the part of the sphere $x^2 + y^2 + z^2 = 1$ in the first octant.

   $\left[\textbf{Ans. } \frac{5\pi}{12} + \frac{2}{15}\right]$

4. Show that $\iint_S (x \cos \alpha + y \cos \beta + z \cos \gamma) \, dS = 3$,

   $\cos \alpha$, $\cos \beta$, $\cos \gamma$ being the direction cosines of the outward drawn normal of the surface S, where S is the positive side of the cube formed by the planes $x = 0$, $y = 0$, $z = 0$ and $x = 1$, $y = 1$, $z = 1$.

   [**Hint.** div $F = 1 + 1 + 1 = 3$, $I = \int_0^1 \int_0^1 \int_0^1 3 \, dx \, dy \, dz = 3$]

5. Use Gauss's theorem to evaluate

   $$\iint_S (y^2 z^2 \, dy \, dz + z^2 x^2 \, dz \, dx + x^2 y^2 \, dx \, dy),$$

S being the surface of the sphere $x^2 + y^2 + z^2 = 1$ above the xy-plane.

[**Ans.** $\pi/24$]

6. Use Gauss's theorem to show that

$$I = \iint_S (4xz\, dy\, dz - y^2\, dz\, dx + yz\, dx\, dy) = \frac{3}{2}.$$

where S is the surface of the cube $x = 0$, $y = 0$, $z = 0$, $x = 1$, $y = 1$, $z = 1$.

[**Hint**. div F = $4z - 2y + y = 4z - y$. By Gauss's theorem.]

$$I = \int_0^1 \int_0^1 \int_0^1 (4z - y)\, dx\, dy\, dz = 2 - \frac{1}{2} = \frac{3}{2}$$

5

SERIES OF ARBITRARY TERMS AND DOUBLE SERIES

5.1 Introduction

In this chapter we will discuss the following terms:

(i) Series of arbitrary terms, (ii) Abel's and Dirichlet's Tests, (iii) Re-arrangement of Series, (iv) Multiplication of Series and (v) Double Series.

The reader is advised to go through the chapter *'Infinite Series'* from the Author's Book entitled **Real Analysis (Brilliant Series)**. The results of Infinite Series will be frequently used in this chapter.

Throughout this chapter we shall use the fact that the series

$1 - \frac{1}{2} + \frac{1}{3} - \frac{1}{4} + \frac{1}{5} = ...$ is convergent and that its sum of log 2.

5.2 Series of Arbitrary Terms

Theorem 1:

Abel's Lemma

If the sequence $< u_n >$ be bounded and if the sequence $< u_n >$ be positive and monotonically decreasing, then the series $\Sigma u_n (v_n - v_{n+1})$ is absolutely convergent.

Proof:

Since the sequence $< u_n >$ is bounded, therefore, there exists a positive number K such that $| u_n | < K$, for all n.

Since $< v_n >$ is monotonically decreasing $v_{n+1} \leq v_n \ \forall$ n.

Now $\sum_{n=1}^{m} |u_n (v_n - v_{n+1})| = \sum_{n=1}^{m} |u_n (v_n - v_{n+1})|$, since $v_n - v_{n+1} \geq 0$.

$$< K \sum_{n=1}^{m} (v_n - v_{n+1})| = K (v_1 - v_{m+1})$$

$$< K v_1 \qquad (\because v_1 = v_{m+1} < v_1, v_{m+1} > 0)$$

Thus the sequence of partial sums of the positive term series $\Sigma \mid u_n (v_n - v_{n+1}) \mid$ is bounded above by Kv_1 and so $\Sigma \mid v_n (V_n - v_{n+1}) \mid$ converges. Hence the series $\Sigma\, v_n (v_n - v_{n+1})$ is absolutely convergent.

Theorem 2:

Abel's Test

If the series $\Sigma\, u_n$ converges and if $< u_n <$ is a positive monotonically creasing sequence, then the series $\Sigma\, u_n v_n$ converges. **(D.U., 1994)**

Proof:

Let $< T_n >$ be the sequence of partial series of $\Sigma\, u_n\, v_n$,

$T_n = u_1 v_1 + u_2 v_2 + ... + u_n v_n$.

We know that

$$T_n = \sum_{k=1}^{n-1} S_k (v_k - v_{k+1}) + S_n v_n. \qquad ...(1)$$

Since $\Sigma\, v_n$ is convergent, so $< S_n >$ is convergent and therefore bounded.

Since $< v_n >$ is positive and monotnically decreasing, therefore by Abel's Lemma.

$$= \sum_{k=1}^{n-1} S_k (v_k - v_{k+1}) \text{ tends to a finite limits as } n \to \infty. \qquad ...(2)$$

It is clear that the sequence $< v_n >$ converges, since it is monotonically decreasing and bounded below by zero. Also $< S_n >$ is convegent, so

$S_n v_n$ tends to a finite limit as $n \to \infty$. ...(3)

Using (2) and (3) in (1), we see that T_n tends to a finite limit as $n \to \infty$. Hence, the series $\Sigma\, u_n\, v_n$ converges.

Theorem 3:

Dirichlet's Test

If for a series $\Sigma\, u_n$, the sequence $< S_n >$ of partial sums of bounded, and $< v_n >$ is a positive monotonically decreasing sequence with lim $v_n = 0$, then the series å $u_n\, v_n$ converges. **(D.U., 1998, 1993)**

Proof:

Let $S_n = u_1 + u_2 + ... u_n$ and

$$T_n = u_1v_1 + u_2v_2 + ... + u_nv_n$$
$$= S_1v_1 + (S_2 - S_1)\, v_2 + ... + (S_n - S_{n-1})\, v_n$$
$$= S_1(v_1 - v_2) + S_2(v_2 - v_3) + ... + S_{n-1}\,(v_{n-1} - v_n) + S_nv_n$$

$$\therefore\ T_n = \sum_{k=1}^{n-1} S_k\,(v_k - v_{k+1}) + S_nv_n. \qquad ...(1)$$

Since the sequence $< S_n >$ is bounded, and the sequence $< v_n >$ is positive and monotonically decreasing, therefore by Abel's Lemma.

$$= \sum_{k=1}^{n-1} S_k\,(v_k - v_{k+1}) \text{ tends to a finite limit (say l) as } n \to \infty. \qquad ...(2)$$

Also, since $v_n \to 0$ as $n \to \infty$ and since $< S_n >$ is bounded, therefore

$$S_nv_n \to 0 \text{ as } n \to \infty. \qquad ...(3)$$

Using (2) and (3) in (1). $\lim_{n\to\infty} T_n = l.$

Thus, the sequence $< T_n >$ of partial sums of the series $\Sigma\, u_nv_n$ converges.

Hence, the series Σu_nv_n converges.

Corollary:

Leibnitz's test is a particular case of Dirichlet's test.

Proof:

The sequence of partial sums of the series $\Sigma(-1)^{n-1}$ is bounded (in fact, $S_n = 0$ if n is even, and $S_n = 1$ if n is odd). Let us take $u_n = (-1)^{n-1}$ in Dirichlet's test.

$$\therefore\ \Sigma u_nv_n = v_1 - v_2 + v_3 - ... + (-1)^{n-1}\, v_n + ...$$

Thus Dirichlet's test can be stated as:

If $< v_n >$ is a positive monotonically decreasing sequence with that $v_n = 0$, then the sereis $v_1 = v_2 + v_3 - ... + (-1)^{n-1}\, v_n + ...$ converges.

This is Leibnitz's test.

Example 1:

Show that the series

$$0 - \frac{1}{2} + \frac{1}{2^2} - \frac{1}{3} + \frac{2}{3^2} - \frac{1}{4} + \frac{3}{4^2} + \ ... + \textit{ is convergem.}$$

Solution:

Let $\quad \Sigma u_n = 1 - 1 + \frac{1}{2} - \frac{1}{2} + \frac{1}{3} - \frac{1}{3} + ...$

and $\quad < v_n > \; = < 0, \frac{1}{2}, \frac{1}{2}, \frac{2}{3}, \frac{2}{3}, \frac{2}{3}, \frac{3}{4}, \frac{3}{4}, ...>$

The the given series can be written as $\Sigma u_n v_n$.

Let $< S_n >$ be the sequence of partial sums of the series Σu_n. Then

$$\lim_{n\to\infty} S_{2n} = 0, \quad \lim_{n\to\infty} S_{2n+1} = \lim_{n\to\infty} \frac{1}{n+1} = 0.$$

$\therefore \lim_{n\to\infty} S_n = 0$ and so the series Σn_u converges.

It is clear that the sequence $< v_n >$ is monotonic and bounded. By Abel's test, $\Sigma u_n v_n$ i.e., the given series is convergent.

Example 2:

Examine the converges of the series:

(i) $\sum \frac{\cos n\theta}{n^\alpha}$ (ii) $\sum \frac{\sin n\theta}{n^\alpha}$.

Solution:

(i) The sequence $< S_n >$ of partial sums of the series $\Sigma \cos n\theta$ is bounded for $\theta \neq 0, 2k\pi$.

For $a > 0$, $\sum \frac{1}{n^\alpha}$ is a monotonically decreasing sequence tending to zero. By Dirichlet's test, the given series converges for $a > 0$ $(\theta \neq 0, 2k\pi)$.

For $\alpha \leq 0$, the series diverges since the nth term does not tend to zero as $n \to \infty$.

For $\theta = 0$ or $2k\pi$, the given series becomes $\Sigma \theta/n^\alpha$ which converges for $\alpha > 1$.

(ii) The sequence $< S_n >$ of partial sums of the series $\Sigma \sin n\theta$ is bounded for $\theta \neq 0, 2k\pi$.

For $a > 0$, $\Sigma 1/n^\alpha$ is a monotonically decreasing sequence tending to zero. By Dirichlet's test, the given series converges for $\alpha > 0$ $(\theta \neq 0, 2k\pi)$.

For $\alpha \leq 0$, the series diverges since the nth term does not tend to zero as $n \to \infty$.

For $\theta = 0, 2k\pi$ the given series become a series of zeros and is therefore conversent for all values of α.

Example 3:

If a series Σu_n is convergent, then show that the series $\sum \left(\frac{n+1}{n}\right) u_n$ *is also convergent.*

Solution:

Let $v_n = \frac{n+1}{n} = 1 + \frac{1}{n}$.

It is clear that $< v_n >$ is a positive monotonically decreasing sequence. By Abel's test, $\Sigma u_n v_n$ is convergent. Hence the series $\sum \left(\frac{n+1}{n}\right) u_n$ is convergent.

Example 4:

Show that the series $\sum \frac{\sin nx}{n}$ *converges for all real values of x.*

Solution:

Let $< S_n >$ be the sequence of partial sum of the series $\Sigma \sin nx$. Then

$$S_n = \sin x + \sin 2x + \ldots + \sin nx = \frac{\cos(x/2) - \cos|(2n+1)x/2|}{2\sin(x/2)}$$

$$u \neq 0.2\, k\pi$$

$$\therefore |S_n| < \frac{1+1}{2|\sin(x/2)|} = \frac{1}{|\sin(x/2)|} \; \forall u.$$

Thus the sequence $< S_n >$ of partial sums of the series $\Sigma \sin nx$ is bounded. Also $< 1/n >$ is a monotonically decreasing sequence tending to zero. By Dirichlet's test, it follows that $\sum \frac{\sin nx}{n}$ converges for $x \neq 0.2\, k\pi$.

For $x = 0$ or $2k\pi$, the given sereis is obviously convergent.

Hence the result follows.

Example 5:

Show that the series

$$1 - \frac{1}{4.3} + \frac{1}{4^2.5} - \frac{1}{4^3.7} + \ldots \textit{ converges.}$$ **(D.U., 1984)**

Solution:

Let $u_n = \left(-\frac{1}{4}\right)^{n=1} . v_n = \frac{1}{2n-1}.$

Then the given series can be written as $\Sigma u_n v_n$.

Since $| u_n | = \left(\frac{1}{4}\right)^{n=1}$, therefore $\Sigma | u_n |$ being a geometric series with common ratio $\frac{1}{4}$ is convergent.

Thus Σu_n is absolutely convergent and so Σu_n is convergent. Since $< v_n >$ is a positive, monotonically decreasing sequence, therefore by Abel's test, the series $\Sigma u_n v_n$ converges.

Ex.: Show that if a series Σu_n is convergent then

(i) $\Sigma \frac{u_n}{n}$ (ii) $\sum_{n=2}^{\infty} \frac{u_n}{\log n}$ are convergent.

Example 6:

Show that the series

$$1 - \frac{1}{3.2^2} + \frac{1}{5.3^2} - \frac{1}{7.4^2} + ...$$

is convergent.

Solution:

Let $\Sigma u_n = \frac{1}{1^2} - \frac{1}{2^2} + \frac{1}{3^2} - \frac{1}{4^2} + ... v_n = \frac{1}{2n-1}.$

The the given series can be written as $\Sigma u_n v_n$. By Leibnitz's Test. Σu_n converges.

Since $< v_n >$ is a positive and monotonically decreasing requence, there.

Example 7:

Examine the convergence of the series

$$\sum_{n=1}^{\infty}\left(1 + \frac{1}{2} + \frac{1}{3} + ... + \frac{1}{n}\right) \frac{\sin n\theta}{n}$$

Solution:

The sequence $< S_n >$ of partial sums of Σu_n is bounded $(\theta \neq 0, 2k\pi)$.

Let $v_n = \frac{1}{n}\left(1 + \frac{1}{2} + \frac{1}{3} + \ldots + \frac{1}{n}\right)$.

Then the given series can be written as $\Sigma\, u_n v_n$.

Let $a_n = \frac{1}{n}$ so that $\lim\limits_{n\to\infty} a_n = 0$.

By Cauchy's First Theorem on limits.

$$\lim_{n\to\infty} \frac{a_1 + a_2 + \ldots. a_n}{n} = \lim_{n\to\infty} \frac{1}{n}\left(1 + \frac{1}{2} + \ldots. + \frac{1}{n}\right) = 0.$$

Thus $\lim\limits_{n\to\infty} v_n = 0$, where $< v_n >$ is a positive monotonically decreasing sequence. Hence by Dirichlet's test, $\Sigma\, u_n v_n$ is convergent for $\theta \neq 0, 2\,k\,\pi$.

For $\theta = 0$ or $2\,k\,\pi$, the given series reduces to the series of zero which is obviously convergent.

Hence the given series is convergent for all values of θ.

Ex: Show that the series

$$\sum_{n=1}^{\infty}\left(1 + \frac{1}{2} + \frac{1}{3} \div \ldots + \frac{1}{n}\right)\frac{\sin n\,\theta}{n}$$

converges, absolutely for $\theta = k\,\pi$, k is any integer, and conditionally for all other real values of θ.

Example 8:

Examine the convergence of the series

$$\Sigma\left[\frac{1}{(n+1)^2} + \frac{1}{(n+2)^2} + \ldots + \frac{1}{(2n)^2}\right]\sin n\,\theta.$$

Solution:

Do your self.

Example 9:

Test the convergence of the series

$$\Sigma\frac{\{(n^3 + 1)^{1/3} - n\}}{\log n}$$ **(D.U., 1981)**

Solution:

Let $u_n = (n^3 + 1)^{1/3} - n$ and $v_n = 1/\log n$.

Then the given series can be written us $\Sigma\, u_n\, v_n$

Now $u_n = n\left(1+\dfrac{1}{n^3}\right)^{1/3} - n = n\left[\left(1+\dfrac{1}{n^3}\right)^{1/3} - 1\right]$

$$= n\left[\left(1+\frac{1}{3n^3}\right)-1\right] = \frac{1}{3n^2}+\ldots\left(-\frac{1}{n^2}\right)$$

Thus, $\Sigma\, u_n$ converges.

Now $< v_n >$ is a positive monotonically decreasing sequence, therefore by Abel's test, $\Sigma\, u_n v_n$ converges.

Example 10:

Test for convergence : $\displaystyle\Sigma\,\frac{(-1)^{n+1}}{n}\left(1+\frac{1}{n}\right)^{-n}$.

Solution:

Do your self.

Hint. $\displaystyle\sum u_n = \sum\frac{(-1)^{n+1}}{n}$ converges, by Leibnitz's Test. $< v_n >$ $= <\left(1+\dfrac{1}{n}\right)^{-n}> = <\dfrac{1}{2}.\dfrac{4}{9}.\dfrac{27}{64},\ldots>$ is a positive and monotonically decreasing sequence. Apply Abel's Test.

Example 11:

Show that the series $\displaystyle\sum_{n=2}^{\infty}\frac{\sin nx}{\log n}$ *converges* $\forall\, x \in R$

Solution:

Do yourself.

Example 12:

Examine the convergence of the series

$$\Sigma\left\{\frac{1}{(n+1)^2}+\frac{1}{(n+2)^2}+\ldots+\frac{1}{(2n)^2}\right\}\cos n\,\theta.$$

Solution:

Let $\Sigma\, u_n = \Sigma\cos n\,\theta$ and

$$v_n = \frac{1}{(n+1)^2} + \frac{1}{(n+2)^2} + ... + \frac{1}{(2n)^2}.$$

Then the given series can be written as $\Sigma\, u_n\, v_n$

Let $< S_n >$ be the sequence of partial sums of $\Sigma\, u_n$. Then

$$S_n = \sum_{r=1}^{n} \cos r\,\theta = \frac{[\cos (n-1)\,\theta/2].\,\sin (n\theta/2)}{\sin(\theta/2)}, \qquad \theta \neq 0,\ 2\,k\pi$$

$$\Rightarrow |S_n| \leq \frac{1}{|\sin(\theta/2)|}\ \forall \qquad (\theta \neq 0,\ 2\,k\pi)$$

Thus $< S_n >$ is bounded. $(\theta \neq 0,\ 2\,k\pi)$

Let $a_n = \frac{n}{(2n)^2} = \frac{n}{(n+n)^2} = \frac{1}{4n}$ so that $\lim_{n\to\infty} a_n = 0.$

By Cauchy's First Theorem on Limits. We have

$$\lim_{n\to\infty} \frac{a_1 + a_2 + ... + a_n}{n} = 0$$

$$\Rightarrow \lim_{n\to\infty}\left[\frac{1}{(n+1)^2} + \frac{1}{(n+2)^2} + ... + \frac{1}{(2n)^2}\right] = 0 \Rightarrow \lim v_n = 0.$$

Now we have

$$v_{n+1} - v_n = \frac{1}{(2n+2)^2} + \frac{1}{(2n+1)^2} - \frac{1}{(n+1)^2} = \frac{-8n^2 - 4n + 1}{4(n+1)^2\,(2n+1)^2} < 0\ \forall\ n$$

$\therefore v_{n+1} \leq v_n\ \forall$ n and so $< v_n >$ is a monotonically decreasing sequence which tends to zero. By Dirichlet's test, it follows that $\Sigma\, u_n\, v_n$ i.e., the given series converges for $\theta \neq 0,\ 2\,k\pi$.

5.3 Rearrangement of Series

A series $\Sigma\, v_n$ is said to be a *rearrangement* of a series $\Sigma\, u_n$ if there exists a one-to-one correspondence between the terms of $\Sigma\, u_n$ and those of $\Sigma\, v_n$ such that the corresponding terms are equal.

For example, the series $1 + \frac{1}{3} - \frac{1}{2} + \frac{1}{5} + \frac{1}{7} - \frac{1}{4} + ...$

is a rearrangement of the series

$$1 - \frac{1}{2} + \frac{1}{3} - \frac{1}{4} + \frac{1}{5} + ...$$

Theorem 4:

Dirichlet's Theorem

Any series obtained from an absolutely convergent series by a rearrangement of terms is absolutely convergent, and has the same sum as the original series. **(D.U., 1994)**

Proof:

Case 1:

We shall first prove the theorem for a series of positive temrs. Let Σu_n be a convergent series of positive terms with sum s and let Σv_n be any rearrangement of Σu_n. Let $< S_n >$ and $< T_n >$ be the sequences of partial sums Σu_n of and Σv_n respectively. If T_m be any member of $< T_n >$, then the terms of T_m consist of a finite number of terms of Σu_n, and therefore form a part of some partial sum S_n of Σu_n. Since the series are a positive temrs, therefore $T_m \leq S_n \leq s$. Thus the sequence $< T_n >$ is bounded above by s and so Σv_n also converges, say to t and $l \leq s$. Similarly Σu_n can also be though of as a rearrangement of Σv_n and so $s \leq l$. Hence $l = s$.

Case 2:

Suppose that Σu_n is any absolutely convergent series and that Σv_n is a rearrangement of Σu_n. We define the sequence

$< P_n >, < q_n >, < r_n >, < S_n >$ as follows:

$$p_n = \begin{cases} u_n, & \text{if } u_n > 0 \\ 0, & \text{if } u_n \leq 0 \end{cases}, \qquad q_n = \begin{cases} -u_n, & \text{if } u_n \leq 0 \\ 0, & \text{if } u_n > 0 \end{cases}$$

$$r_n = \begin{cases} v_n, & \text{if } v_n > 0 \\ 0, & \text{if } v_n \leq 0 \end{cases}, \qquad q_n = \begin{cases} -v_n, & \text{if } v_n \leq 0 \\ 0, & \text{if } v_n > 0 \end{cases}$$

Then $u_n = p_n - q_n \mid u_n \mid = p_n + q_n$, $v_n = r_n - s_n$, $\mid v_n \mid = r_n + s_n$. ...(1)

From (1), we have

$$\left.\begin{aligned} p_n &= \tfrac{1}{2}(u_n + |u_n|, \\ q_n &= \tfrac{1}{2}(|u_n| - u_n \end{aligned}\right\} \qquad \text{...(2)}$$

$$\left.\begin{aligned} r_n &= \tfrac{1}{2}(v_n + |v_n|, \\ s_n &= \tfrac{1}{2}(|v_n| - v_n \end{aligned}\right\} \qquad \text{...(3)}$$

Since Σu_n, $\Sigma \mid u_n \mid$ are both convergent, therefore by (2), Σp_n and Σq_n are also convergent.

Now Σp_n and Σq_n are convergent series of positive terms and Σr_n, Σs_n are rearrangements of these series respectively, therefore, by case 1 Σr_n and Σs_n converge. If p, q, r, s denote the sum of the series Σp_n, Σq_n, Σr_n, Σs_n respectively, then

$$p = r,\ q = s. \qquad ...(4)$$

From (1) and (4), we see that Σv_n and $\Sigma | v_n |$ converge and

$$\sum_{n=1}^{\infty} v_n = r - s = p - q = \sum_{n=1}^{\infty} u_n,$$

$$\sum_{n=1}^{\infty} |v_n| = r + s = p + q = \sum_{n=1}^{\infty} |u_n|.$$

which shows that Σv_n converges absolutely and has the same sum as the series Σu_n.

Remark:

The conclusion of the Dirichtel's theorem may not hold if a series is not absolutely convergent.

Consider the series S : $\sum \frac{(-1)^{n-1}}{n} = 1 - \frac{1}{2} + \frac{1}{3} - \frac{1}{4} + ...$

which is not absolutely convergent. However, by Leibnitz's test the above series is convergent.

Consider the following rearrangement of the given series:

$$S' = 1 - \frac{1}{2} - \frac{1}{4} + \frac{1}{3} - \frac{1}{6} - \frac{1}{8} + \frac{1}{5} - \frac{1}{10} - \frac{1}{12} + ...$$

Suppose we write it as

$$S' = \left(1 - \frac{1}{2}\right) - \frac{1}{4} + \left(\frac{1}{3} - \frac{1}{6}\right) - \frac{1}{8} + \left(\frac{1}{5} - \frac{1}{10}\right) - \frac{1}{12} + ...$$

$$= \frac{1}{2} - \frac{1}{4} + \frac{1}{6} - \frac{1}{8} + \frac{1}{10} - \frac{1}{12} + ...$$

$$S' = \frac{1}{2}\left(1 - \frac{1}{2} + \frac{1}{3} - \frac{1}{4} + \frac{1}{5} - .....\right) = \frac{1}{2} S.$$

Thus S' ≠ S.

It may be noted that the series $\sum \frac{(-1)^{n-1}}{n}$ is conditionally convergent.

Ex: Show that the series $\sum \frac{(-1)^{n-1}}{n}$ *is conditionally convergent. use this series to show that rearrangement of terms and bracketing of terms can lad to a series with a different sum.*

Theorem 5:

(Riemann's Rearrangement Theorem)

If $\sum u_n$ *be a conditionally convergent series, then the terms of this series can be rearranged to give either a divergent series or an oscillating series or a convergent series whose sum is an arbitrary preassigned number.*

The proof is omitted.

Example 13:

Determine the values of x for which the series

$1+\frac{1}{2}(5x-3x^2)+\frac{1}{2^2}(5x-3x^2)^2+\frac{1}{2^3}(5x-3x^2)^3+.$ *converges and can be rearranged in ascending power of x.* . . . (1)

Solution:

The given series being a geometric series with common ratio

$r=\frac{1}{2}(5x-3x^2)$ is convergent if $\left|\frac{1}{2}(5x-3x^2)\right| < 1$

$\Rightarrow -2 < 5x - 3x^2 < 2$

$\Rightarrow \;\; 3x^2 - 5x - 2 < 0$ and $3x^2 - 5x + 2 > 0$

$\Rightarrow (3x - 1)(x - 2) < 0$ and $3x - 2(x - 1) > 0$

Thus the given series converges if

$$\left(-\frac{1}{3}<x<2\right) \text{ and } \left(x<\frac{2}{3} \text{ or } x>1\right)$$

i.e. if $x \in \left]-\frac{1}{3}, \frac{2}{3}\right[$ or ε $[1, 2[$.

The given series can be written as

$$1+\frac{1}{2}(5x-3x^2)+\frac{1}{4}(25x^2-30x^3+9x^4)+....$$

rearranging in ascending powers of x as follows :

$$1+\frac{5}{2}x-\frac{3}{2}x^2+\frac{25}{4}x^2-\frac{30}{4}x^3+\frac{125}{8}x^3+\frac{9}{4}x^4+....$$

if the series (1) is absolutely convergent. The series

$$1+\frac{1}{2}(5|x|+3|x|^2)+\frac{1}{2^2}(5|x|+3|x|^2)^2+....$$

converges if $\frac{1}{2}(5|x|+3|x|^2)<1 \Rightarrow 5|x|+3|x|^2-2<0$

$$\Rightarrow (3|x|-1)(|x|+2)<0$$

This is possible if $|x| < \frac{1}{3}$

Hence the given can be rearranged in ascending powers of x if $|x| < \frac{1}{3}$.

Example 14:

Explain the fallacy:

$$1-\frac{1}{2}+\frac{1}{3}-\frac{1}{4}+\frac{1}{5}-\frac{1}{6}+.....$$

$$=1+\frac{1}{2}+\frac{1}{3}+\frac{1}{4}+\frac{1}{5}+\frac{1}{6}+....-2\left(\frac{1}{2}+\frac{1}{4}+\frac{1}{6}+....\right)$$

$$=1+\frac{1}{2}+\frac{1}{3}+....-\left(1+\frac{1}{2}+\frac{1}{3}+....\right)=0$$

Hence the series $1-\frac{1}{2}+\frac{1}{3}+\frac{1}{4}+....$ *converges to zero!*

Explanation

The given series is conditionally convergent and so, by Riemann's Rearrangement Theorem, can be made to converge to any limit by a rearrangement of the terms.

Example 15:

Explain the fallacy of the following:

$$1-\frac{1}{2}+\frac{1}{3}-\frac{1}{4}+\frac{1}{5}-\frac{1}{6}+.....$$

$$=(2-1)-\frac{1}{2}+\left(\frac{2}{3}-\frac{1}{3}\right)-\frac{1}{4}+\left(\frac{2}{5}-\frac{1}{5}\right)-\frac{1}{6}+....$$

$$=2-1-\frac{1}{2}+\frac{2}{3}-\frac{1}{3}-\frac{1}{4}+\frac{2}{5}-\frac{1}{5}-\frac{1}{6}+....$$

$$= 2 - 1 + \frac{2}{3} - \frac{1}{2} + \frac{2}{5} - \frac{1}{3} + \frac{2}{7} - \frac{1}{4} + \frac{2}{9} - ....$$

$$= 2\left[1 - \frac{1}{2} + \frac{1}{3} - \frac{1}{4} + \frac{1}{5} - \frac{1}{6} + \frac{1}{7} - ....\right]$$

$\therefore$ S = 2S $\Rightarrow$ 1 = 2 ?

Explanation:

The series $1 + \frac{1}{2} + \frac{1}{3} - \frac{1}{4} + ...$ is conditionally convergent.

Therefore by Rieniann's Rearrangement Theorem, a rearrangement of the terms may alter the sum of the series. Thus,

$$2 - 1 - \frac{1}{2} + \frac{2}{3} - \frac{1}{3} - \frac{1}{4} + \frac{2}{5} - \frac{1}{5} - \frac{1}{6} + ...$$ may not be equal to

$$2 - 1 + \frac{2}{3} - \frac{1}{2} + \frac{2}{5} - \frac{1}{3} + \frac{2}{7} - \frac{1}{4} + \frac{2}{9} - ......$$

We know that $1 - \frac{1}{2} + \frac{1}{3} - \frac{1}{4} + ...$ converges and that its sum S = log 2 $\neq$ 0.

Example 16:

If the terms of the series $1 - \frac{1}{2} + \frac{1}{3} - \frac{1}{4} + ...$ *be rearranged, so that the ratio of the number of positive terms to the number of negative terms in succession is k > 0, then the sum of the rearranged series is 1/2 log 4K.*

Solution:

For k = 1, the result is evident.

Let k > 1. Let T_n > be the sequence of partial sums of the rearranged series, where n = m + p such that lim p/m = k gives n $\to \infty \Rightarrow$ m $\to \infty$

and $$T_n = T_{m+p} = \left(1 - \frac{1}{2} + ... + \frac{1}{2m-1} - \frac{1}{2m+1}\right) + \left(\frac{1}{2m+1} + ... + \frac{1}{2p-1}\right). \quad ... (1)$$

$$\text{As } m \to \infty, \left(1 - \frac{1}{2} + \frac{1}{3} - \frac{1}{4} + ... - \frac{1}{2m}\right) \to \log 2 \quad ... (2)$$

$$\text{Now } \lim_{m\to\infty}, \left[\frac{1}{2m+1}+\ldots+\frac{1}{2p-1}\right] \lim_{m\to\infty}, \frac{1}{m}\sum_{r=1}^{p-m}\frac{1}{2+\left(\frac{2r-1}{m}\right)}$$

$$= \int_0^{k-1}\frac{dx}{2+2x}dx = \frac{1}{2}\,|\log(1+x)|_0^{k-1} = \frac{1}{2}\log k \qquad \ldots (3)$$

Hence $\lim_{n\to\infty} T_n = \log 2 + \frac{1}{2}\log k$, using (2) and (3) in (1)

$$= \frac{1}{2}\log 4k.$$

Similarly the result holds for $1 > k > 0$.

Example 17:

Find the sum of the series

$$1-\frac{1}{2}-\frac{1}{4}+\frac{1}{3}-\frac{1}{6}-\frac{1}{8}+\frac{1}{5}-\ldots\ldots$$

Solution:

Let T_{3n} denote the sum of the first 3n terms of the series. Then

$$T_{3n} = \left(1-\frac{1}{2}-\frac{1}{4}\right)+\left(\frac{1}{3}-\frac{1}{6}-\frac{1}{8}\right)+\ldots\ldots\left(\frac{1}{2n-1}-\frac{1}{4n-2}-\frac{1}{4n}\right)$$

$$= \left(\frac{1}{2}-\frac{1}{4}\right)+\left(\frac{1}{6}-\frac{1}{8}\right)+\ldots\left(\frac{1}{4n-2}-\frac{1}{4n}\right)$$

$$= \frac{1}{2}\left(1-\frac{1}{2}+\frac{1}{3}-\frac{1}{4}+\ldots+\frac{1}{2n}\right) = \frac{1}{2}S_{2n}$$

$$\therefore \lim_{n\to\infty} T_{3n} = \frac{1}{2}\lim_{n\to\infty} S_{2n} = \frac{1}{2}\log 2$$

Also $\lim_{n\to\infty} T_{(3n-i)} = \frac{1}{2}\log 2$ for $i = 1, 2$.

Hence $\lim_{n\to\infty} T_n = \frac{1}{2}\log 2$

Example 18:

Find the sum of the series

$$1+\frac{1}{3}+\frac{1}{5}-\frac{1}{2}+\frac{1}{7}+\frac{1}{9}+\frac{1}{11}-\frac{1}{4}+......$$

Solution:

Let T_{4n} denote the sum of the first 4n terms of the series. Then

$$T_{4n}=\left(1-\frac{1}{3}+\frac{1}{5}-\frac{1}{2}\right)+\left(\frac{1}{7}+\frac{1}{9}+\frac{1}{11}-\frac{1}{4}\right)+....$$

$$+\left(\frac{1}{6n-5}+\frac{1}{6n-3}+\frac{1}{6n-1}-\frac{1}{2n}\right)$$

$$=\left(-\frac{1}{2}+\frac{1}{3}+\frac{1}{4}+.....+\frac{1}{2n-1}-\frac{1}{2n}\right)$$

$$+\left(\frac{1}{2n+1}+\frac{1}{2n+3}+.....+\frac{1}{2n+2r-1}+....+\frac{1}{6n-1}\right)$$

$$=S_{2n}+\sum_{r=1}^{2n}\frac{1}{2n+2r-1}$$

$$\lim_{n\to\infty}T_{4n}=\lim_{n\to\infty}S_{2n}+\lim_{n\to\infty}\sum_{r=1}^{2n}\frac{1}{2n+2r-1}$$

$$=\log 2+\lim\sum_{r=1}^{2n}\frac{1/n}{\left(2+\frac{2r}{n}\right)}$$

$$=\log 2+\int_0^2\frac{dx}{2+2x}$$

$$=\log 2+\frac{1}{2}\,|\log(1+x)|_0^2=\log 2+\frac{1}{2}\log 3$$

$$=\frac{1}{2}[2\log 2+\log 3]\frac{1}{2}\log(2^2\times 3)=\frac{1}{2}\log 12$$

Also $\lim_{n\to\infty}T_{(4n-i)}=\frac{1}{2}\log 12$ for $i=1,2,3$

Hence $\lim_{n\to\infty} T_n = \frac{1}{2}\log 12$.

Example 19:

Examine for convergence the following series :

$$1-\frac{1}{2}-\frac{1}{4}-\frac{1}{6}-\frac{1}{8}+\frac{1}{3}-\frac{1}{10}-\frac{1}{12}-\frac{1}{14}-\frac{1}{16}+\frac{1}{5}+\ldots .$$

Solution:

Let $< T_n >$ denote trhe sequence of partial sums of the given series. Then

$$T_{5n} = \left(1-\frac{1}{2}-\frac{1}{4}-\frac{1}{6}-\frac{1}{8}\right)+\left(\frac{1}{3}-\frac{1}{10}-\frac{1}{12}-\frac{1}{14}-\frac{1}{16}\right)+\ldots.$$

$$+\left(\frac{1}{2n-1}-\frac{1}{8n-6}-\frac{1}{8n-4}-\frac{1}{8n-2}-\frac{1}{8n}\right)$$

$$=\left(1-\frac{1}{2}+\frac{1}{3}-\frac{1}{4}+\frac{1}{5}-\ldots..+\frac{1}{2n-1}-\frac{1}{2n}\right)$$

$$-\left(\frac{1}{2n+2}+\frac{1}{2n+4}+\ldots..+\frac{1}{2n+2r}+\ldots.+\frac{1}{8n}\right)$$

$$= S_{2n} - \sum_{r=1}^{3n}\frac{1}{2n+2rn}$$

$$\therefore \lim_{n\to\infty} T_{5n} \lim_{n\to\infty} S_{2n} - \lim_{n\to\infty}\sum_{r=1}^{3n}\frac{1}{2n+2r}$$

$$= \log 2 - \frac{1}{2}\lim_{n\to\infty}\sum_{r=1}^{3n}\frac{1/n}{\left(1+\frac{r}{n}\right)}$$

$$= \log 2 - \frac{1}{2}\int_0^3\frac{dx}{1+x}$$

$$= \log 2 - \frac{1}{2}|\log(1+x)\ |_0^3 = \log 2 - \frac{1}{2}\log 4$$

$$= \frac{1}{2}(2\log 2 - \log 4) = \frac{1}{2}(\log 4 - \log 4) = 0$$

Also $\lim_{n\to\infty}(5n-i) = 0$ of $i = 1, 2, 3, 4$

Hence $\lim_{n\to\infty} T_n = 0.$

Example 20:

Show that the sum of the series

$$1+\frac{1}{3}-\frac{1}{2}+\frac{1}{5}+\frac{1}{7}-\frac{1}{4}+\frac{1}{9}+\frac{1}{11}-\frac{1}{6}+ \ldots\ldots \text{ is } \frac{3}{2}\log 2$$

Solution:

The sum of series is $S = \frac{1}{2}\log(4 \times 2) = \frac{3}{2}\log 2.$

Now we prove it independently. Let $< T_n >$ denote the sequence of partial sums of the series. Then

$$T_{5n} = \left(1+\frac{1}{3}-\frac{1}{2}\right)+\left(\frac{1}{5}+\frac{1}{7}-\frac{1}{4}\right)+\ldots + \left(\frac{1}{4n-3}+\frac{1}{4n-1}-\frac{1}{2n}\right)$$

$$= \left(1-\frac{1}{2}+\frac{1}{3}-\frac{1}{4}+\ldots..+\frac{1}{2n-1}-\frac{1}{2n}\right)$$

$$+\left(\frac{1}{2n+1}+\frac{1}{2n+3}\ldots + \frac{1}{2n+2r-1}+\ldots+\frac{1}{4n-1}\right)$$

$$= S_{2n} + \sum_{r=1}^{n} \frac{1}{2n+2r-1}$$

$$\therefore \lim_{n\to\infty} T_{3n} \lim_{n\to\infty} S_{2n} + \lim_{n\to\infty} \sum_{r=1}^{n} \frac{1}{2n+2r-1}$$

$$= \log 2 + \lim \sum_{r=1}^{n} \frac{1/n}{\left(2+\frac{2r}{n}\right)}$$

$$= \log 2 + \int_0^1 \frac{dx}{2+2x} = \log 2 + \frac{1}{2}\left|\log(1+x)\right|_0^1$$

$$= \log 2 + \frac{1}{2}\log 2 = \frac{3}{2}\log 2$$

Also $\lim T(3n\text{–}e) = \frac{3}{2} \log 2$ for $i = 1, 2$

Hence $\lim_{n\to\infty} T_n = \frac{3}{2}\log 2.$

Example 21:

Examine for convergence the following series :

$$1+\frac{1}{2}-\frac{1}{3}+\frac{1}{4}+\frac{1}{5}-\frac{1}{6}+\ldots\ldots$$ **(D.U., 1995)**

Solution:

Here the number of positive terms is m = 2 and those of negative terms n = 1 (in each succession)

$$\therefore \qquad k=\frac{m}{n}=\frac{2}{1}=2$$

The sum of the series is

$$S=\frac{1}{2}\log 4k=\frac{1}{2}\log(4\times 2)=\frac{1}{2}\log 2^3=\frac{3}{2}\log 2\ .$$

Example 22:

Define rearrangement of a series. Show that any rearrangement of a convergent series of positive terms converges to the same sum. Show by an example that the result may not be true if the series is not that of positive terms. **(D.U., 1996)**

Example 23:

Show that the series

$$1-1+\frac{1}{2}-\frac{1}{2}+\frac{1}{3}-\frac{1}{3}+\frac{1}{4}-\frac{1}{4}+\ldots. \qquad \ldots(A)$$

is convergent. Show that the following rearrangement of this series

$$1+\frac{1}{2}-1+\frac{1}{3}+\frac{1}{4}-\frac{1}{2}+\frac{1}{5}+\frac{1}{6}-\frac{1}{3}+\ldots. \qquad \ldots(B)$$

$$1+\frac{1}{2}+\frac{1}{3}-1+\frac{1}{4}+\frac{1}{5}+\frac{1}{6}-\frac{1}{2}+\ldots. \qquad \ldots(C)$$

converge respective to log 2 and log 3.

Solution:

For the series (A), we have

$$\lim_{n\to\infty} S_{2n}=0,\quad \lim_{n\to\infty} S_{2n-1}=\lim_{n\to\infty}\frac{1}{n}=0.$$

Hence $\lim_{n\to\infty} S_n=0$, which shows that the series (A) is convergent.

Let $< T_n >$ denote the sequence of partial sums of the series (B). Then T_{3n} is the sum of the firt 3n terms of the series so that

$$T_{3n} = \left(1+\frac{1}{2}-1\right)+\left(\frac{1}{3}+\frac{1}{4}-\frac{1}{2}\right)+\ldots.+\left(\frac{1}{2n-1}+\frac{1}{2n}-\frac{1}{n}\right)$$

$$= \left(1-1+\frac{1}{2}-\frac{1}{2}+\ldots.+\frac{1}{n}-\frac{1}{n}\right)+\left(\frac{1}{n+1}+\frac{1}{n+2}+\ldots.+\frac{1}{2n-1}+\frac{1}{2n}\right)$$

$$= 0+\sum_{r=1}^{n}\frac{1}{n+r}$$

$$\therefore \quad \lim_{n\to\infty} T_{3n} = \lim_{n\to\infty}\sum_{r=1}^{n}\frac{1}{n+r} = \lim_{n\to\infty}\sum_{r=1}^{n}\frac{1/n}{1+r/n}$$

$$= \int_0^1 \frac{dx}{1+x} = |\log(1+x)|_0^1 \ \log 2.$$

Also $\lim_{n\to\infty} T_{(3n-i)} = \log 2$ for $i = 1, 2$.

Hence $\lim_{n\to\infty} T_n = \log 2$ and so the series (B) converges to log 2.

Let T_{4n} denote the sume of the first 4n terms of the series (C). Then.

$$T_{4n} = \left(1-1+\frac{1}{2}-\frac{1}{2}+\ldots.+\frac{1}{n}-\frac{1}{n}\right)+\left(\frac{1}{n+1}+\frac{1}{n+2}+\ldots.+\frac{1}{3n}\right)$$

$$= 0+\sum_{r=1}^{2n}\frac{1}{n+r}$$

$$\therefore \quad \lim_{n\to\infty} T_{4n}\int_0^2 \frac{dx}{1+x} = \log 3.$$

Also $\lim_{n\to\infty} T_{(4n-i)} = \log 3$ for $i = 1, 2, 3$.

Hence $\lim_{n\to\infty} T_n = \log 3$.

Example 24:

Find the values of x for which the series

$1 + (x - x^2) + (x - x^2)^2 + (x - x^2)^3 + \ldots.$

(i) converges and (ii) can be rearranged in ascending powers of x.

[Hint. (i) converges if $|x - x^2| < 1$ i.e., $-x - x^2 < x - x^2 < 1$ i.e., $x^2 - x\ 1 < 0$ and , $x^2 - x + 1 > 0$]

[Ans. : $\left]\frac{1}{2}(1-\sqrt{5}), \frac{1}{2}(1+\sqrt{5})\right[$, (ii) $\left]\frac{1}{2}(1-\sqrt{5}), \frac{1}{2}(\sqrt{5}-1)\right[$.]

Example 25:

Show that

(i) $1+\frac{1}{3}+\frac{1}{5}-\frac{1}{2}-\frac{1}{4}+\frac{1}{7}+\frac{1}{9}+\frac{1}{11}-\frac{1}{6}-\frac{1}{8}+.... = \frac{1}{2}\log 6$.

$\left[\textbf{Hint. } T_{5n} = S_{4n} + \sum_{r=1}^{n} \frac{1}{4n+2r-1}\right]$

(ii) $1+\frac{1}{3}-\frac{1}{2}-\frac{1}{4}-\frac{1}{6}-\frac{1}{5}+\frac{1}{7}-\frac{1}{8}-\frac{1}{10}-\frac{1}{12}+.... = \frac{1}{2}\log\frac{8}{3}$.

(iii) $1-\frac{1}{2}-\frac{1}{4}+\frac{1}{5}+\frac{1}{7}-\frac{1}{8}-\frac{1}{10}+\frac{1}{11}+\frac{1}{13}+.... = \frac{2}{3}\log 2$.

Example 26:

Find the values of x for which the series

$$1+\left(\frac{5x}{2}-x^2\right)+\left(\frac{5x}{2}-x^2\right)^2+.....$$

can be arranged as a convergent series in ascending powers of x.

$\left[\textbf{Ans. } -\frac{1}{2}(\sqrt{41}-5) < x < \frac{1}{2}(\sqrt{41}-5).\right]$

[Hint. For absolute convergence, $\frac{5}{2}|x|+|x|^2 < 1$ or $2|x|^2+5$

$|x|-2<0.1]$

(iv) $1-\frac{1}{2}-\frac{1}{4}-\frac{1}{6}+\frac{1}{3}-\frac{1}{8}-\frac{1}{10}-\frac{1}{12}+\frac{1}{5}-\frac{1}{14}-\frac{1}{16}-\frac{1}{18}+\frac{1}{7}-.... = \frac{1}{2}\log\frac{4}{3}$

(v) Examine the convergence of $1-\frac{1}{2}-\frac{1}{3}+\frac{1}{4}+\frac{1}{5}-\frac{1}{6}-\frac{1}{7}+....$

(D.U., 1995)

5.4 Double Sequence

A double sequence is a rectangular array of numbers whose rows and columns continue indefinitely to the right and downward respectively.

$$\begin{array}{cccccc} a_{11} & a_{12} & a_{13} & \ldots\ldots\ldots\ldots & a_{1n} & \ldots\ldots\ldots\ldots \\ a_{21} & a_{22} & a_{23} & \ldots\ldots\ldots\ldots & a_{2n} & \ldots\ldots\ldots\ldots \\ \vdots & \vdots & \vdots & \ldots\ldots\ldots\ldots & \vdots & \ldots\ldots\ldots\ldots \\ a_{m1} & a_{m2} & a_{m3} & \ldots\ldots\ldots\ldots & a_{mn} & \ldots\ldots\ldots\ldots \\ \vdots & \vdots & \vdots & \ldots\ldots\ldots\ldots & \vdots & \ldots\ldots\ldots\ldots \end{array}$$

This double sequence is denoted as $< a_{mn} >$.

Definition:

A double sequence $< a_{mn} >$ converges to a number t as $(m, n) \to \infty$ if for each $\varepsilon > 0$, there exists a positive integer N such that

$| a_{mn} - l | < \varepsilon$ whenever $m > N$ and $n > N$.

We write this limit as

$$\lim_{m,n\to\infty} a_{mn} = l.$$

It can be shown that if such a number l exists, then it must be unique.

5.5 Double Series

If a_{mn} > is a double sequence, then an expression of the form

$a_{11} + a_{12} + \ldots . a_{21} + a_{22} + \ldots . a_{m1} + a_{m2} + \ldots$

is called a double series and is denoted by $\sum_{m,\,n=1}^{\infty} a_{mn}$ or simply $\sum a_{mn}$

Let $S_{mn} \sum_{r=1}^{m} \sum_{s=1}^{n} a_{rs}$

In other words S_{mn} is the sum of all the elements of the $m \times n$ matrix obtained by selecting the *m* rows and *n* coloumns in the upper left-hand corner of the double sequence $< a_{mn} >$. The sequence $< S_{mn} >$ is called the *sequence of partial sums of the series* $\sum a_{mn}$.

If $\lim_{(m,n)\to\infty} S_{mn}$ exists, we say that the double series $\sum a_{mn}$ *converges*; otherwise we say it *diverges*. The convergence of the double series as defined above is sometimes called *convergence by rectangles.* Many of the theorems on double series are direct extensions of those for single series. We state a few theorems on double series.

Theorem 6:

A necessary condition for the convergence of a double series $\sum a_{mn}$ *is that* $\lim\limits_{(m,n)\to\infty} a_{mn} = 0$. *The converse, however, is not true* **(D.U., 1989)**

Proof:

For $m, n \geq 1$, let $S_{m,n} = \sum\limits_{r=1}^{m}\sum\limits_{s=1}^{n} a_{rs}$; $S_{m,0} = S_{0,n} = S_{0,0} = 0$

Then $a_{mn} = S_{m,n} - S_{m-1,\ n} - S_{m,n-1} + S_{m-1,\ ,n-1}$

Since $\sum a_{mn}$ converges, so the sequence $< S_{mn} >$ converges.

Suppose $< S_{mn} >$ converges to S. Therefore.

$$\lim_{m,n\to\infty} a_{nm} = \lim_{m,n\to\infty} (S_{m,n} - S_{m-1,n} - S_{m,n-1} + S_{m-1,n-1})$$

Remark:

If $\lim\limits_{m,n\to\infty} a_{mn} \neq 0$ *then the double series* $\sum a_{mn}$ *does not converge.*

Theorem 7:

Let $\sum a_{mn}$ *and* $\sum b_{mn}$ *be convergent double series and that c and d are constants. Then* $\sum(ca_{nm} + db_{mn}) = c\sum a_{mn} + d\sum b_{mn}$ *is convergent and.*

$\Sigma(ca_{mn} + db_{mn}) = c\ \Sigma a_{mn} + d\Sigma b_{mn}$

Theorem 8:

Suppose that $a_{mn} \geq 0$ *for all and n.*

(i) *If there is a number M such that* $S_{mn} \leq M$ *for all m and n, then* $\sum a_{mn}$ *is convergent.*

(ii) (Comparison Test)

If $0 \leq a_{nm} \leq$ bmn and $\sum b_{mn}$ converges, then $\sum a_{mn}$ also converges.

Theorem 9:

(Cauchy's Principle of Convergence)

A necessary and sufficient condition for a double series Σa_{mn} to be convergent is that, for each $\varepsilon > 0$, there exists a positive integer N such that

$$| S_{m+p,\ n+q} - S_{m,n} | < \varepsilon \ \forall \ m, n \geq N \text{ and } \forall \ p, q \geq 1$$

Example 27:

Show that the double series

$$\sum_{m,n=1}^{\infty} \frac{1}{m^4 + n^4} \textit{ is convergent.}$$

Solution:

We have $(m^2 - n^2)^2 = m^4 + n^4 - 2m^2n^2 \geq 0$

$\Rightarrow \quad m^4 + n^4 \geq 2m^2n^2$

$$\Rightarrow \quad \frac{1}{m^4 + n^4} \leq \frac{1}{2}\frac{1}{m^2n^2} \quad \text{for all m, n.}$$

By Eample 20, $\sum \frac{1}{m^2n^2}$ is convergent. Hence by the comparison test, the given double series is convergent.

Example 28:

Show that $\sum_{m=1}^{\infty}\sum_{n=1}^{\infty} \frac{1}{m^a + n^a}$, *$\alpha > 2$, is convergent.*

Solution:

We have

$$0 \leq \left(m^{\alpha/2} - n^{\alpha/2}\right)^2 = m^{\alpha} + n^{\alpha} - 2\, m^{\alpha/2} n^{\alpha/2}$$

$$\Rightarrow \; m^{\alpha} + n^{\alpha} \geq 2m^{\alpha/2}n^{\alpha/2} \; \Rightarrow \; \frac{1}{m^{\alpha} + m^{\alpha}} \leq \frac{1}{2}.\frac{1}{m^{\alpha/2}}.\frac{1}{n^{\alpha/2}}$$

$$\therefore \quad S_{mn} = \sum_{p,q=1}^{m,n} \frac{1}{p^{\alpha} + q^{\alpha}} \leq \frac{1}{2}\left(\sum_{p=1}^{m} \frac{1}{p^{\alpha/2}}\right)\left(\sum_{q=1}^{n} \frac{1}{q^{\alpha/2}}\right)$$

We know $\sum_{p=1}^{\infty} \frac{1}{p^{\alpha/2}}$ is convergent if $\frac{\alpha}{2}$ > i.e., $\alpha > 2$.

Let $\sum_{p=1}^{\infty} \frac{1}{p^{\alpha/2}}$ $M < \infty$ if $\alpha > 2$.

If follows that, $S_{mn} < \frac{1}{2} M^2$ for all values of *m* and *n* if $\alpha > 2$.

Hence the given double series is convergent for $\alpha > 2$.

Ex: *Show that the double series* $\sum \frac{1}{m^2(1+n^{5/2})}$ *is convergent.*

[Hint. $m^2(1 + n^{5/2}) \geq m^2n^{5/2}$]

Example 29:

Show that the double series $\sum_{m,n=1}^{\infty} \frac{1}{m^2n^2}$ *is convergent.*

Solution:

We have

$$S_{mn} = \sum_{p,q=1}^{m,n} \frac{1}{p^2q^2} = \left(\sum_{p=1}^{m} \frac{1}{p^2}\right)\left(\sum_{q=1}^{n} \frac{1}{q^2}\right)$$

Since $\sum \frac{1}{n^2}$ is convergent, so $\sum_{n=i}^{\infty} \frac{1}{2} = M < \infty$.

If follows that

$S_{mn} < M^2$ for all m, n.

Hence the given double series is convergent.

5.6 Some Modes of Convergence

We defined convergence by rectangles. Now we define some other modes of convergence in the following.

Definition:

We say that a double series $\sum a_{mn}$ converges

1. *By rectangles* if the sequence $< S_{mn} >$, where $S_{mn} = \sum_{r=1}^{m} \sum_{r=1}^{n} a_{rs}$, converges to some number, say S.S is called the *sum by rectangles* or simply the *sum* of the double series $\sum a_{mn}$. Thus

$$S = \lim_{m,n\to\infty} S_{mn} = \sum_{r,s=1}^{\infty} a_{rs} \text{ (simultaneous limit exists)}$$

2. *By squares* if the sequence $< S_n >$, where $S_n = \sum_{r=1}^{n} \sum_{s=1}^{n} a_{rs}$ converges
3. *By triangles (or diagonals)* if the sequence $< T_n >$, where $T_n = a_{11} + (a_{21} + a_{12}) + (a_{31} + a_{22} + a_{13}) + \ldots$

or $\quad T_n = \sum_{k=2}^{n} \left(\sum_{p+q=k} a_{pq} \right) (n \geq 2)$ converges

4. *By rows* if for *m*, the sum R_m of the series $\sum_{n=1}^{\infty} a_{mn}$ exists and the series $\sum_{m=1}^{\infty} R_m$ converges
5. *By columns* if for each n, the sum C_n of the series $\sum_{m=1}^{\infty} a_{mn}$ exists and the series $\sum_{n=1}^{\infty} C_n$ converges.

We shall denote the sum by squares by S $(\sum a_{mn})$, the sum by diagonals by D $(\sum a_{mn})$, the sum by rows by R $(\sum a_{mn})$ or by $\sum_{m=1}^{\infty} \left(\sum_{n=1}^{\infty} a_{mn} \right)$ and the sum by columns by C $(\sum a_{mn})$ or by $\sum_{n=1}^{\infty} \left(\sum_{m=1}^{\infty} a_{mn} \right)$.

Remark:

The sum of a double series depends upon the mode of convergence as seen in the following examples :

Example 30:

For the double series

$$2 + 2 + 2 + 2 + \ldots\ldots$$
$$-2 - 2 - 2 - 2 - \ldots\ldots$$
$$+2 - 2 + 0 + 0 + \ldots\ldots$$
$$+2 - 2 + 0 + 0 + \ldots\ldots$$
$$+2 - 2 + 0 + 0 + \ldots\ldots$$
$$+ \ldots\ldots\ldots\ldots\ldots\ldots$$

Show that the sums by rows and columns do not exist; the sum by rectangle and squares is each zero and the sum by diagonals is 4.

Solution:

We see that the first two rows R_1 and R_2 are not convergent, so the sum by rows does not exist. Similarly the first two columns C_1 and C_2 are not convergent so the sum by columns does not exist. The sums of successive diagonals are

$$2, 0, 2, 0, 0, 0, \ldots..$$

Thus the sum by diagonals is 4. We have $S_{mn} = 0$ if $m, n > 1$ and so the sum by rectangles is 0. Also the sum by squares is 0.

Example 31:

For the double series

$$0 + 2 + 0 + 0 + 0 + \ldots\ldots$$
$$-2 + 0 + 2 + 0 + 0 + \ldots\ldots$$
$$+0 - 2 + 0 + 2 + 0 + \ldots\ldots$$
$$+0 + 0 - 2 + 0 + 2 + \ldots\ldots$$
$$+0 + 0 + 0 - 2 + 0 + \ldots\ldots$$

show that the sums by rows is 2, that by columns –2, that by diagonals 0, and that by squares is 0. Also show that the sum by rectangles does not exist.

Solution:

We see that the sums of successive rows are 2, 0, 0, 0, ... Consequently, the sum by rows is 2.

Again the sums of successive columns are –2, 0, 0, 0, ...

Consequently, the sum by columns is –2.

The sum of successive diagonals are 0, 0, 0, ... and so the sum by liagonals is 0. Similarly the sum by squares is 0. Now

$$S_{mn} = (a_{11} + a_{12}) + (a_{21} + a_{22} + a_{23} + a_{13}) + \ldots + (a_{m1} + a_{mn} + \ldots + a_{1n})$$

$= (0 + 2) + (-2 + 0 + 2 + 0) + 0 + 0 + \ldots + 0 = 2.$

$$\therefore \lim_{(m,n)\to\infty} S_{mn} = 2 \neq 0$$

Hence the sum by rectangles does not exist.

Theorem 10:

(Pringsheim's Theorem)

First we prove the following

Lemma

If the sum R_m of the series $\sum_{n=1}^{\infty} a_{mn}$ exists for all positive integers m, then

$$\lim_{n\to\infty} S_{mn} = R_1 + R_2 + \ldots + R_m$$

Proof:

we know

$$S_{mn} = \sum_{r=1}^{m}\sum_{s=1}^{n} a_{rs}$$

$$= \sum_{s=1}^{n} a_{1s} \sum_{s=1}^{n} a_{2s} + \ldots\ldots + \sum_{s=1}^{n} a_{ms}$$

Hence $\lim_{n\to\infty} S_{mn} = R_1 + R_2 + \ldots.. + R_m.$

Theorem 10:

(Pringsheim's Theorem)

Let $\sum a_{mn}$ be a double series which converges by rows, columns and rectangles; then their sums are equal i.e.,

$$R(\sum a_{mn}) = C(\sum a_{mn}) = \sum a_{mn} \qquad \textbf{(D.U., 1995)}$$

Proof:

Suppose $\sum a_{mn}$ converges by rectangles to S. Since $\sum a_{mn}$ converges by rows, it follows that the sum R_m of the series $\sum_{n=1}^{\infty} a_{mn}$ exist for each positive integer m. Convergence by rectangles implies that for any $\varepsilon > 0$, there exists a positive integer N such that

$$| S_{mn} - S | < \varepsilon \text{ for } m \geq N, n \geq N;$$

Where $S_{mn} = \sum_{r=1}^{m} \sum_{s=1}^{n} a_{mn}$

Thus for $m \geq N$, we obtain

$$\lim_{n\to\infty} | S_{mn} \; S | < \varepsilon \qquad \ldots (1)$$

By above lemma, $\lim_{n\to\infty} | S_{mn} \; R_1 + R_2 + \ldots\ldots + R_m$ ... (2)

$\therefore \quad | R_1 + R_2 + \ldots + R_m - S | < \varepsilon$ for all $m \geq N$, by (1) and (2).

This shows that $\sum_{m=1}^{\infty} R_m$ converges to S, i.e.,

$$R\left(\sum a_{mn}\right) = S$$

Similarly we can show that

$$C\left(\sum a_{mn}\right) = S.$$

Example 32:

Show that the convergence by rows of a double series does not imply convergence by columnus, but if the sum by rows, columns and rectangles all exist, then all three must be equal. Show also that the result may not be true if the convergence by rectangles is not assumed. **(D.U., 1995)**

Solution:

(i) Consider the double series

$$\frac{1}{2} - \frac{1}{5} + \frac{1}{5} - \frac{1}{10} + \frac{1}{10} - \frac{1}{15} + \frac{1}{15} - \ldots\ldots$$

$$+ \frac{1}{2^2} - \frac{4}{5^2} + \frac{4}{5^2} - \frac{9}{10^2} + \frac{9}{10^2} - \frac{14}{15^2} + \frac{14}{15^2} - \ldots\ldots$$

$$+ \frac{1}{2^3} - \frac{4^2}{5^3} + \frac{4^2}{5^3} - \frac{9^2}{10^3} + \frac{9^2}{10^3} - \frac{14^2}{15^3} + \frac{14^2}{15^3} - \ldots\ldots$$

$$+ \frac{1}{2^4} - \frac{4^3}{5^4} + \frac{4^3}{5^4} - \frac{9^3}{10^4} + \frac{9^3}{10^4} - \frac{14^3}{15^4} + \frac{14^3}{15^4} + \ldots\ldots$$

$$+ \ldots\ldots\ldots\ldots\ldots\ldots\ldots\ldots\ldots\ldots\ldots\ldots$$

Thus sums of the successive rows are $\frac{1}{2}, \frac{1}{2^2}, \frac{1}{2^3} \ldots$

The sum by rows = $\frac{\frac{1}{2}}{1-\frac{1}{2}}=1$. Hence the given double series converges by rows. The sums of successive columns are

$$1, -1, 1, -1, 1, -1, \ldots .$$

Thus the sum by columns does not exist and so the doubles series does not converge by columns. For part

(ii) See Theorem 10.

(iii) Let S_{mn} denote the sum of the terms in the rectangle bounded by the first m rows and n columns. Then

$$S_{m,\,2n-1} = 1 - \frac{1}{2^m}$$

and $$S_{m,2n}=\left(1-\frac{1}{2^m}\right)+\left(1-\frac{1}{5^n}\right)^m-1$$

$$=\left(1-\frac{1}{5^n}\right)^m-\frac{1}{2^m}$$

It shows that sums by rectangles and squares do not exist. Thus the sum by rows, columns and rectangle, may not exist (or be equal) if the convergence by rectangles is not assumed.

Note : The double series of Example 23 is not convergent by rectangles, the sum by rows is 2 and the by columns is – 2.

Ex: *Give an example of a double series which is convergent by rectangles bu is not convergent by rows and columns.*

5.7 A Fundamental Theorem for Positive Term Double Series

Theorem 11:

Let $a_{mn} \geq 0$ for all m, n. The following statements are equivalent:

1. $\sum a_{mn}$ *converges by rectangles.*

2. $\sum a_{mn}$ *converges by rows.*

3. $\sum a_{mn}$ *converges by columns.*

4. $\sum a_{mn}$ *converges by squares.*

5. $\sum a_{mn}$ *converges by triangles.*

Further, $\sum a_{mn}$ *has the* same sum under all the five modes of convergence.

Proof:

We show that (1) ⇒ (2).

Suppose $\sum a_{mn}$ converges by rectangles to σ. Then for each m, n ≥ 1, $S_{mn} \le \sigma$. Since $a_{mn} \ge 0$ for each m, n ≥ 1, it follows that for m ≥ 1.

$$a_{m1} + a_{m2} + ... + a_{mn} \le S_{mn} \le \sigma \text{ for each } n \ge 1.$$

Thus, for m ≤ 1, the sequence of partial sums of the series $\sum_{n=1}^{\infty} a_{mn}$ is bounded above by σ and so $\sum_{n=1}^{\infty} a_{mn}$ converges for each m ≥ 1. We denote the sum of $\sum_{n=1}^{\infty} a_{mn}$ by R_m. Again

$S_{mn} \le \sigma \Rightarrow \lim_{n\to\infty} S_{mn} \le \sigma$ for m ≥ 1. But $\lim_{n\to\infty} S_{mn} = R_1 + R_2 + ... + R_m$.

Thus, $R_1 + R_2 + ... + R_m \le \sigma$ for each m ≥ 1. This shows that the sequence of partial sums of $\sum_{m=1}^{\infty} R_m$ is bounded above by σ. Since $R_m \ge 0$ for each m ≥ 1, the series $\sum_{m=1}^{\infty} R_m$ converges. If $\sum_{m=1}^{\infty} R_m$ converges to R, then R ≤σ.

Thus, Σa_{mn} converges by rows to R, R ≤ σ.

Now, we show that (2) ⇒ (3).

Suppose, Σa_{mn} converges by rows to R. For by R_m. For m ≥ 1, we denote the sum of the series $\sum_{n=1}^{\infty} a_{mn}$ a given j ≥ 1, $a_{mj} \le R_m$ so that

$$\sum_{m=1}^{k} a_{mj} \le \sum_{m=1}^{k} R_m \le R.$$

Thus, $\sum_{m=1}^{\infty} a_{mj}$ converges for each j ≥ 1.

We denote the sum of $\sum_{m=1}^{\infty} a_{mj}$ by C_j. For n ≥ 1,

$$C_1 + C_2 + ... + C_n = \sum_{m=1}^{\infty} a_{m1} + \sum_{m=1}^{\infty} a_{m2} + ... + \sum_{m=1}^{\infty} a_{mn}$$

$$= \sum_{m=1}^{\infty} (a_{m1} + a_{m2} + \ldots + a_{mn})$$

$$= \sum_{m=1}^{\infty} (\sum_{k=1}^{m} a_{mj})$$

$$\leq \sum_{m=1}^{\infty} (\sum_{j=1}^{m} a_{mj}) = \sum_{m=1}^{\infty} R_m \leq R.$$

This shows that the sequence of partial sums of $\sum_{n=1}^{\infty} C_n$ is bounded above by R. Thus the series $\sum_{n=1}^{\infty} C_n$ converges, that is, Σa_{mn} converges by columns. If Σa_{mn} converges by columns to C, then $C \leq R$.

Next we show that (3) $\Rightarrow$ (4).

Suppose Σa_{mn} converges by Columns to C.

Let $S_n = = \sum_{r=1}^{n} \sum_{s=1}^{n} a_{rs} = \sum_{s=1}^{n} (\sum_{r=1}^{n} a_{rs}) \leq \sum_{s=1}^{n} (\sum_{r=1}^{\infty} a_{rs})$.

$\therefore S_n \leq \sum_{s=1}^{n} C_s \leq \sum_{s=1}^{\infty} C_s = C$ for each $n \geq 1$. This shows that the sequence $< S_n >$ of partial sums of $\sum_{n=1}^{\infty} a_{mn}$ is bouned above by C and so $< S_n >$ converges.

Thus Σa_{mn} converges by squares. If Σa_{mn} converges by squares to S, then $S \leq C$.

We now prove that (4) $\Rightarrow$ (5).

Suppose Σa_{mn} converges by squares to S. For $n \geq 2$, let $T_n = \sum_{k=2}^{n} (\sum_{r+s=k}^{\infty} a_{rs})$. Then $T_n \leq S_n = \sum_{r=1}^{n} \sum_{s=1}^{\infty} a_{rs} \leq S$ for each $n \geq 1$. Thus $< T_n >$ is bouned above by S. As $< T_n >$ is a monotonically increasing sequence, which is bounded above by S, it follows that $< T_n >$ converges. Hence Σa_{mn} converges by triangles. If Σa_{mn} converges by triangles to T, then $T \leq S$. Lastly we show that (5) $\Rightarrow$ (1).

Suppose Σa_{mn} converges by triangles to T. The $T_n \leq T$ for all $n \geq 1$. Since S_{mn} contains lesser number of terms of the array (a_{ij}) than T_{m+n}. So $S_{mn} \leq T_{m+n} \leq T$ for each m, $n \geq 1$. Thus Σa_{mn} converges by rectangles. If it converges by rectangles to σ, then $\sigma \leq T$. From the above discussion, we have

$$\sigma \geq R\ C \geq S \geq T \geq \sigma.$$

Hence $\sigma = R = C = S = T$.

5.8 Absolute Convergence of a Double Series

Definition:

A double series $\Sigma\ a_{mn}$ as said to converge absolutely if the series $\Sigma \mid a_{mn} \mid$ converges.

It may be noted that the mode of convergence in both the sries must be the same.

Theorem 12:

If $\Sigma\ a_{mn}$ *converges absolutely, then* $\Sigma\ p_{mn}$, $\Sigma\ q_{mn}$ *converges.*

Proof:

Do your self.

Theorem 13:

If $\Sigma\ a_{mn}$ *converges absolutely, then* $\Sigma\ a_{mn}$ *converges.*

Proof:

Since $-1 \mid a_{mn} \mid \leq a_{mn} \leq \mid a_{mn} \mid$ for all m, n ≥ 1. . . . (1)

$\therefore\ 0 \leq a_{mn} + \mid a_{mn} \mid \leq 2 \mid a_{mn} \mid$ for all m, n ≥ 1

Since $\Sigma \mid a_{mn} \mid$ converges, so by the comparison test $\Sigma\ (a_{mn} + \mid a_{mn} \mid)$ converges.

Since $a_{mn} = (a_{mn} + \mid a_{mn} \mid) - \mid a_{mn} \mid$, the result follows by Theorem 6.

Definition:

We define two special sequences $< p_n >$ and $< q_n >$ as follow:

$$p_{mn} = \frac{1}{2}(\mid a_{mn} \mid + a_{mn}),\ q_{mn} = \frac{1}{2}(\mid a_{mn} \mid - a_{mn})$$

From (1), we obtain

$$p_{mn} \geq 0 \text{ and } q_{mn} \geq 0 \text{ for all m, n} \geq 1.$$

Also $\Sigma\ a_{mn} = \Sigma\ (p_{mn} - q_{mn})$ and $\Sigma\ [a_{mn}] = [p_{mn} + q_{mn}]$

Theorem 14:

If $\sum a_{mn}$ be a double series, then the following statements are equivalent:

1. $\sum a_{mn}$ *converges absolutely be rectangles*
2. $\sum a_{mn}$ *converges absolutely by rows.*
3. $\sum a_{mn}$ *converges absolutely by columns.*
4. $\sum a_{mn}$ *converges absolutely by squares.*
5. $\sum a_{mn}$ *converges absolutely by triangles.*

Further $\sum a_{mn}$ has the same sum under all the five modes of convergence.

Proof:

Let Let $p_{mn} = \frac{1}{2}(a_{mn} + |a_{mn}|)$, $q_{mn} = \frac{1}{2}(|a_{mn}| - a_{mn})$

We show that (1) $\Rightarrow$ (2).

Suppose $\sum a_{mn}$ converges absolutely by rectangles. then $\sum |a_{mn}|$ and $\sum a_{mn}$ converge by rectangles and consequently and $\sum p_{mn}$ and $\sum q_{mn}$ converge by rectangles.

We know $p_{mn} \geq q_{mn} \geq 0$, m, n ≥ 1

By Theorem 11, $\sum p_{mn}$ and $\sum q_{mn}$ converge by rows and

$$\sum p_{mn} = R(\sum p_{mn}),\ \sum q_{mn} = R(\sum q_{mn}).$$

By Theorem 6, $\sum |a_{mn}| = \sum (p_{mn} + q_{mn})$ converges by rows and so $\sum a_{mn}$ converges absolutely by rows.

$$\begin{aligned}\text{Now } R(\sum a_{mn}) &= R\left(\sum (p_{mn} - q_{mn})\right)\\ &= R\left(\sum p_{mn}\right) - R\left(\sum q_{mn}\right)\\ &= \sum p_{mn} - \sum q_{mn}\\ &= \sum (p_{mn} - q_{mn}) = \sum a_{mn}\end{aligned}$$

Similarly, we can prove the other parts

Example 33:

Convert the series

$$\frac{x}{1+x} - \frac{2x^2}{1+x^2} + \frac{3x^3}{1+x^3} - \frac{4x^2}{1+x^4} + \ldots\ldots,\quad |x| < 1 \qquad \ldots (1)$$

into a double series and show it is equal to

$$\frac{x}{(1+x)^2} - \frac{x^2}{(1+x^2)^2} + \frac{x^3}{(1+x^3)^2} - \ldots\ldots \quad \ldots (2)$$

(D. U., 1990)

Solution:

For $|x| < 1$, the series in (1) is the sum by rows of the double series:

$$\begin{array}{l} x - x^2 + x^3 - x^4 + x^5 - x^6 + \ldots\ldots \\ -2x^2 + 2x^4 - 2x^6 + 2x^8 - 2x^{10} + 2x^{12} - \ldots\ldots \\ +3x^3 - 3x^6 + 3x^9 - 3x^{12} + 3x^{15} - 3x^{18} + \ldots\ldots \\ - \ldots\ldots\ldots\ldots\ldots\ldots\ldots\ldots\ldots\ldots\ldots\ldots \end{array} \quad \ldots (3)$$

We shall show that the double series in (3) converges absolutely by rows. Replacing x by $|x| = y$ and all negative signs by positive signs in (3), we obtain

$$\begin{array}{l} y + y^2 + y^3 + y^4 + y^5 + y^6 + \ldots\ldots \\ +2y^2 + 2y^4 + 2y^6 + 2y^8 + 2y^{10} + 2y^{12} + \ldots\ldots \\ +3y^3 + 3y^6 + 3y^9 + 3y^{12} + 3y^{15} + 3y^{18} + \ldots\ldots \\ + \ldots\ldots\ldots\ldots\ldots\ldots\ldots\ldots\ldots\ldots\ldots\ldots \end{array} \quad \ldots (4) \quad (0 \le y < 1)$$

For $0 \le y < 1$, the sum by rows of the double series in (4) is

$$\frac{y}{1-y} + \frac{2y^2}{1-y^2} + \frac{3y^2}{1-y^3} + \ldots\ldots \quad \ldots (5)$$

For $y = 0$, the series in (5) converges.

For $0 < y < 1$, we have

$$\frac{a_n}{a_{n+1}} = \frac{ny^n}{1-y^n} \cdot \frac{1-y^{n+1}}{(n+1)\,y^{n+1}}$$

$$\therefore \lim_{n\to\infty} \frac{a_n}{a_{n+1}} = \lim_{n\to\infty} \frac{n}{n+1} \cdot \frac{1}{y} \cdot \frac{1-y^{n+1}}{1-y^n} = \frac{1}{y} > 1 \qquad (\because 0 < y < 1)$$

By D' Alembert's Ratio Test, the series in (5) converges. Thus the double series in (3) converges absolutely by rows and hence columns and the two sums are equal. For $|x| < 1$, the sum by columns of the double series in (3) is given by

$$\frac{x}{(1+x)^2} - \frac{x^2}{(1+x^2)^2} + \frac{x^3}{(1+y^3)^2} - \ldots\ldots$$

Example 34:

Convert the series on L.H.S. into a double series and show it equals on R.H.S. for $| x | < 1$.

$$\frac{x}{1+x^2}+\frac{x^3}{1+x^6}+\frac{x^5}{1+x^{10}}+.....=\frac{x}{1-x^2}-\frac{x^3}{1-x^6}+\frac{x^5}{1-x^{10}}-...... \qquad ...(1)$$

Solution:

For $| x | < 1$, the series on the L.H.S. of (1) is the sum by rows of the double series:

$$\begin{aligned} &x - x^3 + x^5 - x^7 + x^9 - ...... \\ &+x^3 - x^9 + x^{15} - x^{21} + x^{27} - ...... \\ &+x^5 - x^{15} + x^{25} - x^{35} + x^{45} - ...... \\ &.. \end{aligned} \qquad ...(2)$$

We shall show that the double series in (2) converges absolutely by rows. Replacing x by $| x | = y$ and all negative signs by positive signs in (2), the double series becomes

$$\begin{aligned} &y + y^3 + y^5 + y^7 + y^9 + ... \\ &+y^3 + y^9 + y^{15} + y^{21} + y^{27} + ... \\ &+y^5 + y^{15} + y^{25} + y^{35} + y^{45} + ... \\ &+ .... \quad (0 \le y < 1) \end{aligned} \qquad ...(3)$$

For $0 \le y < 1$, the sum by rows of the double series in (3) is

$$\frac{y}{1-y^2}+\frac{y^3}{1-y^6}+\frac{y^5}{1-y^{10}}+..... \qquad ...(4)$$

For $y = 0$, the series in (4) converges.

For $0 < y < 1$, we have

$$\frac{a_n}{a_{n+1}}=\frac{y^{2n-1}}{1-y^{4n}-2}\times\frac{1-y^{4n+2}}{y^{2n+1}}$$

$$\therefore \lim_{n\to\infty}\frac{a_n}{a_{n+1}}=\frac{1}{y^2}>1. \qquad (\because 0 < y < 1)$$

By D' Alembert's Ratio Test, the series in (4) converges for $0 < y < 1$. Ths the double series in (2) converges absolutely by rows and hence by columns and the two sums are equal. for $| x | < 1$, the sum by columns of the double series in (2) is given by

$$\frac{x}{1+x^2} - \frac{x^3}{1-x^6} + \frac{x^5}{1-y^{10}} - ....$$

Hence the two sides in (1) are equal.

Example 35:

Show that for $|x| < 1$,

$$\frac{x}{1+x} + \frac{x^2}{1-x^2} + \frac{x^3}{1-x^3} + ... = \frac{1+x}{1-x} + x^4\frac{1+x^2}{1-x^2} + x^9\frac{1+x^3}{1-x^3} + ... \quad ...(1)$$

Solution:

For $|x| < 1$, the L.H.S. of (1) is the sum by rows of the double series

$$\begin{array}{l} x + x^2 + x^3 + x^4 + x^5 + ... \\ + x^2 + x^4 + x^6 + x^8 + x^{10} + ... \\ + x^3 + x^6 + x^9 + x^{12} + x^{15} + ... \\ + x^4 + x^8 + x^{12} + x^{16} + x^{20} + ... \\ + ... \end{array} \quad ...(2)$$

We shall show that the double series in (2) converges absolutely by rows. Replacing x by $|x| = y$ in (2), we obtain the double series.

$$\begin{array}{l} y + y^2 + y^3 + y^4 + y^5 + ... \\ + y^2 + y^4 + y^6 + y^8 + y^{10} + ... \\ + y^3 + y^6 + y^9 + y^{12} + y^{15} + ... \\ + y^4 + y^8 + y^{12} + y^{16} + y^{20} + ... \\ + ... \quad (0 \le y < 1) \end{array} \quad ...(3)$$

For $0 \le y < 1$, the sum by rows of the double series in (3) is

$$\frac{y}{1-y} + \frac{y^2}{1-y^2} + \frac{y^3}{1-y^3} + \frac{y^4}{1-y^4} + ... \quad ...(4)$$

For $y = 0$, the series in (4) converges.

For $0 < y < 1$, $\frac{a_n}{a_{n+1}} = \frac{y^n}{1-y^n} \cdot \frac{1-y^{n+1}}{y^{n+1}} = \frac{1-y^{n+1}}{y(1-y^n)}$

Therefore $\lim_{n\to\infty} \frac{a_n}{a_{n+1}} = \frac{1}{y} > 1 \quad (\because 0 < y < 1)$

By D' Alembert's Ratio Test, the series in (4) converges. Hence the double series in (2) converges absolutely by rows for $|x| < 1$.

Absolute convergence of the double series implies that it can be summed up in several ways, all having the same sum. We sum up the double series in (2) according to P-shaped blocks as shown in (2).

Thus, the L.H.S. of (1) is

$$= (x + 2x^2 + 2x^3 + 2x^4 + ...) + (x^4 + 2x^6 + 2x^8 + 2x^{10} + ...)$$
$$+ (x^9 + 2x^{12} + 2x^{15} + 2x^{18} + ...) + ...$$

$$= \left[x + 2x^2\left(\frac{1}{1-x}\right)\right] + \left[x^4 + 2x^6\left(\frac{1}{1-x^2}\right)\right] + \left[x^9 + 2x^{12}\left(\frac{1}{1-x^3}\right)\right] + ...$$

$$= x\left(\frac{1+x}{1-x}\right) + x^4\left(\frac{1+x^2}{1-x^2}\right) + \left(\frac{1+x^3}{1-x^3}\right) + ...$$

Example 36:

For $|x| < 1$, prove that

$$1 + \frac{8x}{1-x} + \frac{16x^2}{1+x^2} + \frac{24x^3}{1-x^3} + ... \quad ...(1)$$

$$= 1 + \frac{8x}{(1-x)^2} + \frac{8x^2}{(1+x^2)^2} + \frac{8x^3}{(1-x^3)^2} + ... \quad ...(2)$$ **(D.U., 1986)**

Solution:

For $|x| < 1$, the series in (1) is the sum by rows of the double series:

$$\begin{array}{l} + 1 + 8x + 8x^2 + 8x^3 + 8x^4 + ... \quad ...(3) \\ + 0 + 16x^2 - 16x^4 + 16x^6 - 16x^8 + ... \\ + 0 + 24x^3 + 24x^6 + 24x^9 - 24x^{12} + ... \\ + \text{...} \end{array}$$

We shall show that this double series converges absolutely by rows. Replacing x by $|x| = y$ and all negative sings by positive signs in (3), the double series in (3) becomes

$$\begin{array}{l} 1 + 8y + 8y^2 + 8y^3 + 8y^4 + ... \\ + 0 + 16y^2 + 16x^4 + 16x^6 + 16y^8 + ... \quad ...(4) \\ + 0 + 24y^3 + 24y^6 + 24y^9 + 24y^{12} + ... \\ + \text{...} \quad (0 \le y < 1) \end{array}$$

For $0 \leq y < 1$, the sum by rows of the double series in (4) is

$$1 + \frac{8y}{1-y} + \frac{16y^2}{1-y^2} + \frac{24y^3}{1-y^3} + \ldots\ldots \quad \ldots (5)$$

For $y = 0$, the series in (5) converges. For $0 < y < 1$, we have

$$\frac{a_n}{a_{n+1}} + \frac{8(ny^n)}{1-y^n} + \frac{1-y^{n+1}}{8(n+1)\, y^{n+1}} \quad \text{(neglecting the firs term).}$$

$\lim\limits_{n\to\infty} \frac{a_n}{a_{n+1}} = \frac{1}{y} > 1$ (since $0 < y$, 1). By D' Alembert's Ratio Test, the series in (5) converges. Thus the series in (3) converges absolutely by rows for $|x| < 1$ and hence by columns and the two sums by rows and columns of the double series in (3) are equal.

For $|x|$, 1, the sum by columns of the double series in (3) is

$$1 + \frac{8x}{(1-x)^2} + \frac{8x^2}{(1+x^2)^2} + \frac{8x^3}{(1-x^3)^2} + \ldots..$$

Example 37:

If $\sum_{n=1}^{\infty} a_n$ converges absolutely, then

$$\frac{a_1}{1.2} + \frac{a_1 + 2a_2}{2.3} + \frac{a_1 + 2a_2 + 3a_3}{3.4} + \ldots\ldots = \sum_{n=1}^{\infty} a_n \quad \ldots (1)$$

Solution:

Consider the double series

$$\frac{a_1}{1.2} + \frac{a_1}{2.3} + \frac{a_1}{3.4} + \frac{a_1}{4.5} + \ldots..$$

$$+ 0 + \frac{2a_2}{2.3} + \frac{2a_2}{3.4} + \frac{2a_2}{4.5} + \ldots. \quad \ldots (2)$$

$$+ 0 + 0 + \frac{a_3}{3.4} + \frac{3a_3}{4.5} + \ldots..$$

$+$ ....

The sum of the term in the nth row is a_n. The sum by rows of (2) is

$\sum_{n=1}^{\infty} a_n$. Since $\sum_{n=1}^{\infty} a_n$ converges absolutely, the double series in (2) converges absolutely by rows and hence by columns and the two sums are equal. The sum by columns of (2) is the left-hand side of (1). Hence the result.

Example 38:

Show

$$\sum_{m=1}^{\infty}\sum_{n=1}^{\infty} \frac{(-1)^{n+1}}{(m+n^2)(m+n^2-1)} = \frac{\pi^2}{12} \qquad \textbf{(D.U., 1990)}$$

Solution:

Let the given series be denoted as S.

Let us consider the series

$$T = \sum_{n=1}^{\infty} (-1)^{n+1} \sum_{m=1}^{\infty} \frac{(-1)^{n+1}}{(m+n^2)(m+n^2-1)} \qquad \ldots (1)$$

We have

$$\frac{1}{(m+n^2)(m+n^2-1)} = \frac{1}{m+n^2-1} - \frac{1}{m+n^2}$$

$$\therefore \sum_{m=1}^{k} \frac{1}{(m+n^2)(m+n^2-1)} = \left(\frac{1}{n^2} - \frac{1}{n^2+1}\right) + \left(\frac{1}{n^2+1} - \frac{1}{n^2+2}\right) + \ldots\ldots + \left(\frac{1}{k+n^2-1} - \frac{1}{k+n^2}\right)$$

$$= \left(\frac{1}{n^2} - \frac{1}{k+n^2}\right)$$

Now, $\sum_{m=1}^{\infty} \frac{1}{(m+n^2)(m+n^2-i)} = \frac{1}{n^2}$. from (1), we get

$T = \sum \frac{(-1)^{n+1}}{n^2}$ which converges absolutely be Leibnitz Test.

Consequently,

$$S = T = \sum \frac{(-1)^{n+1}}{n^2} \quad \text{[By Dirichlet's Theorem]}$$

$$= 1 - \frac{1}{2^2} + \frac{1}{3^2} - \frac{1}{4^2} + \frac{1}{5^2} - \frac{1}{6^2} + \ldots.$$

$$= \left(1 + \frac{1}{2^2} + \frac{1}{3^2} - \frac{1}{4^2} + \ldots..\right) - 2\left(\frac{1}{2^2} + \frac{1}{4^2} + \frac{1}{6^2} + \ldots\right)$$

$$= \frac{\pi^2}{6} - \frac{2}{2^2}\left(\frac{\pi^2}{6}\right) = \frac{\pi^2}{12}.$$

Theorem 15:

(Cauchy's Theorem)

Let $\sum a_n$ *and* $\sum b_n$ *converge* absolutely. *Then the Cauchy product* $\sum c_n$ *of* $\sum a_n$ *and* $\sum b_n$ *converges absolutely and*

$$\sum c_n = (\sum a_n)\ (\sum b_n)$$

In other words, if $\sum a_n = A$ *and* $\sum b_n = B$*, then* $\sum c_n = AB$.

Proof:

Consider the double series

$$\begin{aligned} &a_0b_0 + a_0b_1 + a_0b_0 + a_0b_3 + \ldots\ldots. \\ &+ 0 \quad + a_1b_0 + a_1b_1 + a_1b_2 + \ldots\ldots. \qquad \ldots (1) \\ &+ 0 \quad + 0 \quad + a_2b_0 + a_2b_1 + \ldots\ldots. \\ &+ \ldots\ldots\ldots\ldots\ldots\ldots\ldots\ldots\ldots\ldots \end{aligned}$$

Since $\sum a_n$ and $\sum b_n$ converge absolutely, $\sum |a_n|$ and $\sum |b_n|$ converge. Let $\sum |a_n| = a$ and $\sum |b_n| = b$. We have

$$\sum_{m=0}^{\infty}\left(\sum_{n=0}^{\infty} |a_m b_n|\right) = \sum_{m=0}^{\infty} |a_m| \left(\sum_{n=0}^{\infty} |b_n|\right) = \sum_{m=0}^{\infty} b\,|a_m| = ba$$

Thus, the double series in (1) converges absolutely by rows and hence converges by columns and the two sums by rows and columns are equal. Hence the column series given by (1)

$a_0b_0 + (a_0b_1 + a_1b_0) + (a_0b_2 + a_1b_0 + a_2b_0) + \ldots = \sum c_n$ converges absolutely and $\sum c_n = AB = (\sum a_n)\ (\sum b_n)$.

Corollary:

Let $\sum_{n=0}^{\infty} a_n$ *and* $\sum_{n=0}^{\infty} b_n$ *be two positive term convergent series.*

If $C_n = \sum_{k=0}^{n} a_n b_{n-k}$, then $\sum_{n=0}^{\infty} c_n$ *is convergent and its sum is*

$$\left(\sum_{n=0}^{\infty} a_n\right)\left(\sum_{n=0}^{\infty} b_n\right)$$ **(D.U., 1994)**

Proof:

Do your self.

Theorem 16:

(Abel's Theorem)

Let $\sum c_n$ *be the Cauchy rpoduct of two convergent series* $\sum a_n$ *and* $\sum b_n$ *with sums A and B respectively. If* $\sum c_n$ *is convergent, then its sum* is AB.

Proof:

Let $A_n = a_0 + a_1 + ... + a_n$, $B_n = B_0 + b_1 + ... + b_n$. $C_n = c_0 + c_1 + ... + c_n$. As done in the above therem, we have

$C_n = c_0 B_n + a B_{n-1} + ... + a_n B_0$.

$\therefore\ C_0 + C_1 + ... + C_n$

$= a_0 B_0 + (a_0 B_1 + a_1 B_0) + ... + (a_0 B_n + a_1 B_{n-1} + ... + a_n B_0)$

$= (a_0 + a_1 + ... + a_n) B_0 + (a_0 + a_1 + ... + a_{n-1}) B_1 + ... + a_0 B_n$

$= A_n B_0 + A_{n-1} B_1 + ... + A_0 B_n$

If $C_n \to$ (by the convergence of $\sum c_n$), then by Cauchy's First Theorem on limits.

$$\lim_{n\to\infty} \frac{C_0 + C_1 + ... + C_n}{n-1} = C$$

or $$C = \lim_{n\to\infty} \frac{A_n B_0 + A_{n-1} + ... + A_0 B_n}{n+1} \text{ using (1)}$$

$$= \frac{\{AB + (AB)....... + AB\}\ (n+1 \text{ tir}}{n+1}$$

$(\because A_n \to A, B_n \to B)$

Hence $\quad C = AB$

or $\quad \sum c_a = (\sum a_n)(\sum b_n)$

12.9 Multiplication of Series

Definition:

If $\sum_{n=0}^{\infty} a_n$ and $\sum_{n=0}^{\infty} b_n$ are two infinite series, then their product called the Cauchy product, is defined as $\sum_{n=0}^{\infty} c_n$, where

$$c_0 = a_0 b_0,$$
$$c_1 = a_0 b_1 + a_1 b_0,$$
$$c_2 = a_0 b_2 + a_1 b_1 + a_2 b_0$$
$$c_n = a_0 b_n + a_1 b_{n-1} + ... + a_n b_0 = \sum_{n=0}^{\infty} a_k\, b_{n-k}$$

It may be noted that

$$c_n = \sum_{k=0}^{\infty} a_k\, b_{n-k} = \sum_{k=0}^{n} a_{n-k}\, b_k$$

Theorem 17:

(Merten's Theorem)

If a series $\sum_{n=0}^{\infty} a_n$ *is absolutely convergent and a series* $\sum_{n=0}^{\infty} b_n$ *is convergent, then the Cauchy product* $\sum_{n=0}^{\infty} c_n$ *is convergent and*

$$\Sigma C_n = (\Sigma a_n)(\Sigma b_n).$$

Equivalently, $\Sigma C_n = AB$, *where* $A = \Sigma a_n$, $B = \Sigma b_n$.

Proof:

Let $A_n = a_0 + a_1 + ... + a_n$, $B_n = b_0 + b_1 + ... + b_n$,

and $C_n = c_0 + c_1 + ... + c_n$

or $\quad C_n = a_0 b_0 (a_0 b_1 + a_1 b_0) + ... + (a_n b_0 + a_{n-1} b_1 + ... + a_0 b_n)$

$\quad = a_0 (b_0 + b_1 + ... + b_n) + a_1 (b_0 + b_1 + ... + b_{n-1}) + ... + a_n b_0$

$\therefore \quad C_n = a_0 B_n + a_1 B_{n-1} + ... + a_n B_0.$

Put $B_n = B + t_n$, where $t_n \to 0$ as $n \to \infty$. (Here we use fact that Σb_n is convergent and $< B_n >$ is the sequence of its partial sums.)

$$\therefore \quad C_n = a_0 (B + t_n) + a_1 (B + t_{n-1}) + ... + a_n (B_0 + t_0)$$

$$= B (a_0 + a_1 + ... + a_n) + (a_0 t_n + a_1 t_{n-1} + ... a_n t_0)$$

$$= B A_n + (a_n t_1 + a_1 t_{n-1} + ... + a_{n-m} t_m) + (a_{1-m} t_{m-1} + ... + a_n t_0)$$

or $C_n = BA_n + P + Q$. say. . . . (2)

Since Σa_n converges

find some number $k > 0$ such that

$$| a_0 | + | a_1 | + ... | a_n | < k \ \forall \ n. \qquad . . . (3)$$

Since $t_n \to 0$, so for any $\varepsilon > 0$, there exists a positive integer m such that

$$| t_n | < \frac{\varepsilon}{2k} \ \forall \ n \geq m. \qquad . . . (4)$$

Now $p \leq | a_0 | | t_n | | a_1 | | t_{n-1} | + ... + | a_{n-m} | | t_m |$, by (2)

$$< \frac{\varepsilon}{2k} [| a_0 | + | a_1 | + ... + a_{n-m} |], \text{ using (4)}$$

Using (3), we obtain

$$P < \frac{\varepsilon}{2k} . k = \frac{\varepsilon}{2} \qquad . . . (5)$$

Since every absolutely convergent series is convergent, so Σa_n is convergent. Thus each of a_n, ... , $a_{n-m+1} \to 0$ as $n \to \infty$ and so $Q = a_{n-m+1} t_{m-1} + ... + a_n t_0 \to 0$

$$\Rightarrow \qquad Q < \frac{1}{2} \varepsilon \ \forall \ n \geq N > m \qquad . . . (6)$$

Using (5) and (6) in (2) we obtain

$$C_n = B A_n + \varepsilon \ \forall \ n \geq N.$$

Hence, $\lim C_n = \lim B A_n = B \lim A_n = B A = AB.$

Example 39:

Suppose $\sum_{n=0}^{\infty} a_n$ *and* $\sum_{n=0}^{\infty} b_n$ *be two convergent series. A necessary and sufficient condition for the convergence of the Cauchy product series* $\sum_{n=0}^{\infty} c_n$ *of*

$\sum_{n=0}^{\infty} a_n$ *and* $\sum_{n=0}^{\infty} b_n$ *is that the sequence* $< T_n >$*, where* $T_n = \sum_{k=1}^{\infty} a_k\ (b_n + b_{n-1} + \ldots + b_{n-k+1})$ *converges to zero.*

Solution:

Let A_n, B_n, C_n denote the (n + 1)th partial sums of $\Sigma\, a_n$, $\Sigma\, b_n$ and $\Sigma\, c_n$ respectively. Since $\Sigma\, a_n$, $\Sigma\, b_n$ converge, the sequences $< A_n >$ and $< B_n >$ converge. Suppose $A_n \to A$ and $B_n \to B$ as $n \to \infty$.

For $n \leq 0$, we have

$C_n = c_1 + c_2 + \ldots + c_n = a_0 b_0 + (a_0 b_1 + a_1 b_0) + \ldots + (a_0 b_n + \ldots + a_n b_0)$

$= a_0 (b_0 + b_1 + \ldots + b_n) + a_1 (b_0 + \ldots + b_{n-1}) + \ldots + a_{n-1} (b_0 + b_1) + a_n b_0$

$= a_0 B_n + a_1 B_{n-1} + \ldots + a_n B_0$. Now

$T_n = a_1 b_n + a_2 (b_n + b_{n-1}) + \ldots + a_n (b_1 + b_{n-1} + \ldots + b_1)$.

$\therefore\ C_n + T_n = a_0 B_n + a_1 B_n + \ldots + a_n B_n = (a_0 + \ldots + a_n)$

$B_n = A_n B_n$ . . . (1)

Let $T_n \to 0$ as $n \to \infty$. From (1).

$\lim C_n + \lim T_n = \lim A_n B_n \Rightarrow \lim C_n = \lim A_n \lim B_n = AB.$

Hence the series $\Sigma\, c_n$ converges.

Conversely, suppose $\sum_{n=1}^{\infty} c_n$ converges.

By Abel's Theorem, since all the three series $\Sigma\, a_n$, $\Sigma\, b_n$ and $\Sigma\, c_n$ are convergent, so

$$\lim C_n = AB. \quad \ldots (2)$$

From (1), $T_n = A_n B_n - C_n$

$\Rightarrow \lim T_n = \lim A_n \lim B_n - AB = AB - AB = 0$, by (2).

Hence the sequence $< T_n >$ converges to zero.

Example 40:

Prove that for certain values of x and θ which are to be stated, that

$$\sum_{n=1}^{\infty} \frac{x^n}{n} \cos n\theta \sum_{n=1}^{\infty} \frac{x^n}{n} \sin n\theta = \sum_{n=2}^{\infty} \frac{x^n}{n}\left(1 + \frac{1}{2} + \ldots + \frac{1}{n-1}\right) \sin n\theta$$

Solution:

Let $a_n = \dfrac{x^{n+1}}{n+1} \cos (n+1)\theta,\ b_n = \dfrac{x^{n+1}}{n+1} \sin (n+1)\theta$

$\therefore |a_n| \le |x|^{n+1}/(n+1) < |x|^{n+1}$; $|b_n| \le |x|^{n+1}/(n+1) < |x|^{n+1}$.

Thus the series $\Sigma |a_n|$ and $\Sigma |b_n|$ converge if $|x| < 1$ i.e., Σa_n and Σb_n are absolutely convergent if $|x| < 1 \ \forall \ \theta$. Hence the Cauchy product Σc_n is absolutely convergent for $|x| < 1$ and $\forall \ \theta$. Now

$$c_n = \sum_{k=0}^{n} a_k \, b_{n-k} = \sum_{k=0}^{n} \frac{x^{k+1} \cos(k+1)\theta}{k+1} \cdot \frac{x^{n-k+1} \sin(n-k+1)\theta}{n-k+1}$$

$$= x^{n+2} \sum_{k=0}^{n} \frac{\cos(k+1)\,\theta \sin(n-k+1)\,\theta}{(k+1)(n-k+1)}$$

or $$\frac{2(n+2)C_n}{x^{n+2}} = \sum_{k=0}^{n} [\sin(n+2)\,\theta - \sin(n-2k)\,\theta]\left[\frac{1}{k+1} + \frac{1}{n=k+1}\right]$$

$$= \sin(n+2)\,\theta \sum_{k=0}^{n}\left(\frac{1}{k+1} + \frac{1}{n-k+1}\right)$$

$$- \sum_{k=0}^{n} \frac{\sin(n-2k)\,\theta}{k+1} + \frac{\sin(n-2k)\,\theta}{n-k+1}$$

We have

$$\sum_{k=0}^{n} \frac{1}{n-k+1} = \sum_{k=0}^{n} \frac{1}{k+1}; \sum \frac{\sin(n-2k)\,\theta}{n-k+1} = -\sum \frac{\sin(n-2k)\,\theta}{k+1}$$

$$\therefore \frac{2(n+2)c_n}{x^{n+2}} = 2\sin(n+2)\,\theta \sum_{k=0}^{n} \frac{1}{k+1}$$

$$\Rightarrow c_n = \frac{x^{n+2}\sin(n+2)\,\theta}{n+2}\left(1 + \frac{1}{2} + \ldots.. + \frac{1}{n+1}\right)$$

$$\therefore \sum_{n=0}^{\infty} c_n = \sum_{n=2}^{\infty} c_{n-2} = \sum_{n=2}^{\infty} \frac{x^n \sin n\,\theta}{n}\left(1 + \frac{1}{2} + \ldots.. + \frac{1}{n+1}\right)$$

Hence $(\Sigma b_n) = \Sigma c_n$ proves the desired result.

Example 41:

Give an example to show that the Cauchy product of two convergent series may be divergent. **(D.U., 1996)**

Solution:

Let $a_n = b_n = \dfrac{(-1)^n}{\sqrt{n+1}}$

By Leibnitz's Rule, $\Sigma\, a_n$ and $\Sigma\, b_n$ are convergent series.

$$\text{Now } |\, c_n\,| = |\sum_{k=0}^{n} a_k\, b_{n-k}\,| = \sum_{k=0}^{n} \frac{1}{\sqrt{k+1}\,\sqrt{n-k+1}}$$

For $0 \le k \le n,\ \sqrt{k+1} \le \sqrt{n+1}, \le \sqrt{n-k+1} \le \sqrt{n+1}$

$$\Rightarrow \quad \sqrt{k+1}\,\sqrt{n-k+1} \le \sqrt{n+1}\,\sqrt{n+1} = (n+1)$$

$$\Rightarrow \quad |\, c_n\,| \ge \sum_{k=0}^{n} \frac{1}{(n+1)} \ge 1$$

$\Rightarrow \lim_{n\to\infty} |\, c_n\,| \ne 0$ and so $\Sigma\, c_n$ is divergent.

Example 42:

Assuming that log $(1 + x) = x = x - \frac{1}{2}x^2 + \frac{1}{3} - \ldots$ for $|\,x\,|\ 1$, prove

that

$$\frac{1}{2}[\log(1-x)]^2 = \sum_{n=2}^{\infty} (-1)^2 \left[1 + \frac{1}{2} + \frac{1}{3} + \ldots\ldots + \frac{1}{n-1}\right]\frac{x^n}{n}.$$

Solution:

Do your self.

Example 43:

Show that

$$\sum_{n=0}^{\infty} (-1)^n \left[\frac{1}{(n+1)} + \frac{1}{n.2} + \ldots\ldots + \frac{1}{1.(n+1)}\right](\log 2)^2 .$$

Solution:

We know $\sum a_n = -\frac{1}{2} + \frac{1}{3} - \frac{1}{4} + \frac{1}{5} - \ldots\ldots = \log 2$

Thus $(\log 2)^2 = (\Sigma\, a_n)\,(\Sigma\, a_n)$. . . . (1)

For the Cauchy product series $\sum_{n=0}^{\infty} c_n,\ c_0 = 1$ and for $n \ge 1$.

$$c_n = (-1)^n \frac{1}{(n+1).1} + \frac{1}{n.2} + \ldots + \frac{1}{1.(n+1)} \qquad \ldots (2)$$

$$= \frac{(-1)^n}{n+2}\left[\left(1+\frac{1}{(n+1)}\right)+\left(\frac{1}{2}+\frac{1}{n}\right)+ \ldots + \left(\frac{1}{(n+1)}+1\right)\right] \quad \ldots (3)$$

The result will follow from Abel's Theorem, if we can show that Σc_n converges. From (3), we have

$$|c_n| = \frac{2}{n+2}\left(1+\frac{1}{2}+\ldots+\frac{1}{n+1}\right) \to 0 \text{ as } n \to \infty.$$

Now

$$|c_{n+1}|-|c_n| = \left(\frac{2}{n+3}-\frac{2}{n+2}\right)\left(1+\frac{1}{2}+\ldots+\frac{1}{n+1}\right)+\frac{2}{(n+2)(n+3)}$$

$$= \frac{-2}{(n+2)(n+3)}\left[\frac{1}{2}+\frac{1}{3}+ \ldots + \frac{1}{n+1}\right] < 0$$

$\therefore$ $|c_{n+1}| < |c_n|$ $\forall$ n and $|c_n| \to 0$ as $n \to \infty$.

Thus $\Sigma |c_n|$ is convergent and so Σc_n is also convergent.

By Abel's theorem, $(\Sigma a_n)(\Sigma a_n) = \Sigma c_n$. Hence, by (1) and (2).

$$(\log 2)^2 = \sum_{n=0}^{\infty} (-1)^n \left[\frac{1}{(n+1).1}+\frac{1}{n.1}+ \ldots + \frac{1}{1.(n+1)}\right]$$

Example 44:

Give an example to show that the Cauchy product of two divergent series may be convergent.

Solution:

We'll show that if

$$\sum a_n = 1 - \left(-\frac{3}{2}\right)^2 - \left(\frac{3}{2}\right)^3 - \ldots\ldots$$

and $$\sum b_n = 1 + \left[2+\left(\frac{1}{2}\right)^2\right]+\frac{3}{2}\left[2^2+\left(\frac{1}{2}\right)^3\right]+\ldots.$$

then the series Σa_n and Σb_n both diverge but the Cauchy product Σc_n is convergent.

Since Σa_n is a geometric series having the common ratio $\frac{3}{2} > 1$, it is divergent. In Σb_n each term is greater than a fixed number, say 1, and hence its sum tends to infinity as $n \to \infty$. Therefore Σb_n is divergent. We have

$$a_0 = 1, a_n = -(3/2)^n \ ; \ b_0 = 1, b_n = (3/2)^{n-1}\left[2^n + \left(\frac{1}{2}\right)^{n+1}\right], n = 1,2,\ldots\ldots$$

$$c_n = a_0 b_n + a_n b_{0+} \sum_{k=1}^{n-1} a_{n-k} \ b_k$$

$$= r^{n-1}\left[2^n + \left(\frac{1}{2}\right)^{n+1}\right] - r^n - r^{n-1}\sum_{k=1}^{n-1}\left[2^k + \left(\frac{1}{2}\right)^{k+1}\right], \ \left(r = \frac{3}{2}\right)$$

$$= r^{n-1}\left[2^n + \left(\frac{1}{2}\right)^{n+1}\right] - r^2 - r^{n-1}\left\{2\left(\frac{2^{n-1}-1}{2-1}\right) + \frac{1}{4}\left(\frac{1-(1/2)^{n-1}}{1-1/2}\right)\right\}$$

$$= 2^n r^{n-1} + r^{n-1}\left(\frac{1}{2}\right)^{n+1} - r^n - r^{n-1}\left\{2^n - \left(2 - \frac{1}{2}\right) - \frac{1}{2^n}\right\}$$

$$= (r^{n-1}/2^{n+1}) + (r^{n-1}/2^n) = (3/2)(r^{n-1}/2^n) = (r/2)^n = (3/4)^n.$$

Hence, $\Sigma C_n = \Sigma (3/4)^n$ being a geometric series with common ratio $\frac{3}{4} < 1$ is convergent.

Example 45:

Prove that for certain values of x and θ which are to be stated, that

$$2\sum_{n=0}^{\infty} x^n \cos n\theta \sum_{n=0}^{\infty} x^n \sin n\theta \sum_{n=0}^{\infty}(n+1)x^n \sin n\theta.$$ **(D.U., 1984, 81)**

Solution:

Let $a_n = x^n \cos n\,\theta$, $b_n = x^n \sin n\,\theta$. Then

$|a_n| = |x|^n |\cos n\,\theta| \le |x|^n$; $|b_n| = |x|^n |\sin n\,\theta| \le |x|^n \ \forall\ \theta$.

Thus the series $\Sigma |a_n|$ and $\Sigma |b_n|$ converge if $|x| < 1$. Now

$$C_n = \sum_{k=0}^{n} a_{n-k}\ b_n = \sum_{k=0}^{n} x^{n-k} \ \cos(n-k)\ .\ x^k \sin k\theta.$$

$$= \frac{1}{2} x^n \sum_{k=0}^{n} 2\cos(n-k)\,\theta \sin k\theta$$

$$= \frac{1}{2} x^n \sum_{k=0}^{n} [\sin n\,\theta - \sin(n-2k)\,\theta]$$

Now, $$\sum_{k=0}^{n} \sin n\,\theta = (n+1)\sin n\theta,$$

and $\sum_{k=0}^{n} \sin(n-2k) = \sum_{r=n}^{-n} \sin r\theta = 0.$ (Take $n - 2k = r$)

Thus $C_n = \frac{1}{2} x^n (n+1) \sin n\theta$ for $|x| < 1, \forall \theta$.

From $(\Sigma a_n)(\Sigma b_n) = (\Sigma c_n)$, the result follows.

Example 46:

Show that for $-1 < x \leq 1$,

$$\frac{1}{2}\left(x - \frac{x^3}{3} + \frac{x^5}{5} - \frac{x^7}{7} + ...\right)^2 = \sum_{n=1}^{\infty}\left[1 + \frac{1}{3} + \frac{1}{5} + ... + \frac{1}{2n+1}\right]\frac{x^{2n+2}}{2n+n}$$

(D.U., 1990)

Solution:

Let $a_n = b_n = \frac{(-1)^n x^{2n+1}}{2n+1}$.

For $|x| < 1$, the series Σa_n and Σb_n are absolutely convergent. Hence, by Cauchy Theorem, the series $\Sigma c_n = (\Sigma a_n)(\Sigma(b_n)$ is absolutely convergent. Now,

$$c_n = \sum_{k=0}^{n} a_k b_{n-k} = \sum_{k=0}^{n} (-1)^k \frac{x^{2k+1}}{2k+1}(-1)^{n-k}\frac{x^{2(n-k)+1}}{2(n-k)+1}$$

$$= (-1)^n x^{2n+2} \sum_{k=0}^{n} \frac{1}{(2k+1)(2n-1=2k)}$$

$$\therefore \quad c_n = (-1)^n x^{2n+2} a_n \qquad ...(1)$$

where $$a_n = \sum_{k=0}^{n} \frac{1}{(2k+1)(2n+2k-1)}$$

$$= \frac{1}{(2x-1)} + \frac{1}{3(2x-1)} + \frac{1}{5(2x-3)} + ... + \frac{1}{(2n+1)}$$

$$= \frac{1}{(2n-1)}\left[\left(1 + \frac{1}{2n+1}\right) + \left(\frac{1}{3} + \frac{1}{2n-1}\right)\left(\frac{1}{5} + \frac{1}{2n-3}\right) + ...\right.$$

$$\left. + \left(\frac{1}{(2n+1)} + 1\right)\right]$$

$$= \frac{2}{2n+2}\left[1+\frac{1}{3}+\frac{1}{5}\ ....+\frac{1}{2n+1}\right] \qquad ...(2)$$

From (1) and (2), we get

$$\Sigma c_n = \sum_{n=0}^{\infty} \frac{2(-1)^n x^{2n+2}}{(2n+2)}\left[1+\frac{1}{3}+\frac{1}{5}+....+\frac{1}{2n+1}\right]$$

Hence $(\Sigma a_n)(\Sigma b_n) = (\Sigma c_n)$ gives the required result.

At x = 1, Σc_n is convergent as $< \alpha_n >$ is monotonically decreasing and tends to zero as $n \to \infty$.

Example 47:

Show that for all x, y ∈ R

$$\left(\sum_{n=0}^{\infty}\frac{x^n}{n!}\right)\left(\sum_{n=0}^{\infty}\frac{y^n}{n!}\right) = \sum_{n=0}^{\infty}\frac{(x+y)^n}{n!}.$$

Solution:

Let $a_n = x^n/n!$ so that $a_{n+1} = x^{n+1}/(n+1)!$ and

$$\lim_{n\to\infty}\left|\frac{a_n}{a_{n+1}}\right| = \lim_{n\to\infty}\frac{n+1}{|x|} = \infty \quad \text{for all } x \neq 0.$$

By ratio test, $\Sigma\, a_n$ converges absolutely for all $x \neq 0$.

But for x = 0, $\Sigma\, a_n$ is obviously convergent.

Thus $\Sigma a_n = \Sigma x^n/n!$ and $\Sigma b_n = \Sigma y^n/n\,!$ are absolutely convergent for all x, y ∈ R. By Cauchy's Theorem, the Cauchy product Σc_n of Σa_n and Σb_n also converges absolutely for all x, y ∈ R.

$$\text{Further } \left(\sum_{n=0}^{\infty}\frac{x^n}{n!}\right)\left(\sum_{n=0}^{\infty}\frac{y^n}{n!}\right) = \sum_{n=0}^{\infty} c_n, \qquad ...(1)$$

where $c_n = a_n b_0 + a_{n-1} b_1 + ... + a_1 b_{n-1} + a_0 b_n$

$$= \frac{x^n}{n!} + \frac{x^{n-1}\, y}{(n+1)!} + \frac{x^{n-2}}{(n-2)!}\cdot\frac{y^2}{2!} + ... + x.\frac{y^{n-1}}{(n-1)!} + \frac{y^n}{n!}$$

$$= \frac{1}{n!}\left[x^n + nc_1\, x^{n-1}\, y + nc_1\, x^{n-2}\, y^2 + ... + y^n\right] = \frac{(x+y)^n}{n!} \qquad ...(2)$$

Hence the result follows by (1) and (2).

Example 48:

Assuming $\tan^{-1} x = x - \dfrac{x^3}{3} + \dfrac{x^5}{5} - \dfrac{x^7}{7} + \ldots$, *prove that*

$$\frac{1}{2}(\tan^{-1} x)^2 = \sum_{n=0}^{\infty} \frac{(-1)^n\, x^{2n+2}}{(2n+2)}\left[1 + \frac{1}{3} + \frac{1}{5} + \ldots + \frac{1}{2n+1}\right].$$

Also prove that the series converges absolutely for $|x| < 1$ *and converges when* $x = 1$.

Solution:

Similar to Example 39.

Example 49:

For $|x| < 1$, *prove that*

$$\left[x - \frac{x^2}{2} + \frac{x^3}{3} - \ldots\right]^2 = 2\sum_{n=0}^{\infty}(-1)^n\left[1 + \frac{1}{2} + \frac{1}{3} + \ldots + \frac{1}{n-1}\right]\frac{x^n}{n}.$$

(D.U., 1982)

Solution:

Let $a_n = (-1)^n x^{n+1}/(n+1) = b_n$.

The series Σa_n, Σb_n are absolutely convergent for $|x| < 1$. Hence Cauchy product of Σa_n and Σb_n will converge absolutely to Σc_n, where

$$c_n = \sum_{k=0}^{n} a_k\, b_{n-k} = \sum_{k=0}^{n} \frac{(-1)^k\, x^{k+1}}{k+1} \cdot \frac{(-1)^{n-k}\, x^{n-k+1}}{n-k+1}$$

$$= (-1)^n\, x^{n+2} \sum_{k=0}^{n} \frac{1}{(k+1)\,(n-k-1)}$$

Replacing n + 2 by n, we obtain

$$c_{n-2} = (-1)^{n-2}\, x^n \sum_{k=0}^{n-2} \frac{1}{(k+1)\,(n-k-1)} = (-1)^n\, x^n\, a_n, \qquad \ldots(1)$$

$$a_n = \sum_{k=0}^{n-2} \frac{1}{(k+1)(n-k-1)} = \frac{1}{(n-1)} + \frac{1}{2.(n-2)} + \ldots + \frac{1}{(n-1)}$$

$$= \frac{1}{n}\left[\left(1+\frac{1}{n-1}\right)+\left(\frac{1}{2}+\frac{1}{n-2}\right)+\left(\frac{1}{3}+\frac{1}{n-3}\right)+\ldots+\left(\frac{1}{n-1}+1\right)\right]$$

$$= \frac{2}{n}\left[1+\frac{1}{2}+\frac{1}{3}+..+\frac{1}{n-1}\right]. \qquad ...(2)$$

From (1) and (2), $c_{n-2} = 2\,(-1)^n \dfrac{x^n}{n}\left[1+\dfrac{1}{2}+\dfrac{1}{3}+\ldots+\dfrac{1}{n-1}\right].$

The Cauchy product of the series

$$\left(\sum_{n=0}^{\infty} a_n\right)\left(\sum_{n=0}^{\infty} b_n\right) = \sum_{n=2}^{\infty} c_{n-2}$$ prove the required result.

Example 50:

For $|x| < 1$, show that

$$\frac{1}{1-x}\log\left(\frac{1}{1-x}\right) = \sum_{n=1}^{\infty}\left(1+\frac{1}{2}+\frac{1}{3}+....+\frac{1}{n}\right)x^n.$$

Solution:

We know

$$(1-x)^{-1} = \sum_{k=0}^{\infty} x^n,\; -\log(1-x) = \sum_{n=0}^{\infty}\frac{x^{n+1}}{n+1}$$

Both these series are absolutely convergent for $|x| < 1$.

The Cauchy product is

$$c_n = \sum_{k=0}^{n} a_{n-k} b_k = \sum_{k=0}^{n} x^{n-k}\left(\frac{x^{k+1}}{k+1}\right) = x^{n+1}\sum_{k=0}^{n}\frac{1}{k+1}$$

Hence $(\Sigma a_n)(\Sigma b_n) = \Sigma c_n$ gives

$$(1-x)^{-1}\log(1-x)^{-1} = \sum_{n=0}^{\infty} x^{n+1}\left(1+\frac{1}{2}+....+\frac{1}{n+1}\right)$$

$$= \sum_{n=1}^{\infty} x^n\left(1+\frac{1}{2}+\frac{1}{3}+....+\frac{1}{n}\right)$$

Example 51:

Suppose $\sum_{n=1}^{\infty} a_n X^n$ *is absolutely convergent for* $|x| < 1$*, and its sum is* $A(x)$*. Show that for* $|x| < 1$,

$$(1 - x)^{-1} A(x) = \sum_{n=0}^{\infty} S_n X^n, \text{ where } S_n = a_0 + a_1 + \ldots + a_n.$$

Solution:

The series $\sum_{n=0}^{\infty} x^n$ converges absolutely for $|x| < 1$ and its sum is $(1 - x)^{-1}$. By Cauchy's theorem, the Cauchy product $\sum_{n=0}^{\infty} c_n$ of $\sum_{n=0}^{\infty} x^n$ and $\sum_{n=0}^{\infty} a_n X^n$ converges absolutely for $|x| < 1$ and

$$\sum_{n=0}^{\infty} c_n = (1 - x)^{-1} A(x), |x| < 1. \qquad \ldots(1)$$

We have $c_0 = a_0 = 1$. For $n \geq 1$, we have

$$c_n = 1 \,.\, a_n X^n + x(a_{n-1} X^{n-1}) + x^2 (a_{n-2} x^{n-2}) + \ldots + a_0 x^n$$

or $c_n = (a_n + a_{n-1} + \ldots + a_0) x^n = S_n x^n. \qquad \ldots(2)$

From (1) and (2), $(1 - x)^{-1} A(x) = \sum_{n=0}^{\infty} S_n X^n$.

EXERCISES

1. Prove that the Cauchy product of the series $\sum_{n=2}^{\infty} \frac{(-1)^n}{\log n}$ with itself cannot converge.
2. Let $a_0 = 3$, $a_n = 3^n$ $(n \geq 1)$; $b_0 = -2$, $b_n = 2^n$ $(n \geq 1)$. Show that the Cauchy product of the series $\sum a_n$ and $\sum b_n$ is convergent although $\sum a_n$ and $\sum b_n$ are divergent.
3. State at least two conditions ensuring the convergence of the product.

(D.U., 1996)

4. If $S(x) = \sum_{n=0}^{\infty} (-1)^n \frac{x^{2n+1}}{(2n+1)!}$, $C(x) = \sum_{n=0}^{\infty} (-1)^n \frac{x^{2n}}{(2n)!}$

then for all x, y ∈ R, show that

(i) $S(x) C(y) + S(y) C(x) = S(x + y)$,

(ii) $C(x) C(y) - S(x) S(y) = C(x + y)$,

(iii) $2 S(x) C(x) = S(2x)$,

(iv) $\{C(x)\}^2 - \{S(x)\}^2 = C(2x)$.

5. If $0 < b < a$, prove that

$$\sum_{n=1}^{\infty} \frac{nb^n}{a^n - b^n} = = \sum_{m=1}^{\infty} \frac{a^m b^m}{(a^m - b^m)^2}$$

6. Examine for convergence

$$\sum_{m=0}^{\infty} \sum_{n=0}^{\infty} \frac{(-1)^n}{(m+4n^2)(m+4n^2+1)}.$$ **(D.U., 1996)**

7. Prove that

$$\left(\sum_{n=0}^{\infty} \frac{x^n}{(n!)^2}\right)\left(\sum_{n=0}^{\infty} (-1)^n \frac{x^n}{(n!)^2}\right) = \sum_{n=0}^{\infty} \frac{(-1)^n x^{2n}}{(n!)^2 (2n)!}, \ x \in R.$$

8. Check the validity of $(\Sigma a_n)(\Sigma b_n) = \Sigma c_n$.

$c_n = \sum_{k=1}^{n} a_k b_{n-k}$, when

(i) $a_n = b_n = \frac{1}{n^2}$.

(ii) $a_n = b_n = \frac{(-1)}{\sqrt{n}}$,

(iii) $a_n = b_n = \frac{(-1)}{n}$.

9. For $|x| < 1$, show that

$$\frac{x}{1+x^2} - \frac{x^2}{1+x^4} + \frac{x^3}{1+x^5} - \ldots = \frac{x}{1+x} - \frac{x^3}{1+x^3} + \frac{x^5}{1+x^5} - \ldots$$

10. For $|x| < 1$, show that

$$\frac{x}{(1+x)^2} + \frac{x^2}{(1+x^2)^2} + \frac{x^3}{(1+x^3)^2} + \ldots = \frac{x}{1-x} - \frac{2x^2}{1-x^2} + \frac{3x^3}{1-x^3} - \ldots$$

11. Using double series, prove that for $|x| < 1$

$$\log(1+x) = x - \frac{1}{2}x^2 + \frac{1}{3}x^2 - \frac{1}{4}x^4 + \ldots$$

and $\frac{1}{2}[\log(1+x)]^2 = \frac{1}{2}x^2 - \frac{1}{3}x^3\left(1+\frac{1}{2}\right) + \frac{1}{4}x^4\left(1+\frac{1}{2}+\frac{1}{3}\right) - \ldots.$

(D.U., 1989, 88)

6

THE LEBESGUE INTEGRAL

6.1 Introduction

In Riemann's integration, the domain over which the integral is taken is divided into a partition, and the integral is defined as the limit of a Cauchy sum for this partition as the norm of the partition tends to zero. In Lebesgue integration, on the other hand, the domain over which the integral is taken is divided into a number of measurable sets. The integral is then defined as a limit of a certain sum taken for all these measurable sets as the number of measurable sets increases indefinitely.

6.2 Measurable Sets

Let (a, b) be a closed bounded interval in R. By length of (a, b) we mean the real number b – a. If A is any non-empty bounded open subset of (a, b) then the length of A is defined as the sum of the lengths of all its disjoint open intervals I_k, k = 1, 2, 3, ... s.t. $A \subseteq \bigcup_{k=1}^{n} I_k$, and is denoted by $l(A)$. If A_1 and A_2 are two bounded open sets and $A_1 \subseteq A_2$, then $l(A_1) \le l(A_2)$. Again if B is any closed subset of the interval (a, b), the complement C(B) of B relative to any open subset F of (a, b) containing B is $F - B = F \cap C(B)$ and is open in (a, b). The length of B is defined as

$$l(B) = l(F) - l(F - B).$$

6.3 Outer and Inner Measure of a Bounded Set

Let $A \subseteq$ (a, b) be any bounded subset of real numbers. The *outer measure* of A denoted by m^*A, is defined as $m^*A = \inf l(F)$, where the infimum is taken over all open sets F which contain A. F being open, can be expressed as a countable union of open intervals I_n, n = 1, 2 ...., such that $A \subseteq F \subseteq \bigcup_{n=1}^{\infty} I_n$. For each such countable collection containing A, consider the sum of the lengths of the intervals in that collection. The *outer measure* of A is then defined as

$$m^*A = \inf_{A \subseteq \bigcup_n I_n} \sum_n l(I_n)$$

The *inner measure* of A denoted by m_*A is defined as

$$m_*A = \sup l(B)$$

where the supremum is taken over the lengths of all the closed sets B contained in the set A. Thus we observe that the outer measure m^*A is computed by open sets F which contain A and which come closer and closer to A.

$\therefore \quad m^*A \leq l(F)$

whereas the inner measure m_*A is computed by closed sets B which are contained in A.

$\therefore \quad m_*A \geq l(B)$

In general $m^*A \neq m_*A$.

Definition:

A set $A \subseteq (a, b)$ is said to be *measurable* if $m^*A = m_*A$. In this case we define mA, the measure of A as

$$mA = m^*A = m_*A$$

Theorem 1:

For every set A,

$$m_*A \leq m^*A$$

Proof:

It suffices to prove the result if A is bounded.

Let F be any bounded open set containing A.

For any closed subset B contained in A, we have

$$B \subseteq A \subseteq F$$

$\therefore \quad l(B) \leq l(F)$

Taking supremum over all closed subsets B contained in a, we have

$$\sup_{B \subseteq A} l(B) \leq l(F)$$

Again taking infimum over all open subsets F containing A

$$\sup_{B \subseteq A} l(B) \leq \inf_{F \supseteq A} l(F)$$

i.e.,

$$m_*A \leq m^*A.$$

6.4 Measure of an Open Set and a Closed Set

An open set is an enumerable union of non-overlapping open intervals. The measure of an open set is defined as the limiting sum of the lengths of its intervals. If the set G is contained in an interval (a, b), then the measure of G, denoted by m(G), has the property that

$$m(G) \leq b - a.$$

We know that the complement of a closed set w.r.t. an open set is open. Let a closed set F be contained in an open interval (a, b). We defined the measure of F as follows:

$$m(F) = b - a - m(F'), \text{ where } F' = (a, b) - F.$$

6.5 Measure of an Open Interval

The length of any open interval G is defined as its length and is denoted by the symbol m(G).

Examples : (i) $m[(2, 5)] = 5 - 2 = 3$

(ii) $m[(a, b)] = b - a$.

Evidently $\quad m(G) > 0$.

6.6 Measure of a Closed Interval

Let (a, b) be the smallest closed interval containing a closed set F. Then we define $m(F) = b - a - m(F')$.

F' being the complement of F w.r.t. the interval of its length.

To prove that the measure of a closed interval is its length.

Let A = (a, b). Then A is a closed interval and hence a closed set, (a, b) is the smallest closed interval containing A.

Complement of A relative to (a, b) = A' = ϕ. For A = (a, b).

By definition, $m(A) = b - a - m(A') = b - a - m(\phi) = b - a - 0$

$$= b - a$$

$$m([a, b]) = b - a.$$

6.7 Measure of a Rectangle

The area of an open rectangle R $(a < x < b, c < y < d)$ i.e., $(b - a)(d - c)$ is defined as the measure of R.

Thus, $m(R) = (b - a)(d - c)$.

The area of a closed rectangle R $(a \leq x \leq b, c \leq y \leq d)$, i.e., $(b - a)(d - c)$ is defined as the measure of R.

Thus, $m(R) = (b - a)(d - c)$.

6.8 Measure of a Parallelopiped

The volume of an open parallelopiped

$$V(a < x < b, c < y < d, l < z < m)$$

is defined as the measure of V.

Thus, $m(V) = (b - a)(d - c)(m - l)$.

Similarly the measure of a closed parallelopiped

$$V'(a \leq x \leq b, c \leq y \leq d, l \leq z \leq m)$$

is defined as its volume.

Thus, $m(V') = (b - a)(d - c)(m - l)$.

6.9 Exterior and Interior Measure. (Lebesgue Measurable Set)

(Kanpur, 95, 93, 88)

The Lebesgue exterior measure (or simply exterior measure) of any set A, denoted by $m_e(A)$, is defined as follows:

$$m_e(A) = \inf, \{m(G) : G \supset A, G \text{ is open}\} \quad ...(1)$$

From this it follows that $m(G) > m_e(A)$. Hence given $\in < 0$, $\exists$ an open set $G \supset A$ s.t.

$$m(G) < m_e(A) + \in. \quad ...(2)$$

If the set A is contained in an interval (a, b), then

$$0 \leq m_e(A) \leq b - a.$$

Also $A_1 \subset A_2 \Rightarrow m_e(A_1) \leq m_e(A_2)$

and $m_e(\phi) = 0$.

The Lebesgue interior measure (or simply interior measure) of a set A contained in an interval (a, b), denoted by $m_i(A)$, is defined as follows:

$$m_i(A) = \sup \{m(F) : F \text{ is closed}, F \subset A\}.$$

It can also be expressed as

(The equation (2) is of vital importance for further study.)

$$m_i(A) = b - a - m_e(A'), \qquad ...(2)$$

From the definition it is clear that

$$m_i(A) \geq 0 \text{ and } m_i\ (\phi) = 0.$$

A set A is said to be measurable (or Lebesgue measurable) if $m_i(A) = m_e(A)$. The common value of $m_e(A)$ and $m_i(A)$ is called its measure and is denoted by m(A).

Thus, $\quad m_e(A) = m_i(A) = m(A)$, if A is measurable.

Remark:

g.l.b. = inf and l.u.b. = sup.

Theorem 2:

Show that $m_e(A) \geq m_i(A)$ for any set A.

Proof:

Let Any set A be contained in an interval (a, b). Then by definition of exterior measure $\exists$ open sets G_1, G_2 containing A and A' respectively s.t.

$$\left.\begin{aligned} m(G_1) < m_e\,(A) + \in \\ m(G_2) < m_e\,(A) + \in \end{aligned}\right\} \qquad ...(1)$$

This proves that every point of the interval $(a + \in, b - \in)$ is an interior point of an interval contained in G_1 or G_2. By Heine Borel theorem, we can determine a finite set G from these intervals s.t.

$$G \supset (a + \in, b - \in) \qquad ...(2)$$

and $m(G_1) + m(G_2) \geq m(G)$ ...(3)

$(2) \Rightarrow m\ (G) \geq m\ [(a + \in, b - \in)] = b - \in - (a + \in)$

$\Rightarrow m\ (G)\ b - a - 2\in.$ ...(4)

By (1), $\quad m(G_1) + m(G_2) < m_e(A) + m_e(A') + 2\in.$

Using (3),

$$m(G) \leq m(G_1) + m\ (G_2) < m_e(A) + m_e(A') + 2\in$$

or $\quad m(G) < m_e(A) + m_e\ (A') + 2\in.$

Using (4), we get

or $\quad b - a - 2\in\ < m_e\ (A) + m_e(A') + \in$

or $\quad b - a - m_e\ (A') < m_e\ (A) + 4\in$

or $\quad m_i (A) = b - a - m_e (A') < m_e (A) + 4\in$

or $\quad m_i (A) < m_e (A) + 4\in$.

Since $\in$ is arbitrary and hence making $\in \to 0$, we get the required result $m_e (A) \geq m_i (A)$.

Theorem 3:

A set A is measurable (i.e. Lebesgue measurable iff its complement A' is measurable.

Proof:

Let a set A be contained in an interval (a, b). Then we know that

$$m_i(A) = b - a - m_e (A'). \qquad ...(1)$$

(i) Let A be measurable so that

$$m_i (A) = m_e (A) = m(A). \qquad ...(2)$$

To prove that A' is measurable, we have to show that

$$m_e (A') = m_i (A'). \qquad ...(3)$$

Writing (1) with the help of (2),

$$m (A) = b - a - m_e (A')$$

or $\quad m_e (A') = b - a - m (A). \qquad ...(4)$

(1) is true $\forall$ A. Replacing A by A' in (1), we obtain

$$m_i (A') = b - a - m_e (A). \text{ For } (A')' = A.$$

Using (2), we get

$$m_i (A') = b - a - m_e (A). \qquad ...(5)$$

Since, R.H.S. of (4) = R.H.S. of (5)

Hence, L.H.S. of (4) = L.H.S. of (5)

i.e. $\quad m_e (A') = m_i (A'). \qquad ...(6)$

To prove that A is measurable.

By virtue of (1), we have

$$m_i (A) = b - a - m_e (A')$$

and $\quad m_i (A') = b - a - m_e (A)$.

In view of (6), these equations become

$$m_i (A) = b - a - m (A') \qquad ...(7)$$

$$m (A') = b - a - m_e (A) \qquad ...(8)$$

Equating the two values of m(A')

$$b - a - m_i (A) = b - a - m_e (A').$$

or $m_i(A) = m_e(A)$.

This $\Rightarrow$ A is measurable. **Hence Proved.**

Note : Let a measurable set A be contained in (a, b).

Then, $m_i(A) = b - a - m_e(A')$

or $m(A) = b - a - m_e(A')$

or $m(A) + m(A') = b - a$.

For A is measurable $\Rightarrow$ A' is measurable.

Theorem 4:

Let A and B be any two sets such that $A \subset B$.

Then, (i) $m_i(A) \leq m_i(B)$

(ii) $m_e(A) \leq m_e(B)$.

Proof:

Let $A \subset B$.

(i) To prove that $m_i(A) \leq m_i(B)$.

By definition of interior measure,

$m_i(A) = \sup \{m(F) : F \text{ is closed}, F \subset A\}$

$m_i(B) = \sup \{m(F) : F \text{ is closed}, F \subset B\}$.

Let S be the set of numbers consisting of measures of all closed subsets of A. Similarly we define the set T for B. Then

$m_i(A) = \sup S,\ m_i(B) = \sup T.$ ...(1)

F is any closed set s.t. $F \subset A \Rightarrow F \subset A \subset B$. For $A \subset B$.

$\Rightarrow F \subset B$

$\Rightarrow$ F is closed subset of B.

This shows that

any $m(F) \in S \Rightarrow m(F) \in T$.

Hence, $S \subset T$. Consequently $\sup S \leq \sup T$.

For least upper bound, i.e. supremum of a subset of any set can not exceed the least upper bound of the set itself.

$\sup S \leq \sup T \Rightarrow m_i(A) \leq m_i(B)$, by virtue of (1).

(ii) To prove that $m_e(A) \leq m_e(B)$.

By Definition of exterior measure,

$m_e(A) = \inf \{m(G) : G \text{ is open}, G \supset A\}$

$m_e(B) = \inf \{m(G) : G \text{ is open}, G \supset B\}$.

Let S be the set of numbers consisting of measures of all open super sets of A. Similarly we define the set T for B. Then

$$m_e (A) = \inf S, \; m_e (B) = \inf T. \qquad ...(2)$$

G is any open set s.t. $G \supset B \Rightarrow G \supset B \supset A$. For $B \supset A$.

$$\Rightarrow G \supset A.$$

$$\Rightarrow G \text{ is open superset of } A.$$

This proves the fact that

any $m (G) \in T \Rightarrow m (G) \in S$.

Hence $T \subset S$. Consequently $\inf S \leq \inf T$.

For greatest lower bound of a set can not exceed greatest lower bound of a subset of that set

$$\inf S \leq \inf T \Rightarrow m_e (A) \leq m_e (B), \text{ according to (2).}$$

Theorem 5:

If E_1 and E_2 are measurable, (in the sense of Lebesgue), then $E_1 \cup E_2$ is measurable. **(Kanpur, M.Sc. P., 1979)**

Proof:

Let E_k be a measurable set for k = 1, 2 and

$$E = E_1 \cup E_2.$$

To prove that E is measurable.

Consider an arbitrary number $\in > 0$

E_k is a measurable set $\Rightarrow \exists$ a closed set f_k and an open set

$$G_k \text{ s.t. } F_k \subset E_k \subset G_k,$$

$$m(G_k) - m(F_k) < \tfrac{\varepsilon}{2}.$$

Write $\quad F = F_1 \cup F_2, \quad G = G_1 \cup G_2.$

Then F is a closed set and G is an open set so that F and G are measurable sets.

$$F_k \subset E_k \subset G_k \Rightarrow \bigcup_{k=1}^{2} F_k \subset \bigcup_{k=1}^{2} E_k \subset \bigcup_{k=1}^{2} G_k$$

$$\Rightarrow F \subset E \subset G$$

$$\Rightarrow F \subset E, \; E \subset G$$

$$\Rightarrow m_i (F) \leq m_i (E) \leq m_e (E) \leq m (G)$$

$$\Rightarrow m (F) \leq m_i (E) \leq m_e(E) \leq m (G)$$

[For $m_i (F) = m(F)$, $m_e (G) = m(G)$ and $m_i (A) \leq m_e (A) \; \forall \; A$]

$$\Rightarrow m_e(E) - m_i(E) \; m(G) - m(F). \qquad ...(1)$$

$G - F = G \cap F'$ = A finite intersection of open set is

= An open set = A measurable set

$$m(G - F) = m(G) - m(F). \qquad ...(2)$$

Similarly, $m(G_k - F_k) = m(G_K) - m(F_k)$.

Notice that $G - F \subset \sum_{k=1}^{2}(G_k - F_k)$

This $\Rightarrow m(G - F) \le \sum_{k=1}^{2} m(G_k - F_k)$

$$\Rightarrow m(G) - m(F) \le \sum_{k=1}^{2}[m(G_k) - m(F_k)] < \tfrac{\epsilon}{2}. = \epsilon$$

$$\Rightarrow m(G) - m(F) < \epsilon$$

$$\Rightarrow m_e(E) - m_i(E) \le m(G) - m(F) < \epsilon, \text{ by (1)}$$

$$\Rightarrow m_e(E) - m_i(E) < \epsilon.$$

Making $\epsilon \to 0$

$m_e(E) - m_i(E) \le 0$. For ϵ is arbitrary.

This $\Rightarrow$ $m_e(E) \le m_i(E)$.

But $m_e(E) \ge m_i(E)$ is true $\forall$ E.

Hence $m_e(E) = m_i(E)$.

This $\Rightarrow$ E is measurable.

Theorem 6:

If $a \subseteq (a, b)$, then

$$m^*A + m_*C(A) = b - a$$

where $C(A) = (a, b) - A$ is the complement of A relative to (a, b).

Proof:

Let F be any open subset of (a, b) containing A; then $C(F) \subseteq C(A)$. by definition it follows that

$$m_*C(A) \ge l(C(F))$$

Adding $l(F)$ on both sides, we get

$$l(F) + m_*C(A) \ge l(F) + l(C(F)) = (b - a)$$

Taking infimum over all open sets $F \supseteq A$

$$m^*A + m_*C(A) \ge (b - a) \qquad ...(1)$$

Now if B is any closed set such that $B \subseteq C(A)$ and $C(B) \supseteq A$, then by definition we have:

$$m^*A \leq l(C(B))$$

Adding $l(B)$ on both sides, then we have

$$m^*A + l(B) \leq l(C(B) + l(B) = (b - a)$$

Taking supremum over all closed sets $B \subseteq C(A)$

$$m^*A + m_*C(A) \leq (b - a) \qquad ...(2)$$

From (1) and (2), we have

$$m^*A + m_*C\,(A) = (b - a).$$

Corollary:

A subset $A \subseteq (a, b)$ is measurable if and only if

$$m^*A + m^*C\,(A) \leq (b - a)$$

For any set $A \subseteq (a, b)$,

$$m^*A + m_*C(A) = (b - a)$$

Interchanging A with C(A), we have

$$m^*C\,(A) + m_*(A) = (b - a) \qquad ...(1)$$

If A is measurable, then

$$m^*A = m_*A = mA \qquad ...(2)$$

From (1) and (2), we get

$$m^*C(A) + m^*A = (b - a)$$

Hence the condition is necessary.

Given

$$m^*(A) + m^*(C)A \leq (b - a) \qquad ...(3)$$

Subtracting (1) from (3), we get

$$m^*(A) - m_*A \leq 0$$

$$\Rightarrow \; m^*(A) \leq m_*A$$

But $m_*A \leq m^*A$. Hence $m_*A = m^*A \Rightarrow$ A is measurable.

The above result being symmetric in A and CA implies that CA is measurable whenever A is measurable.

Example 1:

If A_1 and A_2 are subsets of (a, b), then

$$m^*A_1 + m^*A_2 \geq m^*(A_1 \cup A_2) + m^*(A_1 \cap A_2)$$

and

$$m_*A_1 + m_*A_2 \leq m_*(A_1 \cup A_2) + m_*(A_1 \cap A_2).$$

Solution:

Do your self.

Hint: If F_1 and F_2 are any two open subsets of (a, b)_ containing A_1 and A_2 respectively, and $F_1 \cup F_2 \supseteq A_1 \cup A_2$. $F_1 \cap F_2 \supseteq A_1 \cap A_2$, then

$$l(F_1) + l(F_2) = l(F_1 \cup F_2) + l(F_1 \cap F_2)$$

and for each $\in > 0$, $m^*A_1 + \in > l(F_1)$.

Theorem 7:

If A_1 and A_2 are measurable subsets of (a, b) the both $A_1 \cup A_2$ and $A_1 \cap A_2$ are measurable and

$$mA_1 + mA_2 = m(A_1 \cup A_2) + m(A_1 \cap A_2).$$

Proof:

Now A_1, A_2 being measurable,

$$m^*A_1 = m_*A_1 \text{ and } m^*A_2 = m_*A_2$$

Thus

$$mA_1 + mA_2 = m^*A_1 + m^*A_2 \geq m^*(A_1 \cup A_2) + m^*(A_1 \cap A_2)$$
$$\geq m_*(A_1 \cup A_2) + m_*(A_1 \cap A_2)$$

and $m_*(A_1 \cup A_2) + m_*(A_1 \cap A_2) \geq mA_1 + mA_2$

This implies that ,

$$m^*(A_1 \cup A_2) + m^*(A_1 \cap A_2) = m_*(A_1 \cup A_2) + m_*(A_1 \cap A_2)$$

or

$$m^*(A_1 \cup A_2) - m_*(A_1 \cup A_2) = m_*(A_1 \cap A_2) - m^*(A_1 \cap A_2) \quad ...(1)$$

But

$$m^*(A_1 \cup A_2) \geq m_*(A_1 \cup A_2)$$

$\therefore$ $m_*(A_1 \cap A_2) - m^*(A_1 \cap A_2) \geq 0$, from (1).

Again

$$m_*(A_1 \cap A_2) \leq m^*(A_1 \cap A_2)$$

$\therefore$ $m_*(A_1 \cap A_2) = m^*(A_1 \cap A_2)$

$\Rightarrow$ $a_1 \cap A_2$ is measurable.

From (1)

$$m^*(A_1 \cup A_2) - m_*(A_1 \cup A_2) = 0$$

$\Rightarrow$ $A_1 \cup A_2$ is measurable.

and

$$mA_1 + mA_2 = m(A_1 \cup A_2) + m(A_1 \cap A_2).$$

Example 2:

If A_1 and A_2 are measurable subsets of the closed integral (a, b) then $A_1 - A_2$ is measurable and if $A_2 \subseteq A_1$ then $m(A_1 - A_2) = mA_1 - mA_2$.

Solution:

Since $A_1 - A_2 = A_1 \cap C(A_2)$ and A_1, A_2 being measurable, $A_1 - A_2$ is also measurable. Now if $A_2 \subset A_1$ then $A_1 = A_2 \cup (A_1 - A_2)$ and $m(A_2 \cap (A_1 - A_2)) = m(\phi) = 0$. Since A_2, $A_1 - A_2$ are disjoint,

$$\therefore \quad mA_1 = m(A_2 \cup (A_1 - A_2)) + m(A_2 \cap (A_1 - A_2))$$

$$mA_1 = mA_2 + m(A_1 - A_2)$$

or

$$mA_1 - mA_2 = m(A_1 - A_2)$$

Theorem 8:

If $A_1, A_2, \ldots\ldots, A_n$ are pairwise disjoint measurable subsets of (a, b), then $\bigcup_{n=1}^{\infty} A_n$ is measurable and

$$m\left(\bigcup_{n=1}^{\infty} A_n\right) = \sum_{n=1}^{\infty} mA_n .$$

Proof:

Let $\in > 0$ be given and F_n, n = 1, 2, 3, ... be a class of open subsets of (a, b) such that $A_n \subseteq F_n$ and $l(F_n) < m^*A_n + \frac{\varepsilon}{2^n}$.

The set $\bigcup_{n=1}^{\infty} F_n$ is an open set containing $\bigcup_{n=1}^{\infty} A_n$.

$$\therefore \quad m^*\left(\bigcup_{n=1}^{\infty} A_n\right) \le l\left(\bigcup_{n=1}^{\infty} F_n\right) \le \sum_{n=1}^{\infty} l(F_n) < \sum_{n=1}^{\infty}\left(m^*A_n + \frac{\in}{2^n}\right)$$

$$= \sum_{n=1}^{\infty} m^*A_n + \in$$

Since $\in$ is arbitrary, then we have

$$m^*\left(\bigcup_{n=1}^{\infty} A_n.\right) \le \sum_{n=1}^{\infty} m^*A_n$$

Also we have

$$m_*\left(\bigcup_{n=1}^{\infty} A_n\right) \geq m_*\left(\bigcup_{n=1}^{N} A_n\right) \geq \sum_{n=1}^{N} m_* A_n - m_*\left(\bigcap_{n=1}^{N} A_n\right)$$

Since A_n, n = 1, 2, ... are pairwise disjoint,

$$\bigcap_{n=1}^{N} A_n = \phi \quad \therefore \; m_*\left(\bigcap_{n=1}^{N} A_n\right) = 0$$

Letting N tend to infinity, then

$$m_*\left(\bigcup_{n=1}^{N} A_n\right) \geq \sum_{n=1}^{\infty} m_* A_n$$

Also

$$m_* A_n = m^* A_n = mA_n \text{ for } n = 1, 2, 3, ...$$

$$\therefore \quad \sum_{n=1}^{\infty} mA_n \leq m_*\left(\bigcup_{n=1}^{\infty} A_n\right) \leq m^*\left(\bigcup_{n=1}^{\infty} A_n\right) \leq \sum_{n=1}^{\infty} mA_n$$

This implies that

$$m^*\left(\bigcup_{n=1}^{\infty} A_n\right) = m^*\left(\bigcup_{n=1}^{\infty} A_n\right) = \sum_{n=1}^{\infty} mA_n .$$

Corollary 1:

If A_n, n = 1, 2, 3, ... are measurable subsets of (a, b) and if $A_l \subseteq A_{i+1}$ for all $i(A_{l+1} \subseteq A_l$ for all i) then $\bigcup_{n=1}^{\infty} A_n \left(\bigcap_{n=1}^{\infty} A_n\right)$ is measurable and

$$m\left(\bigcup_{n=1}^{\infty} A_n\right) = \lim_{n\to\infty} mA_n \left(m\left(\bigcap_{n=1}^{\infty} A_n\right) \lim_{n\to\infty} mA_n \right).$$

Proof:

The sets A_1, $(A_2 - A_1)$, $(A_3 - A_2)$, ... $(A_{n+1} - A_n)$ are pairwise disjoint and measurable. Also

$$\bigcup_{i=1}^{n} A_n = A_1 \cup (A_2 - A_1) \cup (A_3 - A_2) \cup ... \cup (A_n - A_{n-1})$$

$$\therefore \quad m\left(A_1 \cup \bigcup_{i=1}^{n-1} (A_{i+1} - A_i)\right) = mA_1 + \sum_{i=1}^{n-1} m(A_{i+1} - A_i)$$

$$= mA_1 + \sum_{i=1}^{n-1} (mA_{i+1} - mA_i) = mA_n$$

Letting n tend to infinity

$$m\left(\bigcup_{n=1}^{\infty} A_n\right) = \lim_{n\to\infty} mA_n$$

Now for each $n \in N$, $C(A_n)$ is measurable and $mC(A_n) = (b - a) - mA_n$. If $A_{l+1} \subseteq A_i$, then $C(A_{l+1}) \supseteq C(A_i)$, hence

$$m\left(\bigcup_{n=1}^{\infty} C(A_n)\right) = \lim_{n\to\infty} mC(A_n) = b - a - \lim_{n\to\infty} mA_n$$

$$\Rightarrow \bigcup_{n=1}^{\infty} C(A_n) \text{ is measurable}$$

But $\bigcup_{n=1}^{\infty} C(A_n)$ is the complement of $\bigcap_{n=1}^{\infty} A_n$.

Hence $\bigcap_{n=1}^{\infty} A_n$ is also measurable, and

$$m\left(C\left(\bigcap_{n=1}^{\infty} A_n\right)\right) = b - a - m\left(\bigcap_{n=1}^{\infty} A_n\right)$$

$$\Rightarrow m\left(\bigcap_{n=1}^{\infty} A_n\right) = \lim_{n\to\infty} mA_n.$$

Corollary 2:

If $A_1, A_2, \ldots, A_3$ are measurable subsets of (a, b), then $\bigcup_{n=1}^{\infty} (A_n)$ is measurable and

$$m\left(\bigcup_{n=1}^{\infty} (A_n)\right) \leq \sum_{n=1}^{\infty} mA_n$$

Moreover $\bigcap_{n=1}^{\infty} A_n$ is measurable.

Solution:

Do your self.

Hint. $\bigcup_{n=1}^{\infty} A_n = A_1 \cup (A_2 - A_1) \cup (A_2 - (A_1 \cup A_2)) \ldots$

and $\bigcap_{n=1}^{\infty} A_n = C\left(\bigcup_{n=1}^{\infty} C(A_n)\right)$

Theorem 9:

If E_1 and E_2 are measurable then $E_1 \cup E_2$ is measurable.

Proof:

E_1 being measurable, by definition, for each set A, we have

$$m^*A = m^*(A \cap E_1) + m^*(A \cap C(E_1)) \qquad ...(1)$$

Again E_2 is measurable. Thus, taking $A \cap C(E_1)$ in place of A in the condition of measurability, we get

$$m^*(A \cap C(E_1)) = m^*(A \cap C(E_1) \cap E_2) + m^*(A \cap C(E_1) \cap C(E_2)) \;...(2)$$

Now

$$(A \cap E_1) \cup (A \cap C(E_1) \cap E_2) = A \cap (E_1 \cup E_2) \text{ as a disjoint union}$$

$$\therefore \quad m^*(A \cap (E_1 \cup E_2)) = m^*((A \cap E_1) \cup (A \cap C(E_1) \cap E_2))$$

$$\leq m^*(A \cap E_1) + m^*(A \cap C(E_1) \cap E_2) \qquad ...(3)$$

(by Ex. 1)

Using (2) in (3) we get

$$m^*(A \cap (E_1 \cup E_2)) \leq m^*(A \cap E_1) + m^*(A \cap C(E_1)$$
$$- m^*(A \cap C(E_1) \cap C(E_2)$$

or

$$m^* A \geq m^*(A \cap (E_1 \cup E_2)) + m^*(A \cap C(E_1) \cap C(E_2) \text{ (from (1))}$$

Also

$$m^*A \leq m^*(A \cap (E_1 \cup E_2)) + m^*(A \cap C(E_1 \cup E_2))$$

Hence $m^*A = m^*(A \cap (E_1 \cup E_2)) + m^*(A \cap C(E_1 \cup E_2))$.

Theorem 10:

If A is any set and E_1, E_2, ..., E_n is a finite sequence of disjoint measurable sets, then

$$m^*\left(A \cap \left(\bigcup_{i=1}^{n} E_i\right)\right) = \sum_{t=1}^{\infty} m^*(A \cap E_i).$$

Proof:

Do your self.

Hint. This can be proved by induction on n and using

$$A \cap \left(\bigcup_{i=1}^{n} E_t\right) \cap (E_n) = A \cap E_n$$

and

$$A \cap \left(\bigcup_{i=1}^{n} E_t\right) \cap C(E_n) = A \cap \left(\bigcup_{i=1}^{n-1} E_i\right) \quad \because \; E_i \text{ are disjoint.}$$

Theorem 11:

For any sequence of sets (E_i)

$$m^*\left(\bigcup_{i=1}^{n} E_i\right) \leq \sum_{i=1}^{\infty} m^*(E_i).$$

Proof:

The inequality is trivial if any one of the sets has infinite outer measure. Therefore we consider the case i n which the sets have finite outer measures. If E_t's are open then measurable, given $\in > 0$ $\exists$ open sets F_i such that

$F_i \supseteq E_i$, $i = 1, 2, 3, ...$

and

$mF_i = m^*F_i < m^*E_i + \in/2^i$

$mF_i = m^*F_i \geq m^*E_i$, $i = 1, 2, 3, ...$

$$m^*\left(\bigcup_{i=1}^{\infty} E_i\right) \leq m^*\left(\bigcup_{i=1}^{\infty} F_i\right) = m\left(\bigcup_{i=1}^{\infty} F_i\right) \leq \sum_{i=1}^{\infty} mF_i = \sum_{i=1}^{\infty} m^*F_i$$

$$\leq \sum_{i=1}^{\infty}(m^*E_i + \in/2^i) \leq \sum_{i=1}^{\infty} m^*E_i + \in$$

(F_1 being open and hence measurable)

But $\in$ being arbitrary small number.

$$\therefore \quad \left(\bigcup_{i=1}^{\infty} E_i\right) \leq \sum_{i=1}^{\infty} m^*E_i.$$

Theorem 12:

The interval $[a, \infty[$ is measurable.

Proof:

In order to show that $[a, \infty[$ is measurable we have to show that for any set A.

$$m^*A \geq m^*(A \cap]-\infty, a[) + m^*(A \cap [a, \infty[) \qquad ...(1)$$

Let $A_1 = A \cap]-\infty\ a[$, and $A_2 = A \cap [a, \infty[$,

Since $m^*A = \inf_{A \subseteq \bigcup_n I_n} \sum_{n=1}^{\infty} l(I_n)$, therefore for each $\in > 0$ there exist open intervals I_n, $n \in N$, such, that

$$A \subseteq \bigcup_n I_n, \text{ and } m^*A + \in > \sum_{n=1}^{\infty} l(I_n) \qquad ...(2)$$

Define

$$I_n' = I_n \cap]-\infty, a[\text{ and } I_n'' = I_n \cap [a, \infty[,$$

so that

$$l(I_n) = l(I_n') + l(I_n''). \quad [\because]-\infty, a[\text{ and } [a, \infty[\text{ are disjoint}]$$

Then

$$A_1 = A \cap]-\infty, a[\subseteq \bigcup_n I_n \cap]-\infty, a[= \bigcup_n (I_n \cap [-\infty, a[)$$

i.e., $A_1 \subseteq \bigcup_n (I_n')$

Similarly,

$$A_2 \subseteq \bigcup_n (I_n)$$

Now $m^*A_1 + m^*A_2 \leq \sum_{n=1}^{\infty} l(I_n') + \sum_{n=1}^{\infty} l(I_n'') = \sum_{n=1}^{\infty} l(I_n) \leq m^*A + \in$ (by (2))

Since e is arbitrary,

$\therefore \quad m^*A \geq m^*A_1 + m^*A_2.$

Hence $[a, \infty[$ is measurable.

6.10 Sets of Measure Zero

A subset A or R is said to be a *set of measure zero* if for any $\in > 0$ there exists a sequence of bounded open intervals I_1, I_2, ... such that

(i) $A \subseteq \bigcup_{n=1}^{\infty} I_n$

(ii) $\sum_{n=1}^{\infty} l(I_n) \leq \in$.

F· om the definition it is clear that every subset of a set of measure zero has measure zero. Moreover, if $A = \bigcup_{n=1}^{\infty} A_n$, where A_n has measure zero for all n, then A has measure zero.

Theorem 13:

The following statements regarding the set E are equivalent:

(i) E is measurable.

(ii) For all $\in > 0$, $\exists\, 0$, and open set, $0 \supseteq E$ such that $m^*(0 - E) \leq \in$.

(iii) $\exists\, G$, a G_δ-set, $G \supseteq E$ such that $m^*(G - E) = 0$. (A set G is said to be G_δ if $G = \bigcap_{i=1}^{\infty} G_i$, each G_i is an open set).

(iv) For all $\in > 0$, $\exists\, F$, a closed set, $F \subseteq E$, such that $m^*(E - F) \leq \in$.

(v) $\exists\, F$, a F_σ-set, $F \subseteq$ such that $m^*(E - F) = 0$. (A set F is said to be F_σ if $F = \bigcup_{i=1}^{\infty} F_i$, each F_l is a closed set.)

Proof:

(i) $\Rightarrow$ (ii) Let us first consider the case when $mE < \infty$, E being measurable, $\exists$ an open set $0 \supseteq E$ such that

$$m^*0 \leq m^*E + \in$$

$$\therefore \quad m^*(0 - E) = m^*0 - m^*E \leq \in.$$

If $mE = \infty$, let $E \subseteq \bigcup_{n=1}^{\infty} I_n$ where I_n's are disjoint finite intervals. Define $E_n = E \cap I_n$; then $E = \bigcup_{n=1}^{\infty} E_n$. For each n, E_n is measurable, belong intersection of two measurable sets. Moreover $E_n \subseteq I_n$ and $m(I_n) < \infty$. Hence $mE_n < \infty$. From the above result, it follows that, for each $\in > 0$, $\exists$ an open set 0_n such that $0_n \supseteq E_n$ and $m(0_n - E_n) \leq \frac{\varepsilon}{2^n}$, for n = 1, 2, 3, ...

Write $\quad 0 = \bigcup_{n=1}^{\infty} 0_n$; then 0 is an open set. Consider

$$0 - E = \bigcup_{n=1}^{\infty} 0_n - \bigcup_{n=1}^{\infty} E_n \subseteq \bigcup_{n=1}^{\infty} (0_n - E_n)$$

$$m^*(0 - E) \leq m^*\left(\bigcup_{n=1}^{\infty} (0_n - E_n)\right) \leq \sum_{n=1}^{\infty} m^*(0_n - E_n) \leq \sum_{n=1}^{\infty} \frac{\in}{2^n} = \in$$

$\therefore$ (i) $\Rightarrow$ (ii)

(ii) $\Rightarrow$ (iii) For each n, $\exists$ an open set $0_n \supseteq E$ such that $m^*(0_n - E) < \frac{1}{n}$.

Define $G = \bigcap_{n=1}^{\infty} 0_n$; then G is a G_δ-set. Since $E \subseteq 0_n$ for each n, so $E \subseteq \bigcap_{n=1}^{\infty} 0_n = G$. Thus,

$$m^*(G - E) \le m^*(0_n - E) < \frac{1}{n}, \text{ for each } n,$$

$$\therefore \quad m^*(G - E) = 0$$

(iii) $\Rightarrow$ (i) Let G be a G_δ-set containing E such that $m^*(G - E) = 0$, $G - E$ is measurable. G being a G_δ-set is measurable. Hence $E = G - (G - E)$ is measurable.

Similarly equivalence in the remaining statements can be established.

It m^*E is finite the above statements are equivalent to the following;

(vi) Given $\in > 0$, there is disjoint finite union $\bigcup_{i=1}^{n} I_i$ of open intervals such that $m^*\left(\bigcup_{i=1}^{n} I_i \,\Delta E\right) < \in$, where $\bigcup_{i=1}^{n} I_i \,\Delta E = \left(\bigcup_{i=1}^{n} I_i - E\right) \cup \left(E - \bigcup_{i=1}^{n} I_i\right)$.

The intervals I_i can be either open, closed or half open but we will give the proof for open intervals.

Suppose E is measurable; then by (ii), $\forall \in > 0$ $\exists$ an open set 0, $0 \supseteq E$ such that

$$m\,(0 - E) < \frac{\in}{2} \qquad ...(1)$$

mE being finite, so m0 is finite, 0 being the union of disjoint open intervals I_i, $i = 1, 2, 3, ... \infty$, such that

$$m^*\left(\bigcup_{i=1}^{\infty} I_n\right) \le \sum_{i=1}^{\infty} m^* I_i < \infty$$

Hence $\exists$ a positive integer n such that

$$\sum_{i=n+1}^{\infty} mI_i < \in, \text{ i.e., } m\left(\bigcup_{i=n+1}^{\infty} I_i\right) < \frac{\in}{2}$$

[by Cauchy's General principle of convergence for series]

$$\Rightarrow \quad m\left(0 - \bigcup_{i=1}^{n} I_i\right) \frac{\in}{2} \qquad ...(2)$$

Writing $U = \bigcup_{i=1}^{n} I_i$,

$$E\,\Delta U = \left(E - \bigcup_{i=1}^{n} I_i\right) \cup \left(\bigcup_{i=1}^{n} I_i - E\right) \subseteq \left(0 - \bigcup_{i=1}^{n} I_i\right) \cup (0 - E)$$

$$\therefore \quad m^*(E\ DU) \le m^*\left(\left(0 - \bigcup_{i=1}^{n} I_i\right) \cup (0 - E)\right)$$

$$\le m^*\left(0 - \bigcup_{i=1}^{n} I_i\right) + m^*(0 - E) < \tfrac{\in}{2} + \tfrac{\in}{2} = \in$$

[from (1) and (2)]

Conversely, for every given set E and $\in > 0$, $\exists$ an open set 0 containing E such that $m^*0 \le m^*E + \in$. Now we show that $m^*(0 - E) < \in$ whenever

$$m^*\left(E\ \Delta \bigcup_{i=1}^{n} I_i\right) < \tfrac{\in}{3}$$

Let $U = \bigcup_{i=1}^{n} (I_i \cap 0)$, where I_i, i = 1, 2, ..., n, are given open intervals corresponding to the same $\in$.

$$U \subseteq \bigcup_{i=1}^{n} I_i$$

Thus

$$U\ \Delta E \subseteq (E - U) \cup \left(\bigcup_{i=1}^{n} I_i - E\right)$$

But

$$E - U = E - \left(\left(\bigcup_{i=1}^{n} I_i\right) \cap\ 0\right)$$

or

$$E - U - E \cap C\,(0)\ E \cap C\left(\bigcup_{i=1}^{n} I_i\right) = E - \bigcup_{i=1}^{n} I_i$$

As $E \subseteq 0 \quad \therefore \quad E \cap C\,(0) = \phi$

$$\therefore \quad u\ \Delta E \subseteq \left(E - \bigcup_{i=1}^{n} I_i\right) \cup \left(\bigcup_{i=1}^{n} I_i - E\right) = E\ \Delta \left(\bigcup_{i=1}^{n} I_i\right)$$

$$\Rightarrow \quad m^*(U\ \Delta E) \le m^*\left(E\ \Delta \bigcup_{i=1}^{\infty} I_i\right) \le \tfrac{\in}{3} \qquad \text{...(3)}$$

Now,

$$E \subseteq U \cup (U\ \Delta E)$$

$\therefore \quad m^*E \leq m^*U + m^*(U \Delta E) < mU + \frac{\epsilon}{3}$ using (3)

or

$$m^*E < mU + \frac{\epsilon}{3} \qquad \text{...(4)}$$

Also

$$0 - E \subseteq (0 - U) \cup (U \Delta E)$$

$$m^*(0 - E) \leq m^*(0 - U) + m^*(U \Delta E) < m(0 - U) + \frac{\epsilon}{3}$$

(using (3))

or

$$m^*(0 - E) < m0 - mU + \frac{\epsilon}{3}$$

Also 0 and U have finite measure

$\therefore \quad m^*(0 - E) < m^*E - mU + \frac{2\epsilon}{3} \qquad [\because m^*E + \frac{\epsilon}{3} > m0]$

$\therefore \quad m^*(0 - E) < \frac{\epsilon}{3} + \frac{2\epsilon}{3}$ using (4)

$\Rightarrow$ E is measurable.

6.11 Borel Sets

A class $\mathscr{F}$ of subsets of an arbitrary field F is said to be σ-Algebra (sigma algebra) if

(i) $F \in \mathscr{F}$

(ii) $F_1, F_2, \ldots F_n, \ldots \in \mathscr{F}$

implies $\bigcup_{i=1}^{\infty} F_i \in \mathscr{F}$, and $F_1 - F_2 \in \mathscr{F}$ whenever, $F_1, F_2 \in \mathscr{F}$.

From (i) and (ii) it follows that the class $\mathscr{F}$ is closed under countable intersection. The σ-algebra generated by any class of sets is the smallest σ-algebra containing this class of sets.

The *Borel Sets* in R are the members of the σ-algebra generated by the class of intervals of the form [a, b[.

Example 3:

A function f is measurable iff the set {x : f(x) < r} is measurable for every rational number r. **(Kolhapur, 1970)**

Solution:

Suppose a function f is defined over a measurable set E. Let r be an arbitrary rational number.

(i) Suppose f is measurable over E so that E $(f < c)$ is a measurable set, c being any real number.

To prove that $\{x : f(x) < r\}$ is measurable.

E $(f < c)$ is measurable $\forall\ c \in R$

$\Rightarrow$ in particular, E $(f < c)$ is measurable $\forall\ c \in Q$. For $Q \subset R$.

$\Rightarrow$ $E(f < r)$ is measurable for $r \in Q$.

$\Rightarrow$ $\{x : f(x) < r\} = \{x \in E : f(x) < r)$ is measurable.

$\Rightarrow$ $\{ : f(x) < r\}$ is measurable.

(ii) Conversely suppose that $\{x : f(x) < r\}$ is measurable.

To prove that f is measurable.

Observe that

$$E(f < c) = \{x \in E : f(x) < c\}$$
$$= \bigcup_{r \in Q} \{ x \in E : f(x) < r < c\}$$
$$= \bigcup_{r < c} \{ x \in E : f(x) < r,\ r \in Q\}$$

= enumerable union of measurable sets.

[For Q is enumerable]

= measurable set.

$\therefore$ E $(f < c)$ is measurable and so f is measurable over E.

Example 4:

If i is a continuous function and g is a measurable function, then prove that the composite function f o g is measurable.

Solution:

Let G be an open set and g be a measurable function.

Let $A = \{x : g(x) \in G\}$.

To prove A is a measurable set. ...(1)

G is an open set $\Rightarrow$ G is expressible in the form

$$G = \bigcup_n I_n, \text{ where } I_n = (a_n, b_n).$$

Hence $A = \{x : g(x) \in G\} = \bigcup_n \{x : g(x) \in I_n\}$

or $\quad A = \bigcup_n \{ x : g(x) \in (a_n, b_n)\} = \bigcup_n \{ x : a_n < g(x) < b_n\}$

$$= \bigcup_n [\{ x : g(x) > a_n\} \cap \{ x : g(x) < b_n\}]$$

= enumerable union of measurable sets

= measurable set. Hence the result (1).

$$\{x : f(g(x)) < a\} = \{ x : g(x) \in G\} \qquad ...(2)$$

where $G = \{y : f(y) < a\}$

f is continuous $\Rightarrow$ $G = \{y : f(y) < a\}$ is open

$\Rightarrow \{x : g(x) \in G\}$ is measurable, by (1)

$\Rightarrow$ fog is measurable, by (2).

Theorem 14:

(E. Borel) : *Suppose a measurable function f defined on a closed interval (a, b), is finite almost everywhere on the same interval. For all numbers, σ, $\in > 0$, there exists a continuous function ϕ (a, b) such that*

$$m\,[(E \mid f - \phi \mid \geq \sigma)] < \in. \qquad \textbf{(Punjab, 1999)}$$

Proof:

Let f be a measurable function defined on [a, b] s.t., f is finite almost everywhere on [a, b].

Case I. Let f be bounded.

Then $\exists$ a number $k > 0$ s.t. $|f(x)| < k$ (1) $\forall x \in [a, b]$.

Let $\sigma, \in > 0$ be arbitrary numbers. Fixing $\sigma, \in$ by choosing a positive integer m so large that

$$(k/m) < \sigma.$$

Write $$E_r = \left(\frac{r-1}{m} k \leq f < \frac{r}{m} k\right)$$

where $r = 1 - m, 2 - m, ..., m - 1$

and $$E_m = E\left(\frac{m-1}{m} k \leq f < k\right).$$

Evidently $E_i \cap E_j = f$ for $i \neq j$.

where $i, j = 1 - m, 2 - m, ..., m.$

$$[a, b] = \sum_{r=1-m}^{m} E_r.$$

Select a closed interval $F_r \subset E_r$ s.t.

$m(E_r) < m(F_r) + (\in/2m)$.

Take $F = \sum_{r=1-m}^{m} F_r$.

Then $m([a, b]) - (F) < \in$. ....(2)

Define a function ϕ on F by writing

$$\phi(x) = \frac{rk}{m} \ \forall\ x \in F_r,\ r = 1 - m,\ 2 - m, \ldots\ldots, m.$$

Then ϕ is constant on each closed set F_r.

Also $F_r \cap F_r' = \Phi$ for $r \neq r$

Consequently ϕ is continuous on F.

$$|\phi(x)| = \left|\frac{rk}{m}\right| = \frac{rk}{m} \le k,$$

$\therefore$ $|\phi(x)| \le k$.

Finally, if $x \in F$, then

$$|f(x) - \phi(x)| = \left|k - \frac{rk}{m}\right| = \frac{k}{m}|m - r| \le \frac{k}{m} < \sigma.$$

$\therefore$ $|f(x) - \phi(x)| < \sigma\ \forall\ x \in F$.

Now we shall state a Lemma without proof.

Lemma : Let F be a closed set contained in the closed interval [a, b]. If the function $\phi(x)$ is defined and continuous on the set F, then it is possible to find out a function $\psi(x)$ on [a, b] s.t.

(i) $\psi(x)$ is continuous

(ii) $\psi(x) = \phi(x)\ \ \forall\ x \in F$

(iii) max. $[\ \psi(x)\ | =$ max. $|\phi(x)|$

By the lemma, just stated, it i s possible to find a function $\phi(x)$ defined on [a, b] having all the properties stated in the lemma. Moreover

$$E(|f - \psi| \ge \sigma) \subset [a, b] - F,$$

Hence $m[E(|f - \psi| \ge \sigma)] \le m([a, b] - F) < \in$, by (2)

i.e. $m[E(f - \psi) \ge \sigma] < \in$.

Thus we have proved the theorem for a bounded function f.

Case II. Let f be unbounded.

Let us suppose that f(x) is unbounded. Then we can find out a bounded function g such that

$$m[(E\ f \neq g)] < \frac{\in}{2}.$$

Applying Case I, just proved, to the bounded function g, we are able to find a continuous function ϕ s.t.

$$m[(E(|g - \psi| \geq \sigma)] < \frac{\in}{2}.$$

From this we can deduce that

$$E\ (\ |\ f - \psi\ |\ \geq \sigma) \subset E\ (f \neq g) + E\ (\ |\ g - \psi\ |\ \geq \sigma).$$

This $\Rightarrow$

$$m\ [\ E\ (\ |\ f - \psi\ |\ \geq \sigma) \leq m\ [\ E\ (f \neq g)] + m\ [E\ (\ |\ g - \psi\ |\ \geq \sigma)].$$

$$< \frac{\in}{2} + \frac{\in}{2} = \in$$

$$\therefore\quad m\ [\ E\ (\ |\ f - \psi\ |\ \geq \sigma\ )] < \in.$$

The function f satisfies all the requirements.

Hence the theorem is also true in case of unbounded function.

6.12 Borel Measurability Function

A function f is said to be Borel measurable if the set

$$\{x : f(x) < a\}\ \forall\ a \in R$$

is a Borel set. Every Borel measurablee function is (Lebesgue) measurablee function. For every Borel measurable set is a Lebesgue measurable set.

6.13 Little Wood's Three Principles

In connection with the theory of functions of a real variable, J.E. Little Woods says, "The extent of knowledge required is nothing like so great as is sometimes supposed. There are three principles, roughly expressible as follows;

(1) Every (measurable) set is nearly a finite union of intervals;

(2) Every (measurable) function is nearly continuous;

(3) Every convergent sequence of (measurable) functions is nearly uniformly continuous.

Example 5:

If $m^*E = 0$, then E is measurable.

Proof:

Let A be any set, then $A \cap E \subseteq E$, so, $0 \leq m^*(A \cap E) \leq m^*E = 0.$

$\therefore \quad m^*(A \cap E) = 0$

Also

$$A \supseteq A \cap C(E) \quad \therefore \quad m^* A \geq m^*(A \cap C(E))$$

Now, we have

$$m^*(A \cap C(E)) + m^*(A \cap E) = m^*(A \cap C(E)) \leq m^*A \qquad ...(1)$$

But we know that

$$A = (A \cap E) \cup (A \cap C(E))$$

$$\therefore \quad m^*A \leq m^*(A \cap E) + m^*(A \cap C(E)) = m^*(A \cap C(E)) \qquad ...(2)$$

From (1) and (2), we get

$$m^*A = m^*(A \cap E) + m^*(A \cap C(E))$$

Hence E is measurable. **Hence Proved.**

Example 6:

A set consisting of one point is measurable and its measure is zero.

Solution:

Let $\{x\}$ be the given singleton set. Since $x \in I_n = \left]x - \frac{1}{2^{n+1}}, x + \frac{1}{2^{n+1}}\right[$ for all n and $l(I_n) = \frac{1}{2^n}$.

$$\therefore \quad m^*\{x\} = \inf_{\{x\} \subset I_n} l(I_n) = 0$$

$\Rightarrow \quad \{x\}$ is measurable.

Example 7:

A countable set is measurable and its measure is zero.

Solution:

Let $A + \bigcup_{i=1}^{\infty}\{x_i\}$ be a given countable set; then

$$m^*A = m^*\left(\bigcup_{i=1}^{\infty}\{x_i\}\right) \leq \sum_{i=1}^{\infty} m^*\{x_i\} = 0 \quad \text{[by Example 6]}$$

Note : From the above Example, [0, 1] is not countable, since $m^*[0, 1] = 1 \neq 0$, however, one should not conclude that the sets of measure zero consist only of at most countable number of points. Indeed we have the following example.

Example 8:

There exist uncountable sets of measure zero.

Solution:

Let us construct the Cantor sets by dividing the i interval [0, 1] into three equal parts by the points $\frac{1}{3}$, $\frac{2}{3}$ and remove open interal $]\frac{1}{3},\frac{2}{3}[$ from [0, 1]. Then dividing each of the remaining closed intervals $[0, \frac{1}{3}]$; $[\frac{2}{3}, 1]$ into three parts by the points $\frac{1}{9},\frac{2}{9}$ for the first interval $[0, \frac{1}{3}]$ and by the points $\frac{7}{9},\frac{8}{9}$ for the second interval $[\frac{2}{8}, 1]$ and remove the middle open intervals $]\frac{1}{9},\frac{2}{9}[$, $]\frac{7}{9},\frac{8}{9}[$. Next divide each of the remaining four intervals into three equals parts and remove the middle intervals from these. Continue this process indefinitely. By this process we remove an open set G_0 from [0, 1], the set G_0 is the union of a countable family of open intervals, i.e.,

$$G_0 =]\tfrac{1}{3}, \tfrac{2}{3}[\cup]\tfrac{1}{9}, \tfrac{2}{9}[\cup]\tfrac{7}{8}, \tfrac{8}{9}[\ldots$$

The complement of G_0 in [0, 1] is denoted by P_0. Clearly the set P_0 is the set of all those points which are left after deleting the intervals, i.e., P_0 is the intersection of all the closed intervals which are left after each iteration.

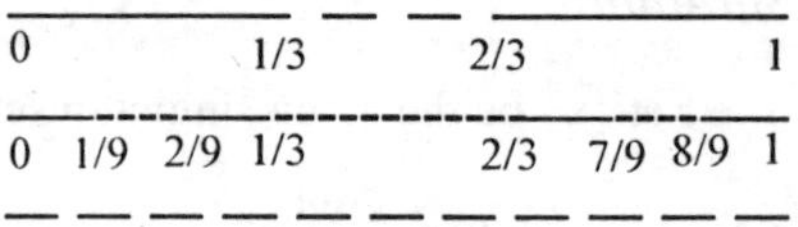

0 1/3 2/3 1

0 1/9 2/9 1/3 2/3 7/9 8/9 1

The set P_0 is *an uncountable set.* Let, if possible, P_0 be countable. We enumerate the points of P_0 by x_i, $i = 1, 2, \ldots \infty$. Let the ternary representation (i.e., expansion inn the scale of three of the form $\sum_{k=1}^{\infty} \frac{a_k}{3^k}$, where each a_k is either zero or two; see also the notes) of the points of P_0 be denoted by

$$x_i = 0.a_{i1}, a_{i2}, a_{i3} \ldots, \; i = 1, 2, 3, \ldots..$$

where $a_{ij} = 0$ to 2.

Now we construct a point

$$y = 0.b_1 b_2 b_3 \ldots$$

$$\text{where } b_i = 0 \text{ if } a_{ij} = 2$$

$$i = 1, 2, 3, \ldots.$$

$$= 2 \text{ if } a_{ij} = 0$$

Since b_i's are either zero or 2, therefore the new point y constructed as above which is different from x_i for any i belongs to P_0 which contradicts the fact that the set P_0 is countable. Hence the set P_0 is uncountable. Now

$$m(G_0) = \frac{1}{3} + \frac{2}{9} + \frac{4}{27} \ldots. = \frac{\frac{1}{3}}{1 - \frac{2}{3}} = 1$$

Hence, the uncountable set P_0 has measure

$m(P_0) = m([0, 1] - G_0) = 0$

The sets G_0 and P_0 are called *Cantor Sets.*

Note : Let $x = \frac{7}{9} \in P_0$. Then by the ternary representation of x (i.e., expansion of x in the scale of 3) we mean

$$\frac{7}{9} = \frac{2.3+1}{3^2} = \frac{2}{3} + \frac{1}{3^2} = .21 = .202222... = \frac{2}{3} + \frac{0}{3^2} + \frac{2}{3^3} + \frac{2}{3^4} + ...$$

Similarly,

$$\frac{8}{9} = \frac{2.3+2}{9} = \frac{2}{3} + \frac{2}{3^2} = .22 = .22000...$$

Example 9:

Show that for any set A, $m^*A = m^*(A + x)$, where

$A + x = \{y + x : y \in A\}$.

Solution:

Let $A \subseteq \bigcup_{n=1}^{\infty} I_n$; then for each e > 0 we have

$$m^*A + \in \leq \sum_{n=1}^{\infty} l(I_n)$$

Also, $A + x \subseteq \bigcup_{n=1}^{\infty} (I_n + x)$. Thus for each $\in < 0$.

$$m^*(A + x) \leq \sum_{n=1}^{\infty} l(I_n + x) = \sum_{n=1}^{\infty} l(I_n) \leq m^*A + \in$$

$\in$ being arbitrary.

$$m^*(A + x) \leq m^*A \qquad ...(1)$$

Writing $A = A + x - x$, from (1), it follows that

$$m^*A = m^*((A + x) - x) \leq m^*(A + x) \qquad ...(2)$$

From (1) and (2) we get the desired result.

Example 10:

If A is measurable, then for each x the set $A + x = \{y + x : y \in A\}$ is measurable, and the measures are the same.

Solution:

As A is measurable, by Theorem 13, for each $\in > 0$, $\exists$ an open set 0, $0 \supseteq A$ such that $m^*(0 - A) < \in$. The set $0 + x$ is open and $0 + x \subseteq A + x$.

Moreover,

$$(0 + x) - (A + x) = (0 - A) + x$$

and

$$m^*(0 - A) = m^*((0 - A) + x) < \in$$

$$\Rightarrow m^*(0 + x - (A + x)) < \in$$

Hence, A + x is measurable.

Example 11:

Show that, for any set A and any $\in > 0$, $\exists$ an open set 0 such that $A \subseteq 0$ and $m^*0 \le m^*A + \in$.

Solution:

Let I_n, n = 1, 2, 3, ... be open intervals such that

$$A \subseteq \bigcup_{n=1}^{\infty} I_n$$

By definition,

$$m^*A = \inf_{A \subseteq \bigcup_n I_n} \sum_{n=1}^{\infty} l(I_n)$$

Hence given $\in > 0$, $\exists$ intervals I_n, n = 1, 2, 3, ... such that

$$m^*A \ge \sum_{n=1}^{\infty} l(I_n) - \frac{\in}{2}$$

If $I_n = [a_n, b_n[$, and $I_n' =]a_n = \in/2^{n+1}, b_n + \in/2^{n+1}[$ [so that $A \subseteq \bigcup_{n=1}^{\infty} I_n'$,

and if $0 = \bigcup_{n=1}^{\infty} I_n'$, then 0 is an open set,

$$m^*0 \le \sum_{n=1}^{\infty} l(I_n') = \sum_{n=1}^{\infty} l(I_n) + \tfrac{1}{2}\in \le m^*A + \tfrac{1}{2}\in + \tfrac{1}{2}\in$$

$$\therefore \quad m^*0 \le m^*A + \in$$

Theorem 15:

The function f on [a, b] is measurable if and only if any one of the following conditions hold:

(i) $\{x : f(x) > \alpha\}$ is α measurable set for every real α.

(ii) $\{x : f(x) \geq \alpha\}$ is α measurable set for every real α.

(iii) $\{x : f(x) < \alpha\}$ is α measurable set for every real α.

(iv) $\{x : f(x) \leq \alpha\}$ is α measurable set for every real α.

Proof:

Suppose f is measurable, then by definition, the set $\{x : f(x) < \alpha\}$ for every $\alpha \in R$ is a measurable set. The set $\{x : f(x) \leq \alpha\}$ is the complement of the set $\{x : f(x) < \alpha\}$ in the space of all measurable sets; and the complement of a measurable set is measurable. Therefore the measurability of the set $\{x : f(x) > \alpha\}$ implies the measurability of the set $\{x : f(x) \leq \alpha\}$ i.e., (i) ⇒ (iv).

Again (i) ⇒ (ii) Since f is measurable, and $\alpha \in R$ then each of the sets $\{x : f(x) < \alpha - 1/n\}$, n = 1, 2, ... is measurable and $\{x : f(x) \geq \alpha\} = \bigcap_{n=1}^{\infty}\{x : f(x) > \alpha - 1/n\}$. This being the arbitrary intersection of measurable sets is measurable.

(i) ⇒ (iii). Because the set $\{x : f(x) < \alpha\}$ is the complement of the set $\{x : f(x) \geq \alpha\}$ and the set $\{x : f(x) \geq \alpha\} = \bigcap_{n=1}^{\infty}\{x : f(x) > \alpha - 1/n\}$. Therefore measurability of (i) implies measurability of (iii).

(iii) ⇒ (iv). Since the set $\{x : f(x) \leq \alpha\} = \bigcap_{n=1}^{\infty}\{x : f(x) < \alpha - 1/n\}$ and each of the sets $\{x : f(x) < \alpha + 1/n\}$ is measurable. The set $\{x : f(x) < \alpha\}$ being the complement of the set $\{x : f(x) \leq \alpha\}$ and $\{x : f(x) \leq \alpha\}$ is measurable because (iii) holds.

(iii) ⇒ (i). Now for each n, the set $\{x : f(x) < \alpha - 1/n)$ is the complement of these et $\{x : f(x) \leq \alpha - 1.n\}$, and $\{x : f(x) \geq \alpha\} = \bigcap_{n=1}^{\infty}\{x : f(x) > \alpha - 1/n\}$ is the complement of $\bigcup_{n=1}^{\infty}\{x : f(x) \leq \alpha - 1/n\} = \{x : f(x) < \alpha\}$.

Hence any of these conditions may be used to define measurability of functions.

Theorem 16:

If f is measurable then $|f|$ is measurable.

Proof:

Since the set $\{x : |f(x)| < \alpha\} = \{x : f(x) < \alpha\} \cap \{x : f(x) > -\alpha\}$ and each of the sets $\{x : f(x) < \alpha\}$, $\{x : f(x) , -\alpha\}$ is measurable. Hence $|f|$ is measurable.

Theorem 17:

If f is a measurable function on $[a, b]$ and if $k \in R$ then $f + k$ and kf are measurable.

Proof:

If $\alpha \in R$ then $\{x : f(x) + k > \alpha\} = \{x : f(x) > \alpha - k\}$. The set on the right is measurable because f is measurable. Hence the set on the left is measurable.

If $k > 0$, $kf(x) < \alpha \Rightarrow f(x) > \frac{\alpha}{k}$. Hence the set

$$\{x : kf(x) > \alpha\} = \left\{x : f(x) > \frac{\alpha}{k}\right\} \quad \text{...(1)}$$

and if $k < 0$, $kf(x) > \alpha \Rightarrow f(x) < \frac{\alpha}{k}$. And the set

$$\{x : kf(x) > \alpha\} = \left\{x : f(x) < \frac{\alpha}{k}\right\}. \quad \text{...(2)}$$

Each of the sets on the right of (1) and (2) is measurable. Hence the set on the left is measurable. Clearly if f is measurable then $-f$ is measurable by taking $k = -1$.

Example 12:

Show that if f is measurable then the set $\{x : f(x) = \alpha\}$ is measurable for each extended real number α.

Solution:

For α to be finite, $\alpha \in R$ the set

$$\{x : f(x) = \alpha\} = \{x : f(x) \geq \alpha\} \cap \{x : f(x) \leq \alpha\}$$

and each of these sets is measurable and so the set $\{x : f(x) = \alpha\}$.

For $a = +\infty$, the set $\{x : f(x) = \infty\} = \bigcap_{n=1}^{\infty} \{x : f(x) > n\}$ is a measurable.

Similarly, for $\alpha = -\infty$.

Example 13:

Constant functions are measurable.

Given $f(x) = C$ for all $x \in R$..

Solution:

Let $\alpha \in R$, if $C \geq \alpha$, then the set $\{x : f(x) \geq \alpha\}$ is R which is measurable.

And if $C < \alpha$, then the set $\{x : f(x) \geq \alpha\}$ is ϕ, which is again measurable, α being arbitrary. Hence f is measurable.

Theorem 18:

If f_1 and f_2 are measurable on [a, b] then so are $f_1 + f_2$, $f_1 - f_2$, f_1f_2 and f_1/f_2 provided $f_2 \neq 0$ on [a, b].

Proof:

Let $q_1, q_2, ..., q_n, ...$ be an enumeration of the set of rational numbers. Let $x \in [a, b]$ and $\alpha \in R$.

$f_1(x) > \alpha - f_2(x)$ if and only if there is a rational number q_n such that

$$f_1(x) > q_n > \alpha - f_2(x)$$

Hence the set

$$\{x : f_1(x) + f_2(x) > \alpha\}$$

$$= \bigcup_{n=1}^{\infty}[\{x : f_1(x) > q_n\} \cap \{x : \alpha - f_2(x) < q_n\}]. \qquad ...(1)$$

For any $n \in N$, the set $\{x : f_1(x) > q_n\}$ is measurable, since f_1 is a measurable function. And the set $\{x : \alpha - f_2(x) < q_n\} = \{x : f_2(x) > \alpha - q_n\}$ is measurable because f_2 is measurable. The set on the right of (1) is measurable which establishes that the sum $f_1(x) + f_2(x)$ is measurable.

Again $f_1(x) - f_2(x) = f_1(x) + (-f_2(x))$ and hence measurable. Now to show that $f_1(x)\, f_2(x)$ is measurable, we will first show that the square of a measurable function is measurable. If g is a measurable function on [a, b] and $\alpha \geq 0$, then

$$\{x : (g(x))^2 > \alpha\} = \{x : g(x) > \sqrt{\alpha}\} \cup \{x : g(x) < -\sqrt{\alpha}\}$$

Each of the sets on the right is measurable and hence their union is measurable.

If $\alpha < 0$ then the set $\{x : (g(x))^2 < \alpha\}$ is the interval [a, b], hence measurable.

Now, if f_1 and f_2 are measurable then $(f_1 + f_2)^2$ and $(f_1 - f_2)^2$ are also measurable.

Since $f_1(x)\, f_2(x) = \frac{1}{4}[f_1(x) + f_2(x))^2 - (f_1(x) - f_2(x))^2]$, it follows that $f_1(x)$, $f_2(x)$ is measurable.

At last to show that $\dfrac{f_1(x)}{f_2(x)}$, $f_2(x) \neq 0$ on [a, b] is measurable, we will first show that $\dfrac{1}{f_2(x)}$, $f_2(x) \neq 0$ on [a, b] is measurable.

As $f_2(x) \neq 0$ on [a, b] we have

$$\left\{x: \frac{1}{(x)} > \alpha\right\}$$

$$= \begin{cases} \{x : f_2(x) > 0\} & \text{if } \alpha = 0 \\ \{x : f_2(x) > 0\} \cap \{x : f_2(x) < 1/\alpha\} & \text{if } \alpha > 0 \\ \{x : f_2(x) > 0\} \cup [\{x : f_2(x) < 0\} \cap \{x : f_2(x) < 1/\alpha\}] & \text{if } \alpha < 0 \end{cases}$$

Each of the sets on the right is measurable and if f_2 is measurable on [a, b] then $1/f_2$, $f_2(x) \neq 0\ \forall\ x$ on [a, b] is measurable.

Now $\dfrac{f_1(x)}{f_2(x)} = f_1(x).\dfrac{1}{f_2(x)}$ provided $f_2(x) \neq 0$, $\forall\ x \in [a, b]$. Hence from the previous result it follows that $\dfrac{f_1(x)}{f_2(x)}$ is measurable, when $f_2(x) \neq 0$ on [a, b].

Theorem 19:

If $\{f_n\}$ is a sequence of measurable functions on [a, b] such that the sequence $\{f_n(x)\}$ is bounded for every $x \in [a, b]$, then the functions

$$G(x) = \text{l.u.b. } \{f_1(x), f_2(x), f_3(x) \ldots\}$$

$$g(x) = \text{g.l.b. } \{f_1(x), f_2(x), f_3(x) \ldots\}$$

$$H(x) = \limsup_{n \to \infty} f_n(x)$$

$$h(x) = \liminf_{n \to \infty} f_n(x)$$

are measurable.

Proof:

Let $\alpha \in R$. Let $x \in [a, b]$ be such that $G(x) > \alpha$, then $\exists$ some n for which $f_n(x) > \alpha$. Also if $f_n(x) > \alpha$ then, $\underset{\pi}{\text{l.u.b.}} f_n(x) < \alpha$, i.e., $G(x) > \alpha$. Thus $G(x) > \alpha$ iff $f_n(x) > \alpha$ for some n.

Let us consider the set

$$\{x : G(x) > \alpha\} = \bigcup_n \{x : f_n(x) > \alpha\}$$

Each f_n, being a measurable function, the set $\{x : f_n(x) > \alpha\}$ is measurable, and hence there countable union, $\bigcup_n \{x : f_n(x) > \alpha\}$ is measurable

$\Rightarrow \{x : G(x) > \alpha\}$ is measurable, α being arbitrary. Hence the function G is measurable.

Similarly from the following relation.

$$\{x : g(x) < \alpha\} = \bigcup_n \{x : f_n(x) < \alpha\}$$

it follows that the function g is measurable.

Now for each $n \in N$ we define the functions $G_n(x)$, $g_n(x)$ as follows:

$$G_n(x) = \text{l.u.b. } \{f_n(x), f_{n+1}(x), f_{n+2}(x), \ldots\}$$

$$g_n(x) = \text{g.l.b. } \{f_n(x), f_{n+1}(x), f_{n+2}(x), \ldots\}$$

The functions $G_n(x)$ and $g_n(x)$ being the l.u.b. and g.l.b. of measurable functions $f_n(x)$ are also measurable.

Also for each $x \in [a, b]$

$$G_1(x) \geq G_2(x) \geq G_3(x) \geq \ldots\, G_n(x) \geq \ldots$$

and

$$g_1(x) \leq g_2(x) \leq g_3(x) \leq \ldots\, g_n(x) \leq \ldots$$

Now

$$H(x) = \limsup_{n\to\infty} f_n(x) = \lim_{n\to\infty} G_n(x)$$

and

$$h(x) = \liminf_{n\to\infty} f_n(x) = \lim_{n\to\infty} g_n(x)$$

The sequences $\{G_n(x)\}$ and $\{g_n(x)\}$ being monotonically decreasing and monotonically increasing, respectively, it follows that $H(x) < G_n(x)$ and $h(x) > g_n(x)$ for all n.

Now, for each $\alpha \in R$ the sets

$$\{x : H(x) < \alpha\} = \bigcup_n \{x : G_n(x) < \alpha\}$$

and

$$\{x : h(x) > \alpha\} = \bigcup_n \{x : g_n(x) > \alpha\}$$

are measurable. Hence the functions H(x) and h(x) are measurable.

Definition 1:

A function f is a *Borel measurable* or a *Borel function* if for each $\alpha \in R$ the set $\{x : f(x) > \alpha\}$ is a *Borel set.*

Definition 2:

A property which holds everywhere except on a set of measure zero is said to hold *almost everywhere,* abbreviated as a.e.

Definition 3:

Any two functions f and g are said to be *equal almost everywhere* if the set $A = \{x : f(x) \neq g(x)\}$ is of measure zero, i.e., f = g a.e. if mA = 0.

Definition 4:

A sequence $\{f_n\}$ of measurable functions defined as finite a.e. on a measurable set E is said to *converge in measure* to a measurable function f which is finite a.e. if $\underset{n\to\infty}{\text{Lt}}\ m(\{x : |f_n - f| \geq \alpha\}) = 0$, for all positive real numbers α.

Theorem 20:

If f = g a.e. and f is a measurable function, then g is also measurable.

Proof:

The set $\{x : f(x) - g(x) > \alpha\} = \{x : f(x) \neq g(x) - g(x) < \alpha\}$ is contained in the sets $\{x : f(x) \neq g(x)\}$ which is of measure zero. Hence $\{x : f(x) - g(x) > \alpha\}$ is of measure zero and hence measurable.

Let $\alpha < 0$, then the set

$$\{x : f(x) - g(x) > \alpha\} = \{x : f(x) = g(x)\}$$
$$\cup \{x : f(x)' \neq g(x),\ f(x) - g(x) > \alpha\}$$

The two sets on the right hand side are measurable, the first set is the complement of the set of measure zero. Thus the set $(x : f(x) - g(x) > \alpha\}$ is measurable for arbitrary real α.

Hence the function f – g is measurable.

Now g = f – (f – g), f and f – g being measurable implies that g is measurable.

Hence the theorem.

Theorem 21:

Every continuous function is measurable.

Proof:

Let $\alpha \in R$, the set $\{x : f(x) < \alpha\}$, being the inverse image of open interval $]\alpha, \infty[$ under a continuous function f is also open, and hence measurable.

The converse of this theorem may or may not be true, i.e., every measurable function may or may not be continuous.

Still then every measurable function which is finite a.e. can be approximated by some continuous function in the following sense.

Theorem 22:

(N.N. Lusin) : *Given f to be a measurable function which is finite a.e. on [a, b]. Then for every $\delta > 0$ there exists a continuous function g such that $m\{x : f \neq g\} < \delta$. Moreover if f is bounded by a constant M then g is also bounded by the same constant.*

Note : (The proof is beyond the scope of the present chapter).

Theorem 23:

(Egoroff): *If $\{f_n\}$ is a sequence of measurable functions which converge to a real valued function f a.e. on a measurable set E of finite measure, then given $\eta > 0$, there is a subset $A \subseteq E$ with $m(E - A) < \eta$ such that $\{f_n\}$ converges to f uniformly on A.*

Proof:

The sequence $\{f_n\}$ being pointwise convergent to f on E a.e., let $F \subseteq E$ be the set of all $x \in E$ for which $\{f_n(x)\}$ converges to f(x). Then

$$m(E - F) = 0 \qquad \ldots(1)$$

Define the sets

$$F_{nm} = \left\{x : | f_N - f | < \frac{1}{n}, \ \forall N \geq m\right\}$$

For a fixed n, f_{nm}'s form an increasing sequence of subsets of E (i.e. $f_{n1} \subseteq F_{n2} \subseteq F_{n3}$ ...) and $F = \bigcup_{m=1}^{\infty} E_{nm}$. The sets F_{nm} being measurable, we have $mF = \underset{m\to\infty}{\text{Lt}} F_{nm}$.

Thus given $\eta > 0$ and for each n, $\exists$ a positive integer m_n depending on η such that

$$m(F - F_{nm_n}) < \frac{\eta}{2^n} \qquad ...(2)$$

If we define $A = \bigcap_{n=1}^{\infty} F_{nm_n}$, then for each $x \in E$, we have

$$| f_N - f | < \frac{1}{n} \quad \forall\, n \geq m_n \quad .$$

Clearly m_n does not depend on the choice of x. Hence $\{f_N\}$ converges to f uniformly on A.

Now, we have to show that $m(E - A) < \eta$.

Consider F – A, where $A = \bigcap_{n=1}^{\infty} F_{nm_n}$. Thus

$$F - A \subseteq \bigcup_n (F - F_{nm_n})$$

and

$$m(F-A) \leq \sum_{n=1}^{\infty} m(F - f_{nm_n}) < \sum_{n=1}^{\infty} \frac{\eta}{2^n} \quad \text{(by (2))}$$

$$\therefore \quad m(E - A) = m(E - F + F - A) \leq m(E - F) + m(F - A) < \eta$$

i.e.,

$$m(E - A) < \eta.$$

Theorem 24:

Prove that the space of measurable functions is closed under the usual operations of Arithmetic. **(Bhagalpur, 1998)**

Proof:

The fundamental operations of Arithmetic are:

addition, subtraction, multiplication, division.

Let f, g be arbitrary members of the space of measurable functions.

If we show that f + g, f – g, fg and $\frac{f}{g}$ (if g vanishes nowhere on E) are measurable functions, the result will follow. Of course the functions f and g are measurable over a measurable set E.

Here write the proof of Theorems 6 and 7.

Example 14:

If f and g are real valued measurable functions on a measurable space X, prove that the $f \cap g$, $f \cup g$ are measurable functions. **(Gujarat, 1990)**

Solution:

Let c be any real constant. Let f and g be real valued measurable functions defined on a measurable set X so that

X (f > c) and X (g > c) are measurable sets.

(i) To prove that $f \cup g$ is measurable.

$$X (f \cup g > c) = [X (f > c)] \cup [X (g > c)]$$
$$= \text{finite union of measurable sets}$$
$$= \text{measurable set.}$$

This proves the result (i).

(ii) To show that $f \cap g$ is measurable.

$$X (f \cap g > c) = [X (f > c)] \cap [X (g > c)]$$
$$= \text{finite intersection of measurable sets}$$
$$= \text{measurable set.}$$

This $\Rightarrow$ the result (ii).

Example 15:

Show that the set of all measurable functions form a vector space over R. **(Kanpur, 1994)**

Solution:

Let E be a measurable set. Let be the set of all measurable functions defined over E. Then

$$V = \{f : f : E \to R \text{ is a measurable function}\}.$$

To prove that V(R) is a vector space.

Firstly we shall prove the following Lemmas.

Lemma 1: If f is measurable, then of is measurable, a being any real number. Also let c be any real number.

$$E(af > c) = \begin{cases} E\left(f > \frac{c}{a}\right) & \text{if } a > 0 \\ E\left(f < \frac{c}{a}\right) & \text{if } a < 0 \end{cases}$$

Both the sets on the R.H.S. are measurable. Hence E(af > c) is measurable so that af is measurable.

Lemma 2: If f and g are measurable, then f + g is also measurable.

$$E(f > g) = \bigcup_{r \in Q} [E(f > r > g)], \quad Q = \text{set of rational numbers}$$

$$= \bigcup_{r \in Q} [E(f > r)] \cap [E(g < r)]$$

= An enumerable union of measurable sets

= measurable set.

g is measurable ⇒ ag is measurable ⇒ – g is measurable

⇒ c – g is measurable.

Now E (f + g > a) = E (f < a – g) = measurable set

For E (f > g) is measurable.

Now we come to the main problem.

R being the set of real numbers.

(i) (V, +) is abelian group

1. f + g ∈ V, by Lemma 2.
2. ∃ 0 ∈ V s.t. f + 0 = 0 + f = f.

Here 0 is zero function so that 0(x) = 0 ∀ x.

It means that 0 is a special case of a constant function and so it is measurable. Also

(f + 0) (x) = f(x) + 0(x) = f(x) + 0 = f(x), this ⇒ f + 0 = f.

3. – f ∈ V s.t. – f + f = 0, by Lemma 1.
4. f + g = g + f as (R, +) is an abelian group.
5. f + (g + h) = (f + g) + h as (R, +) is an abelian group.

(ii) a f ∈ V, by Lemma 1.

6. (a + b) f = af + bf
7. a(f + g) = af + ag.
8. 1. f = f.

(6), (7), (8) follow from the fact that (R, +, .) is a field. From what has been shown it follows that V (R) is a vector space.

Theorem 25:

Let $< f_n >$ be a sequence of measurable functions defined over a measurable set E. Then sup $\{f_1, f_2, f_3, ...\}$ and inf $\{f_1, f_2, f_3, ...\}$ are measurable over E. **(Kanpur, 1996)**

Proof:

Let $< f_n >$ be a sequence of measurable functions defined over a measurable set E so that

E $(f_n > a)$ is a measurable set $\forall$ n, a being any real number.

Write $g(x) = \sup \{f_n(x) : n = 1, 2, 3, ...\}$

$h(x) = \inf \{f_n(x) : n = 1, 2, 3, ...\}$

h(x) can be defined in terms of supremum as follows:

$h(x) = -\sup \{-f_n(x) : n = 1, 2, 3, ...\}$.

To prove that g and h are measurable over E, it is enough to show that g is measurable over E.

$$E(g > a) = \bigcup_{n=1}^{\infty}[E(f_n > a)]$$

$$= \bigcup_{n=1}^{\infty} \{x \in E : f_n(x) > a\}$$

= An enumerable union of measurable sets

= measurable set.

Theorem 26:

Let $< f_n >$ be a sequence of measurable functions defined over a measurable set E. Show that

$\sup \{ f_1, f_2, ..., f_n\}$, $\inf \{ f_1, f_2, ..., f_n\}$,

$\overline{\text{Lim}}\, f_n$, $\underline{\text{Lim}}\, f_n$ are measurable over E. Hence show that Lim f_n is measurable over E if Lim f_n exists.

Proof:

Let $< f_n >$ be a sequence of measurable functions defined over a measurable set E.

Step I. To prove that $\sup \{f_r : 1 \le r \le n\}$ and

$\inf (f_r : 1 \le r \le n)$ are measurable over E.

Define $\quad M(x) = \sup \{f_r(x) : 1 \le r \le n\}$

$$= \bigcup_{r=1}^{n} f_r(x)$$

and $\quad m(x) = \inf \{f_r(x) : 1 \le r \le n\}$

m (x) can also be defined in terms of supremum as follows :

$$m(x) = -\sup \{-f_r(x) : 1 \le r \le n\}.$$

If we show that M (x) is measurable over E, the result will follow.

For proving this, it is enough to prove that E (M > a) is a measurable set, a being any real number.

f_n is a measurable function over E $\Rightarrow$ E ($f_n > a$) is a measurable set $\forall$ n.

Evidently,

$$E[M > a] = \bigcup_{r=1}^{n} E[(f_r > a)]$$

= A finite union of measurable sets

= A measurable set.

This proves that E (M > a) is a measurable set.

Step II. To prove that $\overline{\text{Lim}}\, f_n$ and $\underline{\text{Lim}}\, f_n$ are measurable functions over the set E. **(Banaras, 1995)**

Define $\quad m_K(x) = \sup\limits_{n \ge k} \{f_n(x)\},$

$$m_K(x) = \inf_{n \ge k} \{f_n(x)\},$$

Then $\quad \overline{\text{Lim}}\, f_n(x) = \inf\limits_{k \ge 1} \{M_k(x)\}$

$$\underline{\text{Lim}}\, f_n(x) = \sup_{k \ge 1} \{m_k(x)\}.$$

By case (i), $M_k(x)$ and $m_k(x)$ both are measurable functions defined over E $\forall$ k $\in$ N and again, by case (i),

$$\inf_{k \ge 1} \{M_k(x)\}, \ \sup_{k \ge 1} \{M_k(x)\}$$

are measurable over E, i.e. $\overline{\text{Lim}}\, f_n$, $\underline{\text{Lim}}\, f_n$ both are measurable over E.

Step III. Let Lim f_n exist.

To show that Lim f_n is measurable over E.

By case (ii) $\overline{\text{Lim}}\, f_n$, $\underline{\text{Lim}}\, f_n$ both are measurable over E.

By hypothesis,

$\text{Lim } f_n = \overline{\text{Lim}}\, f_n$, $\underline{\text{Lim}}\, f_n$.

From what has been done, it follows that Lim f_n is measurable over E.

Theorem 27:

A continuous function defined over a measurable set E is measurable.

(Kanpur 76; Meerut 72; Kolhapur 69)

Proof:

Let f be a continuous function defined over a measurable set E.

To prove that f is a measurable function over E.

Let a be any real number.

We claim E $(f \geq a)$ is a closed set.

Let $\quad A = E\ (f \geq a)$. ...(1)

To prove that A is closed, it is enough to prove that

$$D(A) \subset A. \quad ...(2)$$

Let $x_0 \in D(A)$ be arbitrary, then x_0 is a limit point of A so that $\exists$ a sequence $< x_n >$ whose elements $x_n \in A$

s.t. $\quad \lim_{n\to\infty} x_n = x_0$.

f is continuous at x_0, $x_n \to x_0 \Rightarrow f(x_n) \to f(x_0)$. ...(3)

This follows from Heine's definition of continuity.

By (1), $x_n \in A \Rightarrow f(x_n) \geq a \Rightarrow \lim_{n\to\infty} f(x_n) \geq a$

$\Rightarrow f(x_0) \geq a$, according to (3)

$\Rightarrow x_0 \in A$, by (1).

Finally, any $x_0 \in D(A) \Rightarrow x_0 \in A$.

This $\quad \Rightarrow D(A) \subset A$

$\Rightarrow$ A is closed

$\Rightarrow$ A is measurable

$\Rightarrow$ E $(f \geq a)$ is measurable

$\Rightarrow$ f is a measurable function over the set E.

Theorem 28:

If $< f_n >$ is a sequence of measurable functions then show that Lim f_n is measurable.

Proof:

Let $< f_n >$ be a sequence of measurable function.

To prove that Lim f_n is measurable.

$\therefore$ $< f_n >$ is a monotonic sequence and hence it is either monotonic increasing or monotonic decreasing.

If $< f_n >$ is a monotonic increasing sequence, then

$$\text{Lim } f_n = \sup_{n\geq 1} \{f_n\}$$

If $< f_n >$ is a monotonic decreasing sequence, then

$$\text{Lim } f_n = \inf_{n\geq 1} \{f_n\}.$$

Hence if we prove that

$$\sup \{f_1, f_2, ...\} \text{ and } \inf \{f_1, f_2, ...\}$$

are measurable, then the problem will be proved.

Let $g(x) = \sup \{f_1, f_2, ...\} = \inf \{f_1, f_2, ...\}$.

Then $E(g > a) = \bigcup_{n=1}^{\infty} [E(f_n > a)]$ = An enumerable union of measurable sets.

Hence $E(g > a)$ is measurable and so g is measurable.

h(x) is expressible as

$$h(x) = -\sup \{-f_n(x) : n = 1, 2, ...\}.$$

This $\Rightarrow$ h(x) is also measurable as g is measurable.

This completes the proof.

Theorem 29:

The set $E \subseteq [a, b]$ and its characteristic function ψ_E are both measurable or both non-measurable.

Proof:

If ψ_E is measurable then the measurability of E follows from the relation

$$E = \{x : \psi_E > 0\}$$

Conversely if E is a measurable set then the following set

$$[x : \psi_E > \alpha\} = \begin{cases} \phi & \text{if } \alpha \geq 1 \\ E & \text{if } 0 \leq \alpha < 1 \\ [a, b] & \text{if } \alpha < 0 \end{cases}$$

which establishes the measurability of ψ_E.

6.14 Lebesgue Integral

Definition 1:

Let f be any bounded function on [a, b] and let $P = (A_1, A_2, ..., A_n)$ be any partition of [a, b], where $A_1, A_2, ..., A_n$ are measurable subsets of [a, b] (also called the components of P) such that

$$\bigcup_{i=1}^{n} A_i = [a, b] \quad \text{and} \quad m(A_i \cap A_j) = 0, \quad \text{for } i \neq j$$

Such a partition of [a, b] would be called a *measurable partition* of [a, b]. We define it as

$$U(P, f) = \sum_{i=1}^{n} (\sup_{x \in A_i} f(x)) mA_i$$

$$L(P, f) = \sum_{i=1}^{n} (\inf_{x \in A_i} f(x)) mA_i$$

as the upper and lower Lebesgue sums of the function f corresponding to the partition $P(A_1, A_2, ..., A_n)$ of [a, b]. Obviously $U(P, f) \geq L(P, f)$ for every partition P. The infimum of the set of all upper Lebesgue sums is called the *Upper Lebesgue Integral* denoted as

$$L \overline{\int_a^b} f\,dx = \inf U(P, f) \qquad \forall \text{ partitions } P$$

The supremum of the set of all lower Lebesgue sums is called the *Lower Lebesgue Integral* denoted by

$$L \underline{\int_a^b} f\,dx = \sup L(P, f) \qquad \forall \text{ partitions } P$$

Definition 2 :

A bounded function f on [a, b] is said to be *Lebesgue Integrable* if

$$L \underline{\int_a^b} f\,dx = L \overline{\int_a^b} f\,dx$$

In this case we define

$$L \int_a^b f\,dx = L \underline{\int_a^b} f\,dx = L \underline{\int_a^b} f\,dx^*$$

The fact that f is Lebesgue integrable, we express by writing f L ∈ [a ,b].

Lemma 3 : *Let f be a bounded function on [a, b]. Then, for any two measurable partitions P_1, P_2 of [a, b] we have*

$$U(P_1, f) \geq L(P_2, f), \quad \text{and} \quad L\underline{\int_a^b} f\,dx \leq L\overline{\int_a^b} f\,dx$$

Let $P_1 = \{A_1, A_2,..., A_n\}$, $P_2 = \{B_1, B_2,..., B_m\}$ be any two measurable partitions of [a, b]; and P be the measurable partition, called their common refinement whose components are nm subsets $A_i \cap B_j (i = 1, 2, 3,..., n; j = 1, 2,..., m;)$. Now

$$U(P_1, f) \geq U(P, f) \geq L(P, f) \geq L(P_2, f)$$

Taking the infimum over all partitions P_1, we get $\inf_{P_1} U(P_1, f) \geq L(P_2, f)$;

whatever partition P_2 may be

Again taking the supremum over all partitions P_2, we have

$$\inf_{P_1} U(P_1, f) \geq \sup_{P_2} L(P_2, f)$$

i.e.,

$$L\underline{\int_a^b} f\,dx \geq L\overline{\int_a^b} f\,dx.$$

Remark:

Every upper (lower) Riemann integral is greater than (less than) or equal to every upper (lower) Lebesgue integral.

Since $R\overline{\int_a^b} f\,dx = \inf_Q U(Q, f)$, where Q is any partition of [a, b] into intervals, every interval being a measurable set, so every partition can be regarded as a measurable partition. Thus

$$\inf_Q U(Q, f) \geq \inf_P U(P, f) = L\overline{\int_a^b} f\,dx$$

where P is a measurable partition. Similarly

$$L\underline{\int_a^b} f\,dx \geq R\underline{\int_a^b} f\,dx$$

Combining the two results, we have

$$R \underline{\int_a^b} f\,dx \le L \underline{\int_a^b} f\,dx \le L \overline{\int_a^b} f\,dx \le R \overline{\int_a^b} f\,dx$$

Theorem 30:

Every bounded Riemann integrable function over [a, b] is Lebesgue integrable and the two integrals are equal.

Proof:

If f is Riemann integrable, then

$$R \underline{\int_a^b} f\,dx = R \overline{\int_a^b} f\,dx = R \int_a^b f\,dx$$

We know that

$$R \underline{\int_a^b} f\,dx \le L \underline{\int_a^b} f\,dx \le L \overline{\int_a^b} f\,dx \le R \overline{\int_a^b} f\,dx.$$

$$\Rightarrow \quad R \int_a^b f\,dx = L \underline{\int_a^b} f\,dx = L \overline{\int_a^b} f\,dx = L \int_a^b f\,dx.$$

Thus every Riemann integrable function is Lebesgue integrable and the two integrals are the same. However, the converse of this result may or may not be true.

Remark:

A function f is Riemann integrable on [a, b] if and only if the set of discontinuities of f in [a, b] has measure zero, i.e., if f is continuous a.c.

Theorem 31:

Every bounded measurable function on [a, b] is Lebesgue integrable on [a, b].

Proof:

f being bounded, there exist real numbers m, M such that the range of f is a subset of [m, M[. Given $\in > 0$ there exists a finite number of points $y_0, y_1, y_2, ..., y_n$ such that

$$m = y_0 < y_1 < y_2 .... < y_n = M$$

and

$$y_k - y_{k-1} < \frac{\varepsilon}{b-a}$$

For each k = 1, 2, ..., n let $A_k = \{x \in [a, b] : y_{k-1} \le f(x) < y_k\}$. Then each A_k is measurable, since f is measurable.

Thus, $P = \{A_1, A_2, ..., A_n\}$ is measurable partition of [a, b].

Since $\sup_{x \in A_k} f(x) \leq y_k$, and $\inf_{x \in A_k} f(x) \geq y_{k-1}$,

$$U(P, f) = \sum_{k=1}^{n} (\sup_{A_k} f(x))\, mA_k \leq \sum_{k=1}^{n} y_k mA_k$$

$$L(P, f) = \sum_{k=1}^{n} (\inf_{A_k} f(x))\, mA_k \geq \sum_{k=1}^{n} y_k mA_k$$

Now,

$$U(P, f) - L(P, f) \leq \sum_{k=1}^{n} (y_k - y_{k-1})\, mA_k < \frac{\epsilon}{b-a} \sum_{k=1}^{n} mA_k < \epsilon$$

[Since A_k's are pairwise disjoint sets and $\bigcup_{k=1}^{n} A_k = [a, b]$.]

Hence f is Lebesgue integrable on [a, b]. Thus if f is bounded on [a, b], the measurability of f is a sufficient condition for f to be Lebesgue integrable on [a, b].

6.15 Properties of Lebesgue Integral for Bounded Measurable Functions

The following properties hold for a bounded Lebesgue integrable function f on [a, b]:

(i) If $a < c < b$, then f is Lebesgue integrable on [a, c] as well as on [c, b] and

$$\int_a^b f = \int_a^c f + \int_c^b f.$$

(ii) If k is any scalar, $k \in R$, then kf is Lebesgue integrable and

$$\int_a^b kf = k \int_a^b f.$$

(iii) If $f = f_1 + f_2$, where f_1, f_2 are Lebesgue integrable on [a, b], then f is Lebesgue integrable on [a, b], and

$$\int_a^b f = \int_a^b f_1 + \int_a^b f_2.$$

(iv) If $/_k$ is a (finite or infinite) sequence f disjoint measurable subsets of [a, b] whose union A has finite measure, then

$$\int_A f = \sum_k \int_{A_k} f.$$

Example 16:

Let $f(x) = \begin{cases} 1, & \text{if x is an irrational number in } [-4, 4] \\ -2, & \text{if x is a rational number in } [-4, 4] \end{cases}$

Evaluate $\int_{-4}^{4} f(x)\,dx$.

Solution:

As f is bounded by the numbers –2, 1, choose a portion P by subdividing [–2, 1] by points y_n's such that

$$-2 - \in - y_0 < y_1 < y_2 \ldots < y_{n-1} < y_n = 1 + \in, \in > 0$$

and consider the sets A_k's for k = 1, 2, ..., n as follows:

$$A_k = \{x : y_{k-1} \leq f(x) < y_k\}$$

In the figure the bold lines indicate the graph of f in [–4, 4] and the dotted line show the points of sub-division taken along y-axis. It is clear from the diagram that the only non-empty sets are

$$A_1 = \{x : y_0 \leq f(x) < y_1\}$$

and

$$A_n = \{x : y_{n-1} \leq f(x) < y_n\}$$

The set A_1 is the set of rational numbers in [–4, 4] while the set A_n is the set of irrational numbers in [–4, 4]. It follows therefore that $m(A_1) = 0$ and $m(A_n) = 8$, while $mA_k = 0$, for k = 2, 3, 4, ..., (n – 1).

For the partition $P = \{A_1, A_2, A_3, \ldots A_n\}$ of measurable subsets of [–4, 4], the upper and lower Lebesgue sums are defined respectively by

$$U(P, f) = \sum_{k=1}^{n} \sup_{\in A_k} f(x)\, mA_k \leq \sum_{k=1}^{n} y_k mA_k = y_1 mA_1 + y_n mA_n$$

$$= 8y_n = 8(1 + \in)$$

$$L(P, f) = \sum_{k=1}^{n} \inf_{y \in A_k} f(x)\, mA_k \geq \sum_{k=1}^{n} y_{k-1} mA_k = 8y_n = 8(1 + \in)$$

Now by varying the partition, y_n is any number grater than one and y_{n-1} is any number less than one.

$$\therefore \quad \inf_P U(P, f) = \sup_P L(P, f) = 8$$

$$\therefore \quad \int_{-4}^{4} f(x)\,dx = 8.$$

Example 17:

Calculate Lebesgue integral for the function

$$f(x) = \begin{cases} 1 & \text{when x is rational} \\ 2 & \text{when x is irrational} \end{cases}$$

Solution:

As f is bounded on R (the real line) it is bounded on any subset [a, b] of R.

We know that

$$U(P, f) = \sum_{k=1}^{n} \sup_{A_k} f(x)\, mA_k \le \sum_{k=1}^{n} y_k A_k = y_1 mA_1 + y_n mA_n$$

i.e.,

$$U(P, f) \le (2 + \delta)(b - a) \quad \because \quad mA_1 = 0, \ mA_n = b - a$$

$$L(P, f) = \sum_{k=1}^{n} \inf_{A_k} f(x) \ge \sum_{k=1}^{n} y_{k-1} A_k = y_1 mA_1 + y_{n-1} mA_n$$

i.e.,

$$L(P, f) \ge (2 - \delta)(b - a)$$

where $A_k = \{x : y_{k-1} \le f(x) < y_k\}$, $k = 1, 2, 3, \dots n$.

A_1 is the set of rationals in [a, b] which are countable and hence $mA_1 = 0$. A_n is the set of irrationals in [a, b] and $mA_n = b - a$. Taking supremum and infimum over all partitions P, we get

$$2(b - a) \ge \inf_{P} U(P, f) \ge \sup_{P} L(P, f) \ge 2(b - a)$$

$$\therefore \quad \int_a^b f(x)\, dx = 2(b - a)$$

If the interval [a, b] is the whole real line, then

$$\int_R f(x)\, dx = 2mR = \infty.$$

Hence Proved.

Example 18:

Evaluate

$$\int_0^5 f(x)\, dx, \text{ if } f(x) = \begin{cases} 0, & 0 \le x < 1 \\ 1, & \{1 \le x < 2\} \cup \{3 \le x < 4\} \\ 2, & \{2 \le x < 3\} \cup \{4 \le x < 5\} \end{cases}$$

by using Riemann and Lebesgue definitions of the integral.

Solution:

Using Riemann definition of the integral (where the subdivision is taken of the segment [0, 5] by the division points $x_0, x_1, x_2, \dots, x_n$ on x-axis) the upper and lower Riemann sums tend to the common value

$$0(1 - 0) + 1(2 - 1) + 2(3 - 2) + 1(4 - 3) + 2(5 - 4) = 6$$

(since the function is constant on each of the subintervals.)

$\therefore \quad R \int_0^5 f(x)\, dx = 6$

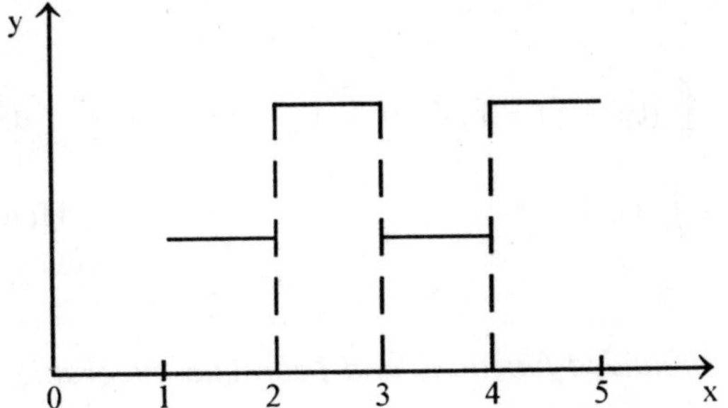

Evaluating the Lebesgue Integral where the subdivision is that of the interval $[0, 2 + \delta[$, $\delta > 0$, we get

$$0(1-0) + 1[(2-1) + (4-3)] + 2(3-2) + (5-4)] = 6$$

$\therefore \quad L \int_0^5 f(x)\, dx = 6.$

Theorem 32:

If f is a bounded and Lebesgue integrable function on [a, b] such that f(x) = g(x) a.e. on [a, b], where g is a bounded function on [a, b] then g is Lebesgue integrable and

$$\int_a^b g\, dx = \int_a^b f\, dx.$$

Proof:

Since f being bounded and Lebesgue integrable, it is measurable. Also $f(x) = g(x)$ a.e. Hence $g(x)$ is measurable.

Moreover g is bounded, hence g is Lebesgue integrable on [a, b].

To show that $\int_a^b g\, dx = \int_a^b f\, dx$, denoted the set A as the set of all points x in [a, b] where $f(x) \neq g(x)$. Clearly $mA = 0$. Now $C(A) = [a, b] - A$ and $f(x) = g(x)$, $x \in C(A)$. Consider the partition $P = \{A, C(A)\}$ of [a, b]; then

$$U(P, g - f) = \sup_A (g - f) \,.\, mA + \sup_{C(A)} (g - f)\, mC(A)$$

$$= \sup_A (g - f) \,.\, 0 + 0 \,.\, mC(A) = 0$$

Similarly, $L(P, g - f) = 0$. Thus

$$0 = L\,(P, g - f) \le \int_a^b (g - f)\,dx \le U(P, g - f) = 0$$

$\therefore \quad \int_a^b (g - f)\,dx = 0$

Now

$$\int_a^b g\,dx = \int_a^b [(g - f) + f]\,dx = \int_a^b (g - f)\,dx + \int_a^b f\,dx = \int_a^b f\,dx$$

$\therefore \quad \int_a^b g\,dx = \int_a^b f\,dx$. **Hence Proved.**

Theorem 33:

If f and g are bounded functions and Lebesgue integrable over [a, b] and if

(i) $f(x) \ge 0$ *a.e. on [a, b], then* $\int_a^b f(x)\,dx \ge 0$

(ii) $f(x) \le g(x)$ *a.e. on [a, b], then* $\int_a^b f\,dx \le \int_a^b g\,dx$.

Proof:

(i) We may assume that $f(x) \ge 0$ for all x in [a, b] as this assumption does not alter the value of $\int_a^b f\,dx$ under given conditions.

Obviously $U(P, f) \ge 0$ for any measurable partition P

$\Rightarrow \quad L\,\overline{\int_a^b} f\,dx \ge 0$, f being Lebesgue integrable

$\Rightarrow \quad L\int_a^b f\,dx = L\,\overline{\int_a^b} f\,dx \ge 0; \qquad \therefore\ L\int_a^b f\,dx \ge 0$

(ii) Since $f(x) \le g(x)$ a.e. on [a, b], so $g(x) - f(x) \ge 0$ a.e. on [a, b]. By (i)

$$\int_a^b (g(x) - f(x))\,dx \ge 0 \Rightarrow \int_a^b g(x)\,dx - \int_a^b f(x)\,dx \ge 0$$

$\Rightarrow \quad \int_a^b f\,dx \le \int_a^b g\,dx$.

Theorem 34:

If a bounded function f is Lebesgue integrable on [a, b] then | f | is Lebesgue integrable over [a, b]. Moreover, if f is Lebesgue integrable then

$$\int_a^b f \,\Big| \le \int_a^b |f|$$

Proof:

f being bounded and Lebesgue integrable on [a, b]

Hence f is measurable.

Define $f^* = \max(f, 0)$, and $f^- = -\min(f, 0)$; then

$$|f| = f^+ + f^- \text{ (where } \max(f, g) = \tfrac{1}{2}(f+g) + \tfrac{1}{2}|f-g|, \text{ and}$$

$$\min(f, g) = \tfrac{1}{2}(f+g) - \tfrac{1}{2}|f-g|\text{)}$$

$|f|$ being the sum of two measurable functions is measurable. Also boundedness of f implies boundedness of $|f|$. Therefore $|f|$ is Lebesgue integrable.

Now,

$$f \le |f|, \text{ and } -f \le |f|$$

$$\therefore \quad \int_a^b f \le \int_a^b |f|, \text{ and} \qquad -\int_a^b f \le \int_a^b |f|$$

$$\Rightarrow \quad \left|\int_a^b f\right| \le \int_a^b |f|.$$

Example 19:

If $f(x) = \frac{1}{2} + \sin x,\ 0 \le x < 2\pi$, *find* $f^{\cdot}$ *and* $f^{\cdot}$.

Solution:

$$f^- = \max(f(x), 0) = \begin{cases} \frac{1}{2} + \sin, x & 0 \le x < \frac{7\pi}{6} \\ 0, & \frac{7\pi}{6} \le x \le \frac{11\pi}{6} \\ \frac{1}{2} + -\sin x, & \frac{11\pi}{6} < x < 2\pi \end{cases}$$

$$f^- = -\min(f(x), 0) = \begin{cases} 0, & 0 \le x < \frac{7\pi}{6} \\ -\frac{1}{2} - \sin x, & \frac{7\pi}{6} \le x \le \frac{11\pi}{6} \\ -0, & \frac{11\pi}{6} < x < 2\pi \end{cases}$$

Thus,

$$f = f^+ - f^-, \text{ and } |f| = f^+ + f^-.$$

Example 20:

If $f(x) = 0$, for every x in the Cantor set P_0 and $f(x) = k$ for x in each of the intervals of length $1/3^k$ in G_0, prove that f is Lebesgue integrable on $[0, 1]$ and that $\int_0^1 f = 3$.

Solution:

Let P_0 and G_0 be Cantor sets as defined in Example 8; then

$$P_0 \cup G_0 = [0, 1]$$

$f(x) = 0, \ \forall x \in P_0$ [the closed intervals which we retain]

$= k \ \ \forall$ x in each of the intervals of length $1/3^k$ in G_0 [the open intervals which we delete]

From Example 8, it follows that the number of intervals of length $1/3^k$ is 2^{k-1}. Thus

$$\int_0^1 f\,dx = 1.\frac{1}{3}.1 + 2.\frac{1}{3^2}.2 + 3.\frac{1}{3^3}.2^2 + \ldots + k.\frac{1}{3^k}.2^{k-1} + \ldots$$

$$= \frac{1}{3}\left(1 + 2.\frac{2}{3} + 3\left(\frac{2}{3}\right)^2 + \ldots. + k\left(\frac{2}{3}\right)^{k-1} + \ldots.\right)$$

$$= 3$$

6.16 Lebesgue Integral of a Bounded Function Over a Set of Finite Measure

Define a function χ on a measurable subset A of [a, b] known as the characteristic function of A as follows:

$$\chi_A(x) = \begin{cases} 1, & x \in A \\ 0, & x \in A \end{cases}$$

A linear combination $\psi(x) = \sum_{i=1}^{n} a_i \chi_{A_i}(x), \bigcup_{i=1}^{n} A_i = A$ of the characteristic function of measurable sets A_i is called a *simple function.* The representation of a simple function is not unique. If A_i', i = 1, 2, ... m is another collection of measurable sets such that $\bigcup_{i=1}^{m} A_i' = A$, then $\sum_{i=1}^{m} a_i' \chi_{A_i}'(x)$ is also a representation of the given simple function ψ. More explicitly, given

$\psi = \sum_{i=1}^{n} a_i \chi_{A_i}$, where A_i's are disjoint and $\bigcup_{i=1}^{n} A_i = A$, then $a_1, a_2 \dots a_n$ are distinct non-zero values of ψ on the sets $A_1, A_2, \dots, A_n$ respectively. The sets A_1 are measurable, if ψ is a measurable function. This representation of ψ is unique and is called the canonical representation of ψ. The integral of the measurable function ψ is defined as follows:

$$\int_A \psi \, dx = \sum_{i=1}^{n} a_i mA_i .$$

Definition:

For any non-negative measurable function f on a measurable set A, its integral is defined as

$$\int_A f \, dx = \sup_{\phi \le f} \int_A \phi \, dx, \text{ for all simple functions } \phi \text{ on } A$$

Integral of f can also be defined as

$$\int_A f \, dx = \inf_{\psi \ge f} \int_A \psi \, dx, \text{for all simple functions } \psi \text{ on } A$$

Since f defined on the set can also be written as $f.x_A$.

$$\therefore \quad \int_A f = \int_A f.\chi_A .$$

Theorem 35:

A necessary and sufficient condition for a bounded function f defined on a measurable set A with finite measure, to be measurable, is that $\inf_{\theta > f} \int_A \psi(x) \, dx = \sup_{\theta > f} \int_A \psi(x) \, dx$, *for all simple functions ϕ and ψ on A.*

Proof:

Necessary Condition: Let f be any bounded measurable function with M as its supremum on the set A. Partitioning the range]–M, M] of the given function f, we define the sets A_i as follows:

$$A_i = \left\{ x : \frac{i-1}{n} M < f(x) > \le \frac{iM}{n}, -n \le i \le n \right\}$$

f being measurable the sets A_i are disjoint measurable sets having A as their union, i.e.,

$$A = \bigcup_{i=-n}^{n} A_i, \text{ and } mA = \sum_{i=-n}^{n} mA_i .$$

The simple functions $\phi_n(x)$ and $\psi_n(x)$ defined by

$$\phi_n(x) = \frac{M}{n}\sum_{i=-n}^{n}(i-1)\chi_{A_i}(x)$$

and

$$\psi_n(x) = \frac{M}{n}\sum_{i=-n}^{n} i\chi_{A_i}(x)$$

Satisfy the relation $\phi_n(x) \le f(x) \le \psi_n(x)$. Thus

$$\inf_{\psi\ge f}\int_A \psi(x)\,dx \le \int_A \psi_n(x)\,dx = \frac{M}{n}\sum_{i=-n}^{n} i\,mA_i$$

and

$$\sup_{\phi\le f}\int_A \phi(x)\,dx \le \int_A \phi_n(x)\,dx = \frac{M}{n}\sum_{i=-n}^{n}(i-1)\,mA_i$$

$$\therefore\ 0 \le \inf_{\psi\ge f}\int_A \psi(x)\,dx - \sup_{\phi\le f}\int_A \phi(x)\,dx \le \frac{M}{n}\sum_{i=-n}^{n} mA_i = \frac{M}{n}mA$$

Taking n to be an arbitrary large number, we have

$$\inf_{\psi\ge f}\int_A \psi(x)\,dx = \sup_{\psi\le f}\int_A \phi_n(x)\,dx$$

Sufficient Condition : Suppose $\inf_{\psi\ge f}\int_A \psi(x)\,dx = \sup_{\phi\ge f}\int_A \phi(x)\,dx.$

$$\therefore\quad \sup_{\phi\le f}\int_A \phi(x)\,dx - \inf_{\psi\ge f}\int_A \psi(x)\,dx = 0$$

$$\Rightarrow\quad \sup_{\phi\le f}\int_A \phi(x)\,dx + \sup_{\psi\ge f}\int_A (-\psi(x))\,dx = 0$$

Thus for each given n, there are simple functions $\phi_n(x)$ and $\psi_n(x)$ such that

$$\int_A \psi(x)\,dx - \int_A \phi_n(x)\,dx < \frac{1}{n}$$

Then the functions inf ψ_n and sup ϕ_n are measurable and

$$\sup \phi_n(x) \le f(x) \le \inf \psi_n(x)$$

Let $F_\alpha = \{x : \sup \phi_n(x) < \inf \psi_n(x) - 1/\alpha\}$; then

$$\bigcup_{\alpha} F_\alpha = F = \{x : \sup \phi_n(x) < \inf \psi_n(x)\}$$

But each F_α is contained in the set $\{x : \phi_n(x) < \psi_n(x) - 1/\alpha\}$ with measure less than α/n. Since n is an arbitrary large number, so $m(F_\alpha) = 0$ and hence $mF = 0$. Thus $\sup \phi_n(x) = \inf \psi_n(x)$ except on a set of measure zero and $\sup \phi_n(x) = f = \inf \psi_n(x)$ a.e. Hence f is measurable.

Theorem 36:

If f and g are non-negative bounded measurable functions defined on set $A \subseteq [a, b]$ of finite measure, then

(i) $\int_A \alpha f + \beta g = \alpha \int_A f + \beta \int_A g$, *where α, β are any real numbers.*

(ii) *If $f = g$ a.e., then* $\int_A f = \int_A g$.

(iii) *If $f \geq 0$ a.e. then* $\int_A f \geq 0$, *and if $f \leq g$ a.e., then* $\int_A f \leq \int_A g$.

(iv) *If $a \leq f(x) \leq B$, then* $amA \leq \int_A f \leq bmA$. *In particular if $\alpha = \beta$ then*

$\int_A f = amA$.

(v) *If A_1 and A_2 are disjoint measurable sets of finite measure, then*

$\int_{a_1 \cup a_2} f = \int_{A_1} f + \int_{A_2} f$.

(vi) *If $mA = 0$, and f is measurable, then* $\int_A f = 0$.

Proof:

(i) Let ψ be any simple function then $\alpha\psi$ is also a simple function. If $\alpha < 0$.

$$\int_A \alpha f = \inf_{\psi \geq f} \int_A \alpha\psi = \alpha \inf_{\psi \geq f} \int_A \psi = \alpha \int_A f$$

If $\alpha < 0$,

$$\int_A \alpha f = \inf_{\phi \leq f} \int_A \alpha\phi = \alpha \sup_{\phi \leq f} \int_A \phi = \alpha \inf_{\psi \geq f} \int_A \psi = \alpha \int_A f$$

Now if ψ_1 and ψ_2 are simple functions greater than or equal to f and g respectively, then $\psi_1 + \psi_2$ is also a simple function greater than or equal to f + g, and

$$\int_A f + g \leq \int_A (\psi_1 + \psi_2) = \int_A \psi_1 + \int_A \psi_2$$

Since infimum on the right hand side is $\int_A f + \int_A g$.

$$\therefore \quad \int_A f + g \le \inf_{\psi_1 \ge f} \int_A \psi_1 + \inf_{\psi_2 \ge g} = \int_A \psi_2 + f + \int_A g$$

On the other hand, if $\phi_1 \le f$ and $\phi_2 \le g$, then $\phi_1 + \phi_2$ is simple function less than or equal to (f + g).

$$\therefore \quad \int_A f + g \ge \int_A \phi_1 + \phi_2 = \int_A \phi_1 + \int_A \phi_2$$

Again supremum on the right hand side is $\int_A f + \int_A g$.

$$\therefore \quad \int_A f + g \ge \sup_{\phi_1 \le f} \int_A \phi_1 + \sup_{\phi_2 \le f} \int_A \phi_2 = \int_A f \int_A g$$

Hence the result follows.

(ii) Since $f - g = 0$ a.e., so if ψ is a simple function greater than or equal to $f - g$, then $\psi \ge 0$ a.e. and hence

$$\int_A \psi \ge 0 \Rightarrow \inf_{\phi \ge (f-g)} \int_A \psi \ge 0 \Rightarrow \int_A (f-g) \ge 0.$$

Similarly, taking $\phi \le (f - g)$, we get

$$\int_A \phi \ge 0 \Rightarrow \sup_{\phi \le (f-g)} \int_A \phi \le 0 \Rightarrow \int_A f - g \le 0$$

$$\therefore \quad \int_A f - g = 0 \Rightarrow \int_A f = \int_A g.$$

(iii) $f \ge 0$ a.e. on A. If ψ is a simple function greater than or equal to f, then

$$\psi \ge 0 \text{ a.e} \Rightarrow \int_A \psi \ge 0 \Rightarrow \inf_{\phi \ge f} \int_A \psi \ge 0 \Rightarrow \int_A f \ge 0$$

$\because \; f \le g$ a.e. $\Rightarrow g - f \ge 0$ a.e. on A.

Let ψ be a simple function greater than or equal to $(g - f)$; then

$$\psi \ge 0 \text{ a.e} \Rightarrow \int_A \psi \ge 0 \Rightarrow \inf_{\psi \ge g-f} \int_A \psi \ge 0$$

$$\Rightarrow \quad \int_A g - f \ge 0 \Rightarrow \int_A g \ge \int_A f.$$

(iv) If $\alpha \le f \le \beta$, then for ψ to be a simple function greater than or equal to f, we have

$$\alpha \le \psi \Rightarrow \int_A \alpha \le \int_A \psi, \text{ or } \alpha mA \le \int_A \psi$$

Taking infimum on the right,

$$\alpha mA \le \inf_{\psi \ge f} \int_A \psi = \int_A f$$

Similarly,

$$\sup_{\phi \le f} \int_A \phi \le \beta mA$$

Thus,

$$\alpha mA \le \inf_{\phi \ge f} \int_A \psi = \sup_{\phi \le f} \int_A \phi \le \beta mA$$

(v) Since A_1 and A_2 are disjoint measurable sets of finite measure, so their characteristic function $\chi_{A_1 \cup A_2}$ of the union $A_1 \cup A_2$ is measurable and is of finite measure. Moreover $\chi_{A_1 \cup A_2} = \chi_{A_1} + \chi_{A_2}$.

Let ψ be a simple function greater than or equal to f on A_1 as well as on A_2. Then ψ will be greater than or equal to f on $A_1 \cup A_2$. Now

$$\begin{aligned}\int_{A_1 \cup A_2} f\,dx = \inf_{\psi \ge f} \int_{A_1 \cup A_2} \psi &= \inf_{\psi \ge f} \int \psi \chi_{A_1 \cup A_2} \\ &= \inf_{\psi \ge f} \int \psi \chi_{A_1} + \inf_{\psi \ge f} \int \psi \chi_{A_2} \\ &= \int_{A_1} f + \int_{A_2} g.\end{aligned}$$

Example 21:

Show that if f is a non-negative measurable function then f = 0 a.e. on a set A iff $\int_A f\,dx = 0$.

Solution:

Let ϕ be any measurable simple function such that $\phi \le f$. Since f = 0 a.e. on A, so $\int_A \phi\,dx \le 0$. Taking supremum over all those measurable simple functions $\phi \le f$, we get $\int_A f\,dx \le 0$. Similarly

$$\inf_{\psi \ge f} \int_A \psi\,dx = \int_A f\,dx \ge 0 \quad \therefore \int_A f\,dx = 0$$

Conversely if $E_n = \left\{x : f(x) > \frac{1}{n}\right\}$.

$$0 = \int_A f\,dx \geq \int \frac{1}{n}\chi_{E_n}\,dx = \frac{1}{n}\,mE_n \Rightarrow mE_n = 0$$

But

$$\{x : f(x) > 0\} = \bigcup_{n=1}^{\infty} E_n. \quad \therefore\ m\left(\bigcup_{n=1}^{\infty} E_n\right) = 0$$

i.e.,

$$m\,\{x : f(x) > 0\} = 0, \qquad \therefore\ f = 0 \text{ a.e. on } A$$

6.17 Lebesgue Integral for Unbounded Functions

We now consider Lebesgue integral for unbounded measurable functions.

Let f be a non-negative measurable function on [a, b].

For each $x \in [a, b]$ and $n \in N$ we define a function F(x, n) as follows:

$$f(x, n) = f(x), \text{ if } 0 \leq f(x) \leq n$$
$$= n, \quad \text{if } f(x) > n$$

Thus, $F(x, n) = \min\,(f(x), n)$

F(x, n) being the minimum of f(x) and n, is bounded and hence measurable which implies that for each $n \in N$, F(x, n) is Lebesgue integrable.

Now if $\underset{n\to\infty}{\text{Lt}} \int_a^b F(x,n)\,dx$ exists finitely then we say that the unbounded function f is Lebesgue integrable and

$$\int_a^b f = \underset{n\to\infty}{\text{Lt}} \int_a^b F(x,n)\,dx$$

If the limit does not exist finitely then f is not Lebesgue integrable. The function F(x, n) is called *truncated* function.

Example 22:

Define

$$f(x) = \frac{1}{x^{2/3}},\ 0 < x \leq 1$$
$$= 0,\ x = 0.$$

Show that f is Lebesgue integrable on [0, 1], and $\int_0^1 \frac{1}{x^{2/3}}\,dx = 3.$ *Find also F(x, 2).*

Solution:

Since $\frac{1}{x^{2/3}} \to \infty$, as $x \to 0$, so f is unbounded in [0, 1]. In order to examine its Lebesgue integrability, define.

$$F(x, n) = \frac{1}{x^{2/3}}, \text{ if } \frac{1}{n^{3/2}} \leq x \leq 1$$

$$= n \qquad \text{if } 0 < x < \frac{1}{n^{3/2}} \qquad ...(1)$$

$$= 0 \qquad \text{if } x = 0$$

For n = 2

$$F(x, 2) = \frac{1}{x^{2/3}}, \text{ if } \frac{1}{2^{3/2}} \leq x \leq 1$$

$$= 2 \qquad \text{if } 0 < x < \frac{1}{2^{3/2}}$$

$$= 0 \qquad \text{if } x = 0$$

Now

$$\int_0^1 F(x, n)\, dx = \int_0^{1/n^{3/2}} F(x, n)\, dx + \int_{1/n^{3/2}}^1 F(x, n)\, dx$$

$$= \int_0^{1/n^{3/2}} n\, dx + \int_{1/n^{3/2}}^1 \frac{1}{x^{2/3}}\, dx$$

$$= \frac{1}{\sqrt{n}} + 3\left[1 - \left(\frac{1}{x^{3/2}}\right)^{1/3}\right] = 3 - \frac{2}{\sqrt{3}}, \quad \forall\, n$$

Thus by the definition of the Lebesgue integral of unbounded functions, we have

$$\int_0^1 f(x)\, dx = \underset{n\to\infty}{\text{Lt}} \int_0^1 F(x, n)\, dx = \underset{n\to\infty}{\text{Lt}} \left(3 - \frac{2}{\sqrt{n}}\right) = 3.$$

Example 23:

If

$$f(x) = \frac{1}{x}, \text{ if } 0 < x \leq 1$$

$$= 9, \text{ if } x = 0$$

then f is not Lebesgue integrable on [0, 1]. Define.

Solution:

$$F(x, n) = \frac{1}{x}, \text{ if } \frac{1}{n} \leq x \leq 1$$

$$= n\ , \quad \text{if } 0 < x < \frac{1}{n}$$

$$= \min(9, n), \text{ if } x = 0$$

$$\int_0^1 F(x, n)\, dx = \int_0^{1/n} n\, dx + \int_{1/n}^1 \frac{1}{x}\, dx = 1 + \log n, \ \forall n$$

$$\therefore \quad \underset{n\to\infty}{\text{Lt}} \int_0^1 F(x, n)\, dx = \underset{n\to\infty}{\text{Lt}} (1 + \log n) = \infty$$

Thus by definition f is not Lebesgue integrable.

6.18 The General Integral

Let f be any real function defined on R and A be any measurable subset of R. The function f^+ and f^- defined as

$f^+(x) = \max(f(x), 0)$, and $f^-(x) = \max(-f(x), 0) = -\min(f(x), 0)$ are such that

(i) $f = f^+ - f^-, |f| = f^+ + f^-;\ f^+, f^- \geq 0$

(ii) f is measurable iff f^+ and f^- are measurable.

Definition:

Let f be a measurable function defined on a measurable set A and $\int_A f^+\, dx < \infty$, $\int_A f^-\, dx < \infty$, then f is Lebesgue integrable on A and its integral is given by

$$\int_A f = \int_A f^+ - \int_A f^- .$$

If $\int_A f^+\, dx = +\infty$, and $\int_A f^- dx = +\infty$, then $\int_A f\, dx$ is undefined. From the definition it follows that $\int_A f$ is finite iff both $\int_A f^+\, dx$ and $\int_A f^-\, dx$ are finite and this will be true iff $\int_A |f| = \int_A f^+ + \int_A f^-$ is finite.

6.19 Some Fundamental Theorems

Theorem 37:

If $A = \bigcup_k A_k$ is the union of a finite or denumerable number of measurable and disjoint subsets $A_1, A_2, a_3, \ldots, A_k, \ldots$ of [a, b], and if $\int_A f\, dx$ exists, then

$$\int_A f\, dx = \sum_k \int_{A_k} f\, dx .$$

Proof:

We will establish the theorem for non-negative functions and extend the results to arbitrary functions by decomposing the function into non-negative and non-positive parts, i.e.,

$$\int_A f\,dx = \int_A f^+\,dx - \int_A f^-\,dx \sum_k\left(\int_{A_k} f^+\,dx - \int_{A_k} f^-\,dx\right)$$

$$= \sum_k \int_{A_k} f\,dx$$

Hence we assume that $f \geq 0$.

Let $E_k = A_1 \cup A_2 \ldots. \cup A_k$.

For bounded functions $F(x, n) = \min(f(x), n)$, we have

$$\int_A F(x, n)\,dx = \sum_k \int_{A_k} F(x, n)\,dx \leq \sum_k \int_{A_k} f(x)\,dx$$

For every n, when $n \to \infty$, it follows that

$$\int_A f\,dx = \operatorname*{Lt}_{n\to\infty} \int_A F(x, n)\,dx \leq \sum_k \int_{A_k} f\,dx \qquad \ldots(1)$$

Also

$$\int_A f\,dx \geq \int_{E_k=\bigcup_{i=1}^{k} A_i} f\,dx \geq \int_{\bigcup_{i=1}^{k} A_i} F(x,n)\,dx = \sum_{i=1}^{k} \int_{A_i} F(x,n)\,dx$$

Taking the limit first with respect to n and then with respect to k (if $\bigcup_k A_k$ is countable) we get

$$\int_A f\,dx \geq \sum_{i=1}^{k} \int_{A_i} f\,dx \text{ or } \int_A f\,dx \geq \sum_{i=1}^{\infty} \int_{A_i} f\,dx \qquad \ldots(2)$$

according as k is finite or infinite.

From (1) and (2) we have

$$\int_A f\,dx = \sum_k \int_{A_k} f\,dx.$$

Remark:

If the sequence of sets A_k is finite then the assumption on the existence of the integral $\int_A f(x)\,dx$ is not necessary. The result of the theorem need not hold if we relax the condition of integrability of f on $A = \bigcup_{k=1}^{\infty} A_k$. This can be illustrated by the following example.

Example 24:

Let

$$f(x) = \begin{cases} n & \text{if } \dfrac{2n}{4n^2-1} < x \le \dfrac{1}{2n-1} \\ -n, & \text{if } \dfrac{1}{2n+1} < x \le \dfrac{2n}{4n^2-1} \end{cases} \qquad n = 1, 2, 3, \ldots..$$

Solution:

For each n we have

$$\int_{1/(2n+1)}^{1/(2n-1)} f(x)\,dx = \int_{1/(2n+1)}^{2n/(4n^2-1)} (-n)\,dx + \int_{2n/(4n^2+1)}^{1/(2n-1)} n\,dx$$

$$= -n\left(\frac{2n}{4n^2-1} - \frac{1}{2n+1}\right) + n\left(\frac{1}{2n-1} - \frac{2n}{4n^2-1}\right)$$

$$= 0$$

and

$$\bigcup_n \left\{\left]\frac{1}{2n+1}, \frac{2n}{4n^2-1}\right] \cup \left]\frac{2n}{4n^2-1}, \frac{1}{2n+1}\right]\right\} =]0, 1]$$

Now,

$$\int_0^1 |f(x)|dx = \sum_{n=1}^{\infty} \int_{1/(2n+1)}^{1/(2n-1)} |f(x)|dx = \sum_{n=1}^{\infty} \int_{1/(2n+1)}^{1/(2n-1)} n\,dx$$

$$= \sum_{n=1}^{\infty} \frac{2n}{4n^2-1} \ge \sum_{n=1}^{\infty} \frac{1}{4n}$$

But the series $\frac{1}{4}\sum_{n=1}^{\infty}\frac{1}{n}$ is divergent.

$\therefore$ $\int_0^1 |f(x)|\,dx$ tends to infinity, and hence $\int_0^1 f(x)\,dx$ does not exist.

Theorem 38:

Let f be a measurable function over the interval [a, b], then f is Lebesgue integrable over [a, b] iff | f | is Lebesgue integrable. Moreover if f is Lebesgue integrable over [a, b], then

$$\left|\int_a^b f\right| \le \int_a^b |f|$$

Proof:

Suppose f is measurable and $|f|$ is Lebesgue integrable over [a, b]. By definition of f^+ and f^- we have for $x \in [a, b]$

$$0 \le f^+(x) \le |f(x)|$$

$$0 \le f^-(x) \le |f(x)|$$

which implies that f^+ and f^- are Lebesgue integrable over [a, b]. Hence f is Lebesgue integrable on [a, b]. Again if f is Lebesgue integrable over [a, b] then both f^+ and f^- are Lebesgue integrable.

But $|f| = f^+ + f^-$. Hence $|f|$ is Lebesgue integrable. Now

$$f \le |f|, \text{ and } -f \le |f|$$

$$\therefore \quad \int_a^b f \le \int_a^b |f| \text{ and } -\int_a^b f \le \int_a^b |f|$$

$$\Rightarrow \quad \left|\int_a^b f\right| \le \int_a^b |f|.$$

Example 25:

If f is Lebesgue integrable on [a, b], and $\left|\int_a^b f\,dx\right| \le \int_a^b |f|\,dx$, *when does the equality hold?*

Hint. $|f| - f \ge 0$ so, $\int |f|\,dx \ge \int f\,dx$. Also $|f| + f \ge 0$ and hence $\int_a^b |f|\,dx \ge -\int_a^b f\,dx$. Hence $\int_a^b |f|\,dx \ge \left|\int_a^b f\,dx\right|$.

If $\int_a^b f\,dx \ge 0$, then $\int_a^b |f|\,dx = \int_a^b |f|\,dx$, i.e., $\int_a^b (|f| - f)\,dx = 0$ $\Rightarrow |f| = f$ a.e., i.e. $f \ge 0$ a.e.

If $\int_a^b f\,dx < 0$, then $\int_a^b |f|\,dx = -\int_a^b f\,dx$, i.e., $\int_a^b (|f| + f)\,dx = 0$ and hence $|f| = -f$ a.e., i.e. $f \le 0$ a.e.

Hence, $f \ge 0$ a.e., or $f \le 0$ a.e. is a necessary condition for the equality of $\left|\int_a^b f\,dx\right|$ and $\int_a^b |f|\,dx$.

Theorem 39:

Let f be a Lebesgue integrable function on [a, b], then given $\in > 0 \; \exists$

a $\delta > 0$ *such that* $\left|\int_4 f\right| < \in$ *whenever A is a measurable subset of [a, b] with mA <* δ.

Proof:

We will first prove the result for non-negative functions. Assuming f to be a non-negative function we have

$$\underset{n\to\infty}{\text{Lt}} \int_a^b F(x,n) = \int_a^b f(x)\,dx, \text{ where } F(x,n) = \begin{cases} f(x), & \text{if } 0 \le f(x) \le n \\ n, & \text{if } f(x) > n \end{cases}$$

The given $\in > 0$, $\exists\ N \in N$ such that

$$\int_a^b f(x)\,dx - \int_a^b F(x, N)\,dx < \frac{\in}{2}$$

$\left[\because \left\{\int_a^b F(x, n)\,dx\right\}\right.$ is a non-decreasing sequence converging to $\left.\int_a^b f(x)\,dx\right]$ i.e.,

$$\int_a^b f(x) - F(x, N))\,dx < \frac{\in}{2} \qquad \text{...(1)}$$

Choose any $\delta > 0$, with $d < \frac{\in}{2N}$. If A is a measurable subset of [a, b] and $mA < \delta$, then we have

$$\int_A F(x, N)\,dx \le \int_A N\,dx = N\,mA < \frac{\in}{7} \qquad \text{...(2)}$$

(using $mA < \in/2N$)

From (1) and (2) we get

$$\int_A f(x)\,dx = \int_A (f(x) - F(x, N))\,dx + \int_A F(x, N)\,dx < \frac{\in}{2} + \frac{\in}{2}$$

For an arbitrary Lebesgue integrable function f over [a, b] we have

$$f = f^+ - f^-$$

By the first part, for a given $\in > 0$, $\exists$ a $\delta_1 > 0$ such that $\int_A f + dx < \frac{\in}{2}$ whenever $mA < \delta_1$.

Similarly, $\exists$ a $\delta_2 > 0$ such that $\int_A f - dx < \frac{\in}{2}$ whenever $mA < \delta_2$.

Thus, if $mA < \delta = \min(\delta_1 . \delta_2)$ we get

$$\left|\int_A f\right| \leq \int_A |f| = \int_A f^+ + \int_A f^- < \frac{\epsilon}{2} + \frac{\epsilon}{2} = \epsilon$$

Hence $\left|\int_A f\right| < \epsilon$, whenever $m < \delta$.

6.20 Lebesgue Theorem on Bounded Convergence

Let $\{f_n\}$ be a sequence of functions measurable on a measurable subset $A \subseteq [a, b]$ such that $\underset{n\to\infty}{\text{Lt}} f_n(x) = f(x)$. Then, if there exists a constant M such that $|f_n(x)| \leq M$, for all n and for all x, we have

$$\underset{n\to\infty}{\text{Lt}} \int_A f_n(x)\, dx = \int_A f_n(x)\, dx$$

Since $\underset{n\to\infty}{\text{Lt}} f_n(x) = f(x)$, and $|f_n(x)| \leq M \Rightarrow |f(x)| \leq M$, the function f is bounded and measurable (being the limit of bounded and measurable function), hence Lebesgue integrable. We shall show that

$$\underset{n\to\infty}{\text{Lt}} \int_A |f_n(x) - f(x)\, dx = 0$$

For a given $\epsilon > 0$ we define a partition A into disjoint measurable sets A_k's as follows:

$$A_k = \{x : |f_{k-1} - f| \geq \epsilon, |f_n - f| < \epsilon, \forall\, n \geq k\},\ k = 1, 2, 3, \ldots..$$

In particular,

$$A_1 = \{x : |f_n - f| < \epsilon, n = 1, 2, \ldots\}$$

$$A_2 = \{x : |f_1 - f| \geq \epsilon, |f_n - f| < \epsilon, n = 2, 3, 4 \ldots.\}$$

Clearly

$$A = \bigcup_{k=1}^{\infty} A_k = \left(\bigcup_{k=1}^{n} A_k\right) \cup \left(\bigcup_{k=n+1}^{\infty} A_k\right) = P_n \cup Q_n \text{ (say)}$$

and

$$mA = m(P_n \cup Q_n) = mP_n + mQ_n$$

Now

$$\int_A |f_n - f|\, dx = \int_{P_n} |f_n - f|\, dx + \int_{Q_n} |f_n - f|\, dx \qquad \ldots(1)$$

For each n we have

$$|f_n - f| < \epsilon \text{ on } P_n, \text{ and } |f_n - f| \leq |f_n| + |f| \leq 2M \text{ on } Q_n \qquad \ldots(2)$$

Thus

$$\int_A |f_n - f|\,dx < \varepsilon m P_n + 2MmQ_n$$

As $n \to \infty$, $\underset{n\to\infty}{\text{Lt}}\ mP_n = mA$ and, $\underset{n\to\infty}{\text{Lt}}\ mQ_n = 0$. Thus

$$\int_A |f_n - f|\,dx < \varepsilon mA$$

$$\therefore \quad \left|\int_A (f_n - f)\right| \le \int_A (f_n - f) < \varepsilon mA \text{ , } \in \text{ being arbitrary}$$

$$\therefore \quad \underset{n\to\infty}{\text{Lt}} \int_A f_n(x)\,dx = \int_A f(x)\,dx$$

Example 26:

Verify Bounded Convergence Theorem for the sequence of function

$f_n(x) = \dfrac{1}{(1+x/n)^n}$, $0 \le x \le 1$, $n \in N$.

Solution:

$$|f_n(x)| = \left|\frac{1}{(1+x/n)^n}\right| \le 1, \ \forall\ n \text{ and } \forall\ x.$$

Each f_n being bounded and measurable, the limit function

$$\underset{n\to\infty}{\text{Lt}}\ f_n(x) = \underset{n\to\infty}{\text{Lt}} \frac{1}{(1+x/n)^n} = \frac{1}{e^x}$$

is also bounded and measurable. Now

$$\int_0^1 \frac{dx}{(1+x/n)^n} = \left.\frac{n(1+x/n)^{-n+1}}{(-n+1)}\right|_0^1 = \frac{n}{n-1}\left(1 - \frac{(1+1/n)}{(1+1/n)^n}\right)$$

$$\therefore \quad \underset{n\to\infty}{\text{Lt}} \int_0^1 \frac{dx}{(1+x/n)^n} = \underset{n\to\infty}{\text{Lt}} \frac{n}{n-1}\left(1 - \frac{(1+1/n)}{(1+1/n)^n}\right)$$

$$= \left(1 - \frac{1}{e}\right) = \frac{e-1}{e}$$

Also

$$\int_0^1 \underset{n\to\infty}{\text{Lt}} \frac{1}{(1+x/n)^n}\,dx = \int_0^1 \frac{1}{e^x}\,dx = \left(1 - \frac{1}{e}\right) = \frac{e-1}{e}$$

Hence Bounded Convergence Theorem is applicable.

Theorem 40:

(Monotone Convergence Theorem) : *Let A be any measurable subset of [a, b] with finite measure. Let $\{f_n\}$ be a sequence of measurable functions such that for $x \in A$*

$$0 \le f_1(x) \le f_2(x) \le \ldots, \le \ldots f_n(x) \le \ldots$$

Proof:

If $\underset{n\to\infty}{\text{Lt}} f_n(x) = f(x)$, then $\underset{n\to\infty}{\text{Lt}} \int_A f_n = \int_A f$.

For a given x, the sequence $\{f_n(x)\}$ is monotonically increasing and $\underset{n\to\infty}{\text{Lt}} f_n(x) = f(x)$. Assume that $\underset{n\to\infty}{\text{Lt}} \int_A f_n = \alpha$, for some α.

Since $\int_A f_n \le \int_A f$, so taking the limit as $n \to \infty$, we get

$$\alpha \le \int_A f. \qquad \ldots(1)$$

Let β be any real number lying between 0 and 1, i.e., $0 < \beta < 1$ and ϕ be any simple measurable function such that $0 \le \phi \le f$.

Define the sets $A_n = \{x : f_n(x) \ge \beta\, \phi(x)\}$, n = 1, 2, 3 ....

By definition of $\{f_n\}$, we have

$$A_i \subseteq A_{i+1},\ i = 1, 2, \ldots, n, \ldots \text{ and } A = \bigcup_{i=1}^{\infty} A_i$$

Also for any n,

$$\int_A f_n \ge \int_{A_n} f_n \ge \beta \int_{A_n} \phi$$

Taking the limit as n tends to infinity, we get

$$\alpha \ge \beta \int_A \phi$$

Letting $\beta \to 1$, we get $\alpha \ge \int_A \phi$ Since ϕ is arbitrary, so the result holds for all simple function $\phi \le f$.

$$\therefore \quad \alpha \ge \sup_{\phi \le f} \int_A \phi = \int_A f \qquad \ldots(2)$$

The equality follows from (1) and (2).

Theorem 41:

(Classical Lebesgue Dominated Convergence Theorem). Let $\{f_n\}$ be a sequence of measurable functions on [a, b] such that

$$\underset{n\to\infty}{\text{Lt}}\ f_n(x) = f(x) \text{ a.e. on } [a, b].$$

Proof:

If there exists a Lebesgue integrable function g on [a, b] such that for each $n \in N$

$$|\ f_n(x)\ | \le g\ (x) \text{ a.e. on } [a, b]$$

Then f is Lebesgue integrable and

$$\underset{n\to\infty}{\text{Lt}} \int_a^b f_n = \int_a^b f$$

f being the limit of a sequence of measurable functions is also measurable. For $n \in I$

$$|\ f_n\ (x)\ | \le g(x) \text{ a.e. on } [a, b] \qquad \text{...(1)}$$

$$\Rightarrow \underset{n\to\infty}{\text{Lt}}\ |\ f_n(x)\ | \le g\ (x) \text{ a.e.} \Rightarrow |\ \underset{n\to\infty}{\text{Lt}}\ f_n\ (x)\ | \le g(x) \text{ a.e.}$$

$$\Rightarrow\ |\ f(x)\ | \le g(x) \text{ a.e. on } [a, b]. \qquad \text{...(2)}$$

Since g is Lebesgue integrable on [a, b], so $|\ f\ |$ is Lebesgue integrable on [a, b], which in turn implies that f is Lebesgue integrable on [a, b].

Corresponding to a given $\in > 0$ and $N \in N$, define

$$A_N = \left\{x : |f_n(x) - f(x)| < \frac{\in}{2(b-a)},\ n \ge N,\ x \in [a, b]\right\}$$

Here,

$$A_N \subset A_{n+1} \quad \forall\ \ N \in N.$$

Since $\underset{n\to\infty}{\text{Lt}}\ f_n(x) = f(x)$ a.e.,so every point of [a, b] either lies in the union $\bigcup_{N=1}^{\infty} A_N$ or in a subset A of the interval [a, b] whose measure is zero, i.e.

$$[a, b] = \left(\bigcup_{N=1}^{\infty} A_N\right) \cup A, \text{ where } mA = 0$$

$$\therefore\ m\left(\bigcup_{N=1}^{\infty} A_N\right) = b - a$$

Also,

$$m\left(\bigcup_{N=1}^{\infty} A_N\right) = \underset{N\to\infty}{\text{Lt}}\ m\left(\bigcup_{n=1}^{N} A_n\right), \text{ i.e., } m\left(\bigcup_{N=1}^{\infty} A_N\right) = \underset{N\to\infty}{\text{Lt}}\ mA_N$$

$$\therefore \qquad b - a = \underset{N\to\infty}{\text{Lt}} \; mA_N \qquad \text{...(3)}$$

It is given that g is Lebesgue integrable; therefore for every $\in > 0$ there exists a $\delta > 0$ such that $\int_A g < \frac{\in}{4}$, for $mA < \delta$

From (1) and (2) it follows that

$$\int_A |f_n| \le \int_A g < \frac{\in}{4}, \text{ for } mA < \delta \qquad \text{...(4)}$$

and

$$\int_A |f| \le \int_A g < \frac{\in}{4}, \text{ for } mA < \delta \qquad \text{...(5)}$$

From (3) there exists a positive integer M, such that

$$b - a - mA_N < \delta, \quad \forall \; N \ge M \qquad \text{...(6)}$$

i.e.,

$$mC(A_N) < \delta, \quad \forall \; N \ge M$$

Now

$$\int_a^b | f_n - f | = \int_{A_M} |f_n - f| + \int_{C(A_M)} |f_n - f| \qquad \text{(by def. of } A_M)$$

$$< \frac{\in}{2} \frac{m(A_M)}{2(b-a)} + \int_{C(A_M)} |f_n| + \int_{C(A_M)} |f_n|. \; \forall \; n \ge M$$

or

$$\int_a^b |f_n - f| < \frac{emA_M}{2(b-a)} + \frac{\in}{4} + \frac{\in}{4} = \in \; \forall \; n \ge M \qquad \text{(using (4) and (5))}$$

$$\therefore \qquad \left| \int_a^b f_n - f \right| \le \int_a^b |f_n - f| < \in \qquad \forall \; n \ge M$$

or

$$\left| \int_a^b f_n - \int_a^b f \right| < \in \qquad \forall \; n \ge M$$

$$\Rightarrow \quad \underset{n\leftarrow\infty}{\text{Lt}} \int_a^b f_n = \int_a^b f.$$

Example 27:

Define the functions $f_n(x)$ on [0, 1] as follows:

$$f_n(x) = -2xn^2 + 2n$$

$$for \quad 0 \le x \le \frac{1}{n}$$

$$= 0 \; for \; \frac{1}{n} < x \le 1$$

Solution:

$$\int_0^1 f_n(x)\,dx = \int_0^{1/n} (-2xn^2 + 2n)\,dx$$

$$+ \int_{1/n}^1 0\,dx$$

$$= 1, \; n = 1, 2, 3, ...$$

$$\therefore \quad \underset{n\to\infty}{Lt} \int_0^1 f_n(x)\,dx = 1$$

Also $f_n(x) \to 0$ for every x. Hence $\int_0^1 f(x)\,dx = 0$.

Fatou's Lemma : Let $\{f_n\}$ be a sequence of non-negative measurable functions on [a, b]. Let $\lim\limits_{n\to\infty} f_n(x) = f(x)$ a.e. on [a, b]. Then

$$\liminf_{n\to\infty} \int_a^b f_n \ge \int_a^b f$$

If f is Lebesgue integrable on [a, b]; otherwise

$$\liminf_{n\to\infty} \int_a^b f_n = \infty$$

For any m ∈ N, we have

$$f_n(x, m) = f_n(x), \;\; \text{if } 0 \le f_n(x) \le m \qquad x \in [a,b]$$

$$= m, \qquad \text{if } f_n(x) > m$$

Thus $F_n(x,m) = \min(f_n(x), m)$. Each $F_n(x, m)$ is bounded by m. Also

$$\underset{n\to\infty}{Lt} F_n(x, m) = \underset{n\to\infty}{Lt} \min(f_n(x), m) = \min(\underset{n\to\infty}{Lt} f_n(x), m)$$

$$= \min(f(x), m) \text{ a.e. on } [a, b]$$

$$= F(x, m) \text{ a.e. on}[a, b]$$

For each m, $F_n(x, m)$ being bounded and measurable, is Lebesgue integrable on [a, b]. Thus by the Dominated Convergence Theorem (for g(x) = m) its limit viz. F(x, m) is also Lebesgue integrable. Moreover.

$$\underset{n\to\infty}{Lt} \int_a^b F_n(x, mx\,dx = \int_a^b \underset{n\to\infty}{Lt} F_n(x, m)\,dx = \int_a^b F(x, m)\,dx.$$

$F_n(x, m)$ is min $(f_n(x), m)$. Hence $F_n(x, m) \le f_n(x)$ for all n

$$\int_a^b F(x, m)\, dx = \underset{n\to\infty}{\text{Lt}} \int_a^b F_n(x, m)\, dx$$

$$= \liminf_{n\to\infty} \int_a^b F_n(x, m)\, dx \le \liminf_{n\to\infty} \int_a^b f_n$$

$$\therefore \quad \liminf_{n\to\infty} \int_a^b f_n \ge \int_a^b F(x, m)\, dx \quad \text{for each } m \in I$$

$F(x, m)$ tends to $f(x)$ when $m \to \infty$

$$\therefore \quad \liminf_{n\to\infty} \int_a^b f_n \ge \int_a^b f.$$

6.21 An Important Result for Integrable Functions

Every integrable function f defined on [a, b] can always be *approximated by some continuous function* g for which the following condition holds:

$$\int_a^b |f(x) - g(x)|\, dx < \in, \text{ for any arbitrary positive number } \in$$

This can be illustrated by the following example.

Define

$$f(x) = \begin{cases} 1, & x \in [a_1, b_1] \subset [a, b] \\ 0, & x \notin [a_1, b_1] \end{cases}$$

Show that given e > 0 there exists a continuous function g defined on [a, b] such that

$$\int_a^b |f(x) - g(x)|\, dx < \in$$

Let the graph of g be represented by PLABMQ in the Figure; i.e.,

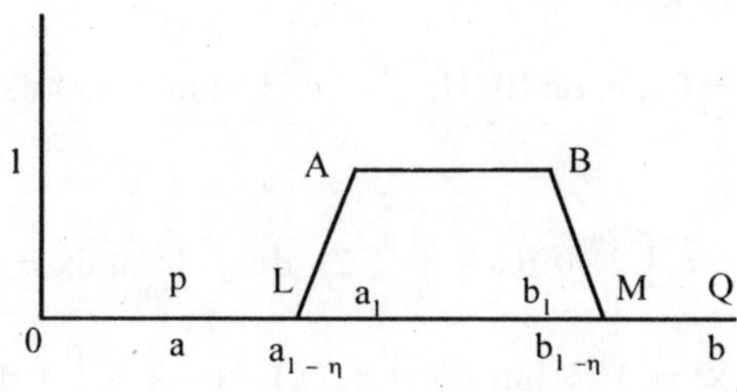

$$g(x) = \begin{cases} 0, & a \le x \le a_1 - \eta \\ 1 + (x - a_1)/\eta, & a_1 - \eta \le x \le a_1 \\ 1, & a_1 \le x \le b_1 \\ 1 - (x - b_1)/\eta, & b_1 \le x\ b_1 + \eta \\ 0, & b_1 + \eta \le x \le b \end{cases}$$

where η is a +ve small number,

Clearly g is continuous on [a, b]. Also

$$\int_a^b |f(x) - g(x)|\,dx = \int_{a_1-\eta}^{a_1} \left(1 + \frac{x - a_1}{\eta}\right) dx$$

$$+ \int_{b_1}^{b_1+\eta} \left(1 - \frac{x - b_1}{\eta}\right) dx = 2\eta < \epsilon.$$

Example 28:

For $n \in N$ let

$$f_n(x) = 2n, \text{ for } x \in \left]\frac{1}{2n}, \frac{1}{n}\right[$$

$$= 0, \text{ for } \in \left]0, \frac{1}{2n}\right[\cup \left]\frac{1}{n}, 1\right[.$$

Calculate $\int_0^1 \lim_{n\to\infty} f_n(x)\,dx$, and $\lim_{n\to\infty} \int_0^1 f_n(x)\,dx$

Show also that Fatou's Lemma holds but that Lebesgue dominated convergence theorem does not.

Solution:

By definition, we have

$$\lim_{n\to\infty} f_n(x) = 0 \text{ a.e. on } [0, 1], \qquad \therefore \int_0^1 \lim_{n\to\infty} f_n(x)\,dx = 0$$

Also,

$$\int_0^1 f_n(x)\,dx = \int_0^{1/2n} 0.dx + \int_{1/2n}^{1/n} 2n\,dx + \int_{1/n}^1 0\,dx = 0$$

$$\therefore \quad \lim_{n\to\infty} \int_0^1 f_n(x) = 1 \Rightarrow \liminf_{n\to\infty} \int_0^1 f_n(x).dx = 1 > \int_0^1 f\,dx = 0$$

Hence strict inequality holds in Fatou's Lemma. In order that the Lebesgue dominated convergence theorem is applicable, the functions $f_n(x)$ should be bounded by some Lebesgue integrable function g(x). But the functions $f_n(x)$ as defined above are unbounded.

Example 29:

Put

$$g(x) = \begin{cases} 0, & 0 \le x \le \frac{1}{2} \\ 1, & \frac{1}{2} < x \le 1 \end{cases}$$

$$f_{2k}(x) = g(x), \qquad 0 \le x \le 1$$

$$f_{2k+1}(x) = g(1-x), \qquad 0 \le x \le 1$$

Then show that

$$\liminf_{n\to\infty} \int_0^1 f_n(x)\,dx > \int_0^1 \liminf_{n\to\infty} f_n(x)\,dx$$

Solution:

Since $f_n(x) = g(x)$, $0 \le x \le 1$, if n is even,

$$\therefore \quad f_{2n}(x) = 0, \text{ if } \quad 0 \le x \le \frac{1}{2}$$

$$= 1, \text{ if } \quad \frac{1}{2} < x \le 1$$

$$f_{2n-1}(x) = g(1-x), \quad 0 \le x \le x\ 1$$

i.e.

$$f_{2n-1}(x) = 1, \qquad 0 \le x < \frac{1}{2}$$

$$= 0, \qquad \frac{1}{2} \le x \le 1$$

$$\liminf_{n\to\infty} f_n(x) = \sup\{\inf(f_n, f_{n+1}, f_{n+2}, \ldots.\} = 0, \quad 0 \le x \le 1$$

$$\therefore \quad \int_0^1 \liminf_{n\to\infty} f_n(x)\,dx = 0 \qquad \ldots(1)$$

Now,

$$\int_0^1 f_{2n}(x) = \int_0^{1/2} 0\,dx + \int_{1/2}^1 1.\,dx = \frac{1}{2}$$

and

$$\int_0^1 f_{2n-1}(x)\, dx = \int_0^{1/2} 1.dx + \int_{1/2}^1 0.\, dx = \tfrac{1}{2}$$

$$\therefore \quad \liminf_{n\to\infty} \int_0^1 f_n(x)\, dx = \tfrac{1}{2} \qquad \text{...(2)}$$

From (1) and (2),

$$\liminf_{n\to\infty} \int_0^1 f_n(x)\, dx > \int_0^1 \liminf f_n(x)\, dx.$$

Example 30:

Let $f_n(x) = \begin{cases} 1/n, & |x| \le n \\ 0, & |x| > n \end{cases}$

Then, $f_n(x) \to 0$ *uniformly on R but* $\int_{-\infty}^{\infty} f_n\, dx = 2,\ n = 1, 2, 3.$

Solution:

$$\underset{n\to\infty}{\text{Lt}} f_n(x) = \underset{n\to\infty}{\text{Lt}} \frac{1}{n} = 0. \text{ when } |x| \le n$$

$$= 0, \qquad \text{when } |x| > n$$

$\therefore \quad \underset{n\to\infty}{\text{Lt}} f_n(x) = 0,$ uniformly on the whole real line

Now

$$|f_{2m}(x) - f_m(x)| = \left|\frac{1}{2m} - \frac{1}{m}\right| = \frac{1}{2m} < \in, \text{ whenever } m > \frac{1}{2\in}$$

and

$$\int_{-\infty}^{\infty} f_n(x)\, dx = \int_{-\infty}^{n} 0\, dx + \int_{-n}^{n} \frac{1}{n} dx + \int_{n}^{\infty} 0\, dx = 2.$$

This implies that uniform convergence of $\{f_n(x)\}$ is not enough for $\underset{n\to\infty}{\text{Lt}} \int f_n = \int \underset{n\to\infty}{\text{Lt}} f_n$. This equality in Lebesgue integration, in general, is only due to dominated convergence of the sequence $\{f_n(x)\}$. However, on the set of finite measure, uniformly convergent sequences of bounded functions are boundedly convergent.

Example 31:

The function f is defined on the interval [0, 1] by

$$f(x) = \begin{cases} 0, & x \text{ is rational} \\ [1/x], & x \text{ is irrational} \end{cases}$$

where [1/x] is the integral part of 1/x ≤ 1/x for x to be irrational. Show that

$$\int_0^1 f\, dx = \infty.$$

Solution:

Define a function g(x) = [1/x], for x ∈]0, 1[. Clearly f ≤ g on]0, 1[. Moreover f = g a.e. Since f < g at the rational points of]0, 1[which are of measure zero, thus f = g a.e.

$$\therefore \quad \int_0^1 (g - f)\, dx = 0 \qquad ...(1)$$

Consider

$$\int_0^1 g\, dx < \int_{1/(m+1)}^1 g\, dx = \sum_{n=1}^{m} \frac{1}{n+1}$$

The series $\sum_{n=1}^{\infty} \frac{1}{n+1}$ diverges to ∞.

$$\therefore \quad \int_0^1 g\, dx = \infty \qquad ...(2)$$

But $\int_0^1 g\, dx = \int_0^1 [(g-f)+f]\, dx = \int_0^1 (g - f)\, dx + \int_0^1 f\, dx$

$$= \int_0^1 f\, dx \quad \text{[using (1)]}$$

$$\therefore \quad \int_0^1 f\, dx = \infty.$$

Example 32:

The function f is defined on [0, 1] by

f(x) = 0, when x is rational

= n, when x is irrational

where n denotes the number of zeros immediately after the decimal point in the representation of x on the decimal scale. Show that f is measurable and find $\int f\, dx$.

Solution:

Let g(x) = n, if $\frac{1}{10^{n+1}} \le x < \frac{1}{10^n}$, n = 0, 1, 2 .... and x ∈]0, 1].

Clearly, $f \le g$ over $]0, 1[$. Moreover $f = g$ a.e.

Now the set $\{x : g(x) = n\}$ is measurable (since g(x) is constant on each of the intervals $\left[\frac{1}{10^{n+1}}, \frac{1}{10^n}\right[$, n = 1, 2, 3 and each interval is measurable).

Hence the function f is measurable and

$$\int_0^1 f\,dx = \int_0^1 g\,dx$$

But,

$$\int_0^1 g\,dx = \sum_{n=0}^{\infty} n\left(\frac{1}{10^n} - \frac{1}{10^{n+1}}\right) = \sum_{n=1}^{\infty} \frac{9n}{10^{n+1}} = \frac{1}{9}$$

$$\therefore \quad \int_0^1 f\,dx = \frac{1}{9}.$$

Example 33:

If f is integrable, show that f is finite-value a.e.

Solution:

Let A be any measurable, subset of R.

As $\int_A f$ is finite, so $\int_A |f|$ is also finite.

Let if possible $|f| = \infty$ on a set $E \subseteq A$ with $mE > 0$; then

$$\int_A |f| > \int_E |f| > mE, \forall\ n\ [\because\ |f|, \forall\ n]$$

$$\Rightarrow \quad \int_A |f| \to \infty \text{ an } n \to \infty$$

which is a contradiction to the fact that $\int_A f$ is finite. Thus $mE = 0$. Hence f is finite valued a.e.

Example 34:

If f is measurable but g is integrable on [a, b] and M and M' are real numbers such that $M \le f \le M'$ a.e., then $\exists$ a real number ξ. $M \le \xi \le M'$ such that

$$\int_a^b f|g|\,dx = \xi \int_a^b |g|dx$$

Solution:

Since $|fg| \le |M'|\,|g| \le (|M| + |M'|)\,|g|$ a.e. and g is integrable, so $(|M| + |M'|)\,|g|$ is integrable $\Rightarrow$ fg is integrable. Also

$$M\,|g| \leq f\,|g| \leq M'\,|g| \text{ a.e.}$$

$$\therefore \quad M\int_a^b |g|\,dx \leq \int_a^b f\,|g|\,dx \leq M'\int_a^b |g|\,dx$$

If $\int_a^b |g|\,dx = 0$ then g = 0 a.e. and the result is trivial. If $\int_a^b |g|\,dx \neq 0$, by taking ξ to be a real number

$$\xi = \frac{\int_a^b f\,|g|\,dx}{\int_a^b |g|\,dx}$$

we get the required result.

Example 35:

If the function f is constant on the measurable set E, say f(x) = c, then

$$\int_E f(x)\,dx = c\,m\,(E).$$

Solution:

Let f be real valued function defined over a measurable st E s.t. f(x) = c (constant) $\forall\ x \in E$.

To prove that

$$\int_E f(x)\,dx = c.m\,(E).$$

$$f(x) = c\ \forall\ x \in E \Rightarrow c \leq f(x) \leq c\ \forall\ x \in E.$$

Using first mean value Theorem (2),

$$c\,m\,(E) \leq \int_E f(x)\,dx \leq c.m\,(E).$$

$$\Rightarrow \int_E f(x)\,dx = c.m\,(E).$$

Example 36:

Let f be a bounded measurable function defined over a measurable set E s.t. m(E) = 0.

Show that $\int_E f(x)\,dx = 0$.

Solution:

Let f be a bounded measurable function defined over a measurable set E s.t. m (E) = 0.

Show that $\int_E f(x)\,dx = 0$.

We claim $\int_E f(x)\,dx = 0$.

We can take $\alpha < f(x) \leq \beta$ on the set E.

using first mean value Theorem, we get

$$a.m(E) \leq \int_E f(x)\,dx \leq \beta, m(E)$$

Using the fact that m(E) = 0, we get

$$0 \leq \int_E f(x)\,dx = 0$$

or $$\int_E f(x)\,dx = 0$$

This completes the proof.

Example 37:

Let f be a bounded measurable function defined over a measurable set E s.t. (i) f (x) ≥ 0

(ii) $\int_E f(x)\,dx = 0.$

Prove that f(x) = 0 almost everywhere in E, i.e., f(x) is equivalent to zero function on E. **(Bhagalpur, 1998)**

Solution:

f be a bounded measurable function defined over a measurable set E s.t.

(i) $f(x) \geq 0 \quad \forall\, x \in E$

(ii) $\int_E f(x)\,dx = 0.$

To prove that f(x) = 0 almost everywhere in E.

Let $E_0 = E\,(f(x) = 0)$,

$$E_n = E\left(\frac{M}{n+1} < f(x) \leq \frac{M}{n}\right) \qquad \forall\, x \in N.$$

M being an upper bound of f(x) in E.

Aim. $f(x) = 0 \ \forall\, x \in E_0$

and $m(E - E_0] = 0$

Then, $E = \bigcup_{n=0}^{\infty} E_n,\ E_k \cap E_k' = \varnothing$ for $k \neq k'$.

Since f is measurable over E and hence the measurability of the sets E_n. It means that

$$m(E - E_0) = \sum_{k=1}^{\infty} m(E_k). \qquad ...(1)$$

We have $\frac{M}{n+1} < f(x)$ on the set E_n $\forall$ $n \in N$.

Using first mean value theorem, we get

$$\frac{M}{n+1}.m(E_n) < \int_{E_n} f(x)\,dx \le \int_E f\,dx = 0$$

Finally, $\frac{M}{n+1}.m(E_n) \le 0 \quad \forall\ n \in N$

or $\quad m(E_n) \le 0 \quad \forall\ n \in N$

But $\quad m(E_n) \ge 0 \quad \forall\ n \in N$ is true for any set E_n.

Combining the last two inequalities, we get

$$m(E_n) = 0 \quad \forall\ n \in N$$

This $\Rightarrow \sum_{n=1}^{\infty} m(E_n) = 0.$

Making use of (1), we get $m(E - E_0) = 0$,

Thus we have shown that

$$f(x) = 0 \qquad \forall\ x \in E_0 \subset E \text{ s.t. } m(E - E_0) = 0.$$

This proves that

$f(x) = 0$ a.e. (almost everywhere) on the set E.

Hence the result.

Remark 1:

Show that if f is integrable and

$$\int_a^x f(t)\,dt = 0 \quad \forall x \in [a, b],$$

then $\quad f = 0$ *a.e. in* (a, b).

Remark 2:

Let $f(x) = -1 \qquad \forall x \in (0, ½)$

$\quad f(x) = 1 \qquad \forall x \in (½, 1).$

Evidently f(x) is defined over (0, 1)

$$\int_0^1 f(x)\,dx = \int_0^{1/2} f(x)\,dx + \int_{1/2}^1 f(x)\,dx$$

$$= -\int_0^{1/2} dx + \int_{1/2}^1 dx$$

$$= -\frac{1}{2} + \frac{1}{2} = 0.$$

Thus, $$\int_0^1 f(x) - 0,$$

but $f(x) \neq 0$ for any $x \in (0, 1)$.

This proves that the integral of the function vanishes although the function is not equivalent to zero. Contrary to the result, just proved.

Example 38:

Let a function f be defined on the interval (0, 1) as follows:

$$f(x) = \begin{cases} 0 & x \text{ is irrational} \\ 1 & x \text{ is rational.} \end{cases}$$

Is f integrable in Riemann sense? Is it integrable in the Lebesgue sense?

(Kanpur, 1994)

Solution:

Here notice that

$$\begin{aligned} \Omega(f, E_1) &= \{(x, y) : x \in E_1, 0 \le y \le f(x)\} \\ &= \{x : x \in E_1\} \times [y : 0 \le y \le 0\} \\ &= E_1 \times \varnothing. \end{aligned}$$

Hence $m\,\Omega(f, E_1) = m(E_1) \,.\, m(\varnothing) = 0$ as $m(\varnothing) = 0$,

$$m[\Omega(f, E_2)] = m(E_2) \,.\, m([0, 1]) = 1 \times 1 = 1.$$

Example 39:

Let $p > 0$, $q > 0$, then show that

(i) $\displaystyle\int_0^1 \frac{x^{p-1}}{1+x^q} dx = \frac{1}{p} - \frac{1}{p+q} + \frac{1}{p+2q} - \frac{1}{p+3q} + \cdots$ *and deduce that*

(ii) $\displaystyle \text{Log } 2 = 1 - \frac{1}{2} + \frac{1}{3} - \frac{1}{4} + \ldots.$

(ii) $\displaystyle \frac{\pi}{4} = 1 - \frac{1}{3} + \frac{1}{5} - \frac{1}{7} + \ldots$

Solution:

We have $x^{p-1}(1 + x^q)^{-1} = x^{p-1}(1 - x^q + x^{2q} - x^{3q} - x^{4q} - \ldots.)$

$$= x^{p-1} \sum_{n=0}^{\infty} (x^2 n^q - x^{(2n+1)q})$$

$$= \sum_{n=0}^{\infty} [x^{2nq+p-1} - x^{(2n+1)q+p-1})$$

Integration yields the result

$$\int_0^1 \frac{x^{p-1}}{1+x^q} dx = \sum_{n=0}^{\infty} \left[\frac{x^{2nq+p}}{2nq+p} - \frac{x^{(2n+1)q+p}}{(2+1)\, q+p} \right]_{x=0}^{1}$$

$$= \sum_{n=0}^{\infty} \left[\frac{1}{2n+p} - \frac{1}{(2n+1)\, q+p} \right]$$

$$= \frac{1}{p} - \frac{1}{p+q} + \frac{1}{p+2q} - \frac{1}{p+3q} + \ldots.$$

$$\int_0^1 \frac{x^{p-1}}{1+x^q} dx = \frac{1}{p} - \frac{1}{p+q} + \frac{1}{p+2q} - \frac{1}{p+3q} + \ldots. \qquad \ldots(1)$$

Hence the result (i)

Put, p= q = 1 and observe that

$$\int_0^1 \frac{1}{1+x} dx = [\log(1+x)]_0^1 = \log 2,$$

we get $\log 2 = 1 - \frac{1}{2} + \frac{1}{3} - \frac{1}{4} + \ldots.$ Hence the result (ii).

Taking p = 1, q = 2 in (1) and observing that

$$\int_0^1 \frac{1}{1+x} dx = [\tan^{-1} x]_0^1 = \frac{\pi}{4},$$

we get $\frac{\pi}{4} = 1 - \frac{1}{3} + \frac{1}{5} - \frac{1}{7} + \ldots.$ Hence the result (iii).

Example 40:

Prove that the function $\frac{\sin x}{x}$ *is not Lebesgue integrable over [0, ∞[.*

(Banaras, 1991)

Solution:

To show that the function $\frac{\sin x}{x}$ is not L-integrable. If we show that $\frac{\sin x}{x}$ is not summable over [0, ∞[, we can conclude the result.

For this we must show that $\int_0^{\infty} \frac{|\sin x|}{x} = \infty.$

Consider the integral

$$\int_0^{n\pi} \frac{|\sin x|}{x}\,dx = \sum_{r=1}^{n} \int_{(r-1)\pi}^{n\pi} \frac{|\sin x|}{x}\,dx$$

$$= \sum_{r=1}^{n} \int_0^{\pi} \frac{|\sin \{y+(r-1)\pi|}{y+(r-1)\pi}\,dy$$

$$\geq \sum_{r=1}^{n} \int_0^{\pi} \frac{|\sin \{y+(r-1)\pi|}{r\pi}\,dy$$

$$= \sum_{r=1}^{n} \frac{1}{r\pi} \int_0^{\pi} |\sin y|\,dy = \sum_{r=1}^{n} \frac{2}{r\pi}.$$

$$\lim_{n\to\infty} \int_0^{n\pi} \frac{|\sin x|}{x}\,dx \geq \frac{2}{\pi} \sum_{r=1}^{n} \frac{1}{r} = \frac{2}{\pi} \times \infty = \infty$$

i.e., $$\int_0^{\infty} \frac{|\sin x|}{x}\,dx = \infty.$$

Example 41:

Give an example of a function which is not integrable in the sense of Lebesgue. **(Kanpur, 1994, 95)**

Solution:

Do your self.

Example 42:

Prove that the function

$$f(x) = \frac{d}{dx}\left(x^2 \sin \frac{1}{x^2}\right) = 2x \sin \frac{1}{x^2} - \frac{2}{x} \cos \frac{1}{x^2}$$

is not L-integrable over (0, 1).

Solution:

Let $f(x) = 2x \sin \frac{1}{x^2} - \frac{2}{x} \cos \frac{1}{x^2}.$

The function $2x \sin \frac{1}{x^2}$ is bounded and continuous over [0, 1]. Continuity of $2x \sin \frac{1}{x^2}$ implies that it is measurable over [0, 1]. Being bounded and measurable over [0, 1], 2x in $\frac{1}{x^2}$ is L-integrable over [0, 1]. To show that f(x) is not L-integrable over [0, 1], it suffices to show that $-\frac{2}{x}\cos\frac{1}{x^2}$ is not summable over [0, 1[.

For this we must show that

$$\int_0^1 \frac{1}{x}\left|\cos\frac{1}{x^2}\right| dx = \infty.$$

Let $a_n = \{(2n + \frac{1}{3})\pi\}^{-1/2}$, $b_n = \{(2n - \frac{1}{3})\pi\}^{-1/2}$,

$$a_n \le x \le b_n \Rightarrow \frac{1}{a_n} \ge \frac{1}{x} \ge \frac{1}{b_n} \Rightarrow \frac{1}{b_n^2} \le \frac{1}{x^2} \le \frac{1}{a_n^2}$$

$$\Rightarrow (2n - \tfrac{1}{3})\pi \le \frac{1}{x^2} \le (2n + \tfrac{1}{3})\pi$$

$$\Rightarrow \left|\cos\frac{1}{x^2}\right| \ge \frac{1}{2}$$

$$\int_0^1 \frac{1}{x}\left|\cos\frac{1}{x^2}\right| dx = \sum_{n=1}^{\infty}\int_{a_n}^{b_n} \frac{1}{x}\left|\cos\frac{1}{x^2}\right| dx$$

$$\ge \sum_{n=1}^{\infty}\int_{a_n}^{b_n} \frac{1}{2x}\, dx = \sum_{n=1}^{\infty}\frac{1}{4}\log\frac{b_n^2}{a_n^2} = \frac{1}{4}\sum_{n=1}^{\infty}\log\frac{6n+1}{6n-1} = \infty$$

$$\therefore \quad \int_0^1 \frac{1}{x}\left|\cos\frac{1}{x^2}\right| dx = \infty$$

This proves the required result.

Example 43:

Answer the following questions in 'Yes' or 'No',

(i) Do the Lebesgue integrable functions over (0, 1) from a vector space?

(ii) Is a function equal to a measurable function almost everywhere itself measurable?

(iii) Is the Lebesgue measure of the union of two measurable sets equal to the sum of their measures?

(iv) Is a function, which is differentiable over (0, 1), Lebesgue integrable over (0, 1)?

(v) Is a continuous function over (0, 1) Lebesgue integrable?

Solution:

(i) Yes (ii) Yes (iii) No

(iv) Yes (v) Yes

Example 44:

Show that if

$$f(x) = \frac{1}{x} \ (0 < x \leq 1),\ f(0) = 19,$$

then f is not integrable on [0, 1].

Solution:

Let n be a natural number. Evidently f is an unbounded function. Recall that

$$[f(x)]_n = \begin{cases} f(x) \text{ if } 0 \leq f(x) \leq n \\ n \text{ if } f(x) > n. \end{cases}$$

Here we have

$$[f(x)]_n = \frac{1}{x} \text{if } \frac{1}{n} \leq x \leq 1\ [f(x) \leq n \Rightarrow \frac{1}{x} \leq n \Rightarrow \frac{1}{n} \leq x$$

$$[f(x)]_n = n \text{ if } 0 < x < \frac{1}{n}$$

$$[f(0)]_n = 19.$$

The value of $[f(x)]_n$ at x = 0 does not affect our calculations.

$$\int_0^1 [f]_n\, dx = \int_0^{1/n} n\, dx + \int_{1/n}^1 \frac{1}{n} dx$$

$$= [\log x]_{1/n}^1 + [nx]_0^{1/n}$$

$$= -\log \frac{1}{n} + 1 = 1 + \log n$$

$$\lim_{n\to\infty} \int_0^1 [f]_n dx = 1 + \lim_{n\to\infty} \log n.$$

Since the value of R.H.S. does not exist, hence f is not Lebesgue integrable over [0, 1].

Example 45:

Show that the function f defined by

$$f(x) = \frac{1}{x^{1/3}}, \; 0 < x \le 1.$$

and f(0) = 0

is Lebesgue integrable over [0, 1].

Solution:

Let n be a real number. Recall that

$$[f(x)]_n = \begin{cases} f(x) & \text{if } 0 \le f(x) \le n \\ n & \text{if } f(x) > n \end{cases}$$

Here we have

$$[f(x)]_n = \frac{1}{x^{1/2}}, \; \frac{1}{n^3} \le x \le 1$$

$$\left[f(x) \le n \Rightarrow \frac{1}{x^{1/3}} \le n \Rightarrow \frac{1}{n^3} \le x \right]$$

$$[f(x)]_n = n \text{ if } 0 < x < \frac{1}{n^3}$$

$$[f(0)]_n = 0$$

$$\int_0^1 [f]_n \, dx = \int_{1/n^3}^1 \frac{1}{x^{1/3}} \, dx + \int_0^{1/n^3} n \, dx$$

$$= \frac{3}{2}\left(x^{2/3}\right)_{1/n^3}^1 + n\,(x)_0^{1/n^3}$$

$$= \frac{3}{2}\left(1 - \frac{1}{n^2}\right) + \frac{1}{n^2} = \frac{3}{2} - \frac{1}{n^2}$$

$$\therefore \quad \lim_{n \to \infty} \int_0^1 [f]_n \, dx = \frac{3}{2}.$$

This $\Rightarrow$ f is Lebesgue integrable over [0, 1].

Hence Proved.

7

FUNCTION OF TWO AND THREE VARIABLE

7.1 Explicit and Implicit Functions

If we consider a set of n independent variables x, y, z, ..., t and one dependent variable u, the equation

$$u = f(x, y, z, ..., t) \qquad ...(1)$$

denotes the functional relation. In this case if x_1, y_1, z_1, ..., t_1, are the n arbitrarily assigned values of the independent variables, the corresponding values of the dependent variable u are determined by the function relation.

The function represented by equation (1) is an *explicit* function most of the functions of more than one variable are *implicit* functions, that is to say we are given a functional relation

$$\phi(x, y, z, ..., t) = 0 \qquad ...(2)$$

7.2 An Explicit Function of Two Variables

If x, y are two independent variables and a variable z depends for its values on the values of x, y by a functional relation

$$z = f(x, y) \qquad ...(3)$$

then we say z is a *function of* x, y. The ordered pair of numbers (x, y) is called a *point* and the aggregate of the pairs of numbers (x, y) is said to be the *domain* (or region) *of definition* of the function.

When the domain of definition is bounded by a closed curve C, it is said to be *closed if* f is defined for all points within and on the curve C; but *open or unclosed* when the function is defined for points within but not on the curve C.

7.3 The Neighbourhood of a Point

The set of values x_1, y_1 other than a, b that satisfy the conditions

$$|x_1 - a| < \delta, \; |y_1 - b| < \delta$$

where δ is an arbitrarily small positive number, is said to form a *neighbourhood* of the point (a, b). Thus a neighbourhood is the square

$$(a - \delta, a + \delta; b - \delta, b + \delta)$$

where x takes any value from a – δ to a + δ except a, and y from b – δ to b + δ except b.

7.4 Limit Point

A point (ξ, η) is called a *limit point* or *a point of condensation* of a set of points S, if every neighbourhood of (ξ, η) contains an infinite number of points of S.

7.5 The Limit of a Function

A function f is said to tend to a limit *l* as a point (x, y) tends to the point (a, b) if to every arbitrarily small positive number ε, there corresponds a positive number δ, such that

$$|f(x, y) - l| < \varepsilon,$$

for every point (x, y), [different from (a, b)] which satisfies

$$|x - a| < \delta, \; |y - b| < \delta.$$

In other words, a *function f tends to a limit l, when (x, y) tends to (a, b) if to every positive number ε, there corresponds a neighbourhood N of (a, b) such that*

$$|f(x, y) - l| < \varepsilon,$$

for every point (x, y) other than (a, b) of the neighbourhood N.

Symbolically we write

$$\lim_{(x,y)\to(a,b)} f(x, y) = l,$$

l is the limit (the *double limit* or the *simultaneous limit*) of f when x, y tend to a, b simultaneously.

7.6 Non-existence of Limit

We know that if $\lim\limits_{(x,y)\to(a,b)} f(x, y) = l$ and if $y = \phi(x)$ is any function such that $\phi(x) \to b$, when $x \to a$, then $\lim\limits_{x\to a} f(x,\phi(x))$, must exist and be equal to l.

Thus, if we can find two functions $\phi_1(x)$, $\phi_2(x)$ such that the limits of $\phi(x, \phi_1(x))$ and $f(x, \phi_2(x))$ are different, then the simultaneous limit in question does not exist.

Example 1:

Let

$$f(x, y) = \begin{cases} \dfrac{x^2y}{x^4+y^2}, & \text{if } x^4 + y^4 \neq 0 \\ 0 \quad , & \text{if } x = y = 0 \end{cases}$$

Solution:

If we approach the origin along any axis, $f(x, y) = 0$.

If we approach (0, 0) along any line $y = mx$, then

$$f(x, y) = f(x, mx) = \frac{mx^3}{x^4 + m^2x^2}$$

$$= \frac{mx^3}{x^4 + m^2x^2} \to 0, \quad \text{as } x \to 0$$

So any straight line approach gives,

$$\lim_{(x,y)\to(a,b)} f(x, y) = 0$$

But putting $\quad y = mx^2$,

$$\lim_{(x,y)\to(0,0)} f(x, y) = \lim_{x\to 0} f(x, mx^2) = \frac{m}{1+m^2}$$

which is different for the different m selected.

Hence, $\lim\limits_{(x,y)\to(0,0)} f(x, y)$ does not exist.

Thus, the function possesses no limit at the origin, but a straight line approach gives the limit zero.

Example 2:

Show that

$$\lim_{(x,y)\to(0,0)} \frac{2xy^2}{x^2+y^4} \text{ does not exist.}$$

Solution:

If we put $x = my^2$ and let $y \to 0$, we get

$$\lim_{y\to 0} \frac{2my^4}{(m^2+1)y^4} = \frac{2m}{1+m^2}$$

which is different for different values of m.

Hence the limit does not exist.

Example 3:

Show that

$$\lim_{(x,y)\to(0,0)} xy\frac{x^2-y^2}{x^2+y^2} = 0.$$

Solution:

Put $x = r\cos\theta$, $y = r\sin\theta$

$$\left|xy\frac{x^2-y^2}{x^2+y^2}\right| = |r^2 \sin\theta\cos\theta\cos 2\theta|$$

if

$$\frac{x^2}{4} < \frac{\varepsilon}{2}, \frac{y^2}{4} < \frac{\varepsilon}{2}$$

$$\text{or if } = \left|\frac{r^2}{4}\sin 4\theta\right| \le \frac{r^2}{4} = \frac{x^2+y^2}{4}\ \varepsilon$$

$$|x| < \sqrt{2\varepsilon} = \delta, |y| < \sqrt{2\varepsilon} = \delta$$

Thus, for $\varepsilon > 0$, $\exists\ \delta > 0$ such that

$$\left|xy\frac{x^2-y^2}{x^2+y^2} - 0\right| < \varepsilon, \text{ when } |x| < \delta, |y| < \delta$$

$$\Rightarrow \lim_{(x,y)\to(0,0)} xy\frac{x^2-y^2}{x^2+y^2} = 0$$

Example 4:

Show that

$$\lim_{(x,y)\to(0,0)} \frac{\sqrt{x^2y^2+1}-1}{x^2+y^2} = 0.$$

Solution:

Since x, y are small

$$\frac{\sqrt{x^2y^2+1}-1}{x^2+y^2} = \frac{(1+x^2y^2)^{1/2}-1}{x^2+y^2} \simeq \frac{\frac{1}{2}x^2y^2}{x^2+y^2}.$$

Now changing to polars, we can show, as in the above example, that

$$\lim_{(x,y)\to(0,0)} \frac{\frac{1}{2}x^2y^2}{x^2+y^2} = 0$$

Hence the required result.

Example 5:

Show that for the function f defined by setting

$$f(x, y) = \frac{xy}{x^2+y^2} \; (x^2+y^2 \neq 0),\; f(0, 0) = 0$$

$\lim\limits_{x\to 0} \lim\limits_{y\to 0} f(x, y) = \lim\limits_{y\to 0} \lim\limits_{x\to 0} f(x, y) = 0$, but $\lim\limits_{(x, y)\to(0, 0)} f(x, y)$ *does not exist.*

Solution:

$$\lim_{x\to 0} \lim_{y\to 0} \frac{xy}{x^2+y^2} = \lim_{x\to 0} (0) = 0$$

$$\lim_{y\to 0} \lim_{x\to 0} \frac{xy}{x^2+y^2} = \lim_{y\to 0} (0) = 0$$

Suppose $(x, y) \to (0, 0)$ along the path $y = mx$. Then

$$f(x, mx) = \frac{mx^2}{m^2+m^2x^2}.$$

$$\therefore \lim_{x\to 0} f(x, mx) = \frac{m}{1+m^2},$$ which depends on m.

Thus different limits are obtained as we consider the path y = mx for different values of *m*

$$\therefore \lim_{(x,y)\to(0,0)} \frac{xy}{x^2+y^2} \text{ does not exist.}$$

Example 6:

Show that for the function f defined by setting

$f(x, y) = |x| + |y| \quad \forall (x, y) \in R^2,$

$\lim_{(x,y)\to(0,0)} f(x, y), \lim_{x\to 0}\lim_{y\to 0} f(x, y), \lim_{y\to 0}\lim_{x\to 0} f(x,y)$ *all exist and are equal*

Solution:

(i) Let $\varepsilon > 0$ be arbitrary. Let $\delta = \frac{1}{2}\varepsilon$. Then

$|f(x, y) - 0| = |x| + |y| < \varepsilon$ if $|x| < \delta, |y| < \delta$

$\therefore |f(x, y) - 0| < \varepsilon$ if $|x - 0| < \delta, |y - 0| < \delta.$

Hence $\lim_{(x,y)\to(0,0)} f(x, y) = 0$

(ii) $\lim_{x\to 0}\lim_{y\to 0}(|x| + |y|) = \lim_{x\to 0} |x| = 0$

(iii) $\lim_{y\to 0}\lim_{x\to 0}(|x| + |y|) = \lim_{y\to 0} |y| = 0$

Example 7:

Show that for the function f defined by setting

$$f(x, y) = \begin{cases} x \sin(1/y) + y \sin(1/x), & \text{if } xy \neq 0 \\ 0, & \text{if } xy = 0 \end{cases}$$

the two repeated limits do not exist, but $\lim_{(x,y)\to(0,0)} f(x, y)$ exists.

Solution:

(i) $|f(x, y) - 0| = |x \sin(1/y) + y \sin(1/x)|$

$\leq |x| \, |\sin(1/y)| + |y| \, |\sin(1/x)|$

$\leq |x| .1 + |y| .1 \quad$ if $xy \neq 0.$

Let $\varepsilon > 0$ be given. Choose $\delta = \frac{1}{2}\varepsilon$. Then

$| f(x, y) - 0 \leq | x | + | y | < \varepsilon$ if $| x | < \delta, | y | < \delta$

$$\therefore \lim_{(x, y)\to(0, 0)} f(x, y) = 0$$

(ii) Since $\lim_{y\to 0} \sin(1/y)$ does not exist and $\lim_{y\to 0} y \sin(1/x) = 0$,

$\therefore \lim_{x\to 0} \lim_{y\to 0} f(x, y)$ does not exist.

(iii) Since $\lim_{x\to 0} \sin(1/x)$ does not exist and $\lim_{x\to 0} x \sin(1/y) = 0$,

$\therefore \lim_{y\to 0} \lim_{x\to 0} f(x, y)$ does not exist.

Example 8:

Show that for the function f defined by setting

$$f(x, y) = \frac{x^2 - y^2}{x^2 + y^2}, \quad x^2 + y^2 \neq 0,$$

$\lim_{x\to 0} \lim_{y\to 0} f(x, y)$ and $\lim_{y\to 0} \lim_{x\to 0} f(x, y)$ both exist but are unequal. Also show that $\lim_{(x, y)\to(0, 0)} f(x, y)$ does not exist.

Solution:

$$\lim_{x\to 0} \frac{x^2 - y^2}{x^2 + y^2} = -1, \quad \lim_{y\to 0} \frac{x^2 - y^2}{x^2 + y^2} = 1.$$

$$\therefore \lim_{y\to 0} \lim_{x\to 0} \frac{x^2 - y^2}{x^2 + y^2} = -1, \quad \lim_{x\to 0} \lim_{y\to 0} \frac{x^2 - y^2}{x^2 + y^2} = 1.$$

Thus the two repeated limits exist but are unequal.

Suppose $(x, y) \to (0, 0)$ along the path $y = mx$.

Now $f(x, mx) = \dfrac{x^2 - m^2x^2}{x^2 + m^2y^2}$

$$\Rightarrow \quad \lim_{x\to 0} f(x, mx) = \frac{1 - m^2}{1 + m^2}$$

which is different for different values of m.

Hence $\lim\limits_{(x, y) \to (0, 0)} = \dfrac{x^2 - y^2}{x^2 + y^3}$ does not exist.

7.7 Continuity of a Function at a Point

Definition:

Let $f : D \to R$, where *D* is an open subset of R²". Then f is said to be continuous at a point $(a, b) \in D$ if

$$\lim_{(x, y) \to (a, b)} f(x, y) = f(a, b)$$

In other words, f is *continuous* at a point $(a, b) \in D$ if given any $\varepsilon > 0$, there exists some $\delta > 0$ such that

$$| f(x, y) - f(a, b) | < \varepsilon, \text{ when } \sqrt{[(x-a)^2 + (y-b)^2]} < \delta.$$

OR

$$| f(x, y) - f(a, b) | < \varepsilon, \text{ when } | x - a | < \delta, | y - b | < \delta.$$

OR

Note : The definition of continuity of a function f at a point (a, b) requires that besides (a, b). f is defined in a certain neighbourhood of (a, b) and moreover the limit of f when $(x, y) \to (a, b)$ exists and equals the value f(a, b).

A function which is not continuous at a point is said to be *discontinuous*

Remark:

A point to be particularly noticed is that if a function of more than one variable is continuous at a point, it is continuous at that point when considered as a function of a single variable. To be more specific if a function f of two variables x, y is continuous at (a, b) then f(x, b) is a continuous function of x at x = a and f(a, y) that of y at y = b.

The converse however is not true, i.e., a function may be a continuous function of one variable when the others remain constant and yet not be a continuous function of all the variables.

For instance, consider a function f, where

$$f(x, y) = \begin{cases} \dfrac{2xy}{x^2 + y}, & (x, y) \neq (0, 0) \\ 0, & \text{at } (0, 0) \end{cases}$$

The function is not continuous at (0, 0) for $\lim\limits_{(x, y)\to(0, 0)} f(x, y)$ does not exist. But

$$\lim_{x\to 0} f(x, 0) = 0 = f(0, 0), \text{ and } \lim_{y\to 0} f(0, y) = 0 = f(0, 0)$$

so that f is continuous at (0, 0), when considered as a function of a single variable x or that of y.

A function is said to be continuous in a region if it is continuous at every point of the same.

As in limits, it can be easily proved that the sum, difference, product and quotient (provided the denominator does not vanish) of two continuous functions are also continuous.

The theorems on continuity for functions of a single variable can be easily extended to functions of several variables; the proofs for some of them, except for verbal changes, are the same while for others the method is not quite the same. However, within the scope of the present work, it is not possible to discuss all of them here.

Example 9:

Investigate the continuity at (0, 0) of

$$f(x, y) = \begin{cases} \dfrac{x^2 - y^2}{x^2 + y^2}, & (x, y) \neq (0, 0) \\ 0, & (x, y) = (0, 0) \end{cases}$$

Solution:

Since $\lim\limits_{(x, y)\to(0, 0)} f(x, y)$ does not exist, therefore the function is not continuous at (0, 0).

Example 10:

Investigate for continuity at (1, 2)

$$f(x, y) = \begin{cases} x^2 + 2y, & (x, y) \neq (1, 2) \\ 0, & (x, y) = (1, 2) \cdot \end{cases}$$

Solution:

Here $\lim\limits_{(x, y)\to(1, 2)} f(x, y) = 5 \neq f(1, 2).$

Hence, the function is not continuous at (1, 2).

The point (1, 2) is a *point of discontinuity* of the function.

However, if the function had the value 5 at (1, 2), it was then continuous at the point.

Remark:

If, as in the above example, it is possible to so redefine the value of the function at a point of discontinuity that the new function is continuous, we say that the point is a *removable discontinuity* of the original function.

Example 11:

Show that the function

$$f(x, y) = \begin{cases} \dfrac{xy}{\sqrt{x^2 + y^2}}, & (x, y) \neq (0, 0) \\ 0, & (x, y) = (0, 0) \end{cases}$$

is continuous at the origin.

Solution:

Let $x = r \cos \theta$, $y = r \sin \theta$.

$$\therefore \quad \left| \frac{xy}{\sqrt{x^2 + y^2}} \right| = r \mid \cos \theta \sin \theta \mid \leq r = \sqrt{x^2 + y^2} < \varepsilon, \text{ if}$$

$$x^2 < \frac{\varepsilon^2}{2}, \quad y^2 < \frac{\varepsilon^2}{2}$$

or, if

$$| x | < \frac{\varepsilon}{\sqrt{2}}, \quad | y | < \frac{\varepsilon}{\sqrt{2}},$$

Thus,

$$\left| \frac{xy}{\sqrt{x^2 + y^2}} - 0 \right| < \varepsilon, \text{ when } | x | < \frac{\varepsilon}{\sqrt{2}}, | y | < \frac{\varepsilon}{\sqrt{2}}$$

$$\Rightarrow \lim_{(x, y) \to (0, 0)} \frac{xy}{\sqrt{x^2 + y^2}} = 0$$

$$\therefore \lim_{(x, y) \to (0, 0)} f(x, y) = f(0, 0)$$

Hence, f is continuous at (0, 0).

Example 12:

Show that the function f defined by

$$f(x, y) = xy\left(\frac{x^2 - y^2}{x^2 + y^2}\right), \text{ if } (x, y) \neq (0, 0) \text{ and } f(0, 0) = 0 \text{ is continuous}$$

at (0, 0)

Solution:

For $(x, y) \neq (0, 0)$, we have

$$|f(x, y) - f(0, 0)| = \left|xy\left(\frac{x^2 - y^2}{x^2 + y^2}\right) - 0\right|$$

$$= |xy|\left|\frac{x^2 - y^2}{x^2 + y^2}\right|$$

$$\leq |xy| .1 \leq \frac{1}{2}(x^2 + y^2)$$

Let $\varepsilon > 0$ be given. Then

$$|f(x, y) - f(0, 0)| < \varepsilon, \text{ if } \sqrt{x^2 + y^2} < \delta = \sqrt{2\varepsilon}.$$

Hence, the given function is continuous at (0, 0)

Example 13:

Show that the function: $f(x, y) = |x| + |y|$ *is continuous at the origin.*

Solution:

$$\lim_{(x, y)\to(0, 0)} f(x, y) = 0. \text{ Also } f(0, 0) = 0 + 0 = 0.$$

$$\therefore \lim_{(x, y)\to(0, 0)} f(x, y) = f(0, 0). \text{ Hence f is continuous at } (0, 0)$$

Example 14:

Show that the function f defined as

$$f(x, y) = \frac{xy}{\sqrt{x^2 + y^2}}, \text{ when } (x, y) \neq (0, 0)$$

$$= 0, \qquad \text{when } (x, y) = (0, 0)$$

is continuous of the origin. **(D.U., Physics (H) 1996)**

Solution:

Putting $x = r\cos\theta$, $y = r\sin\theta$; we see that

$$|f(x, y) - f(0, 0)| = \left|\frac{r\cos\theta .\, r\sin\theta}{r} = 0\right|$$

$$= r\,|\cos\theta|\,|\sin\theta| \le r, \text{ 1.1.} = r$$

$\therefore |f(x, y) - f(0, 0| \le \sqrt{x^2 + y^2}$

Let $\varepsilon > 0$ be given. Choose $\delta = \varepsilon$. Then

or $|f(x, y) - f(0, 0)| < \varepsilon$ if $\sqrt{x^2 + y^2} < \varepsilon$

or $|f(x, y) - f(0, 0)| < \varepsilon$ if $\sqrt{[(x - 0)^2 + (y - 0)^2]} < \delta$.

Hence the given function is continuous at (0, 0)

Example 15:

Show that the function f defined by setting

$$f(x, y) = \frac{x^3 y}{x^6 + y^2}, \text{ when } (x, v) \ne (0, 0) \text{ and } f(0, 0) = 0,$$

is not continuous of the origin.

Solution:

Do your self.

Example 16:

Let $f : R^2 \to R$ *be defined by setting*

$$f(x, y) = \frac{x^2 y}{x^4 + y^2}, \text{ if } (x, y) \ne (0, 0) \text{ and } f(0, 0) = 0.$$

Show that f is not continuous at the origin.

Solution:

Suppose $(x, y) \to (0, 0)$ along the path $y = mx$. Then

$$\lim_{x\to 0} f(x, mx) = \lim_{x\to 0} \frac{mx^3}{x^4 + m^2x^2} = \lim_{x\to 0} \frac{mx}{x^2 + m^2} = 0,$$

for all values of m.

Suppose now $(x, y) \to (0, 0)$ along the curve $y = x^2$. Then

$$\lim_{x\to 0} f(x, x^2) = \lim_{x\to 0} \frac{x^4}{x^4 + x^4} = \frac{1}{2}.$$

Thus, f(x, y) tends to different limits as $(x, y) \to (0, 0)$ along different paths and so $\lim_{(x,y)\to(0,0)} f(x, y)$ does not exist. Hence the given function is not continuous at the origin.

Example 17:

Show that the following function is discontinuous at (0, 0) :

$$f(x, y) = \begin{cases} \dfrac{x^3 + y^3}{x - y}, & x \neq y, \\ 0, & x = y. \end{cases}$$

Solution:

Suppose $(x, y) \to (0, 0)$ along the curve $y = x - mx^3$. Then

$$f(x, x - mx^3) = \frac{x^3 + (x - mx^3)^3}{x - (x - mx^3)} = \frac{x^3[1+(1-mx^2)^3]}{mx^3}$$

$$\therefore \lim_{x\to 0} f(x, x - mx^3) = \frac{2}{m},$$

which is different for different values of m.

Thus, $\lim_{(x, y)\to(0,0)} f(x, y)$ does not exist and so the given function is discontinuous at (0, 0).

Example 18:

If f(x, y) = x sin (1/x) + y sin (1/y), when xy ≠ 0 and f(0, 0) = 0; then prove that f is continuous at (0, 0).

Solution:

$$| f(x, y) - f(0, 0) | = | x \sin 1/x + y \sin 1/y - 0 |$$

$$\leq | x | \left|\sin \frac{1}{x}\right| + | y | \left|\sin \frac{1}{y}\right| \leq | x | + | y |.$$

Let $\varepsilon > 0$ be given. Choose $\delta = \frac{1}{2}\varepsilon$. Then

$$| f(x, y) - f(0, 0) | < \varepsilon \text{ if } | x | < \delta, | y | < \delta.$$

Hence, the given function is continuous at (0, 0).

Example 19:

Show that the function

$f(x, y) = xy/(x^2 + y^2)$, $(x, y) \neq (0, 0)$ *and* $f(0, 0) = 0$

is not continuous at (0, 0).

Solution:

Do your self.

[Hint. $\lim_{x \to 0} f(x, mx) = \frac{m}{1+m^2}$, which does not exist.]

Example 20:

Show that the function $f(x, y) = xy^2/(x^2 + y^2)$, *when* $(x, y) \neq (0, 0)$ *and* $f(0, 0)$ *is continuous at the origin.*

Solution:

Putting $x = r \cos \theta$, $y = r \sin \theta$; we obtain

$$| f(x, y) - f(0, 0) | < \left| \frac{r \cos \theta . r^2 \sin^2 \theta}{r^2} - 0 \right|$$

$$= r \mid \cos \theta \mid \mid \sin \theta \mid \mid \sin \theta \mid$$

$$\leq r = \sqrt{x^2 + y^2}.$$

Let $\varepsilon > 0$ be given. Choose $\delta = \varepsilon$. Then

$$| f(x, y) - f(0, 0) | < \varepsilon \text{ if } \sqrt{x^2 + y^2} < \delta = \varepsilon$$

Hence, the given function is continuous at the origin.

Example 21:

Show that the function f defined by

$$f(x, y) = \begin{cases} \dfrac{x^3 - y^3}{x^2 + y^2}, & (x, y) \neq (0, 0) \\ 0, & (x, y) = (0, 0) \end{cases}$$

is continuous at (0, 0).

Solution:

Putting $x = r \cos \theta$, $y = r \sin \theta$; we obtain

$$| f(x, y) - f(0, 0) | = r \mid (\cos^3 \theta - \sin^3 \theta) \mid$$
$$\leq r \mid (\cos^3 \theta \mid + \mid \sin^3 \theta \mid)$$
$$\leq 2 = 2 \sqrt{x^2 + y^2}.$$

Let $\varepsilon > 0$ be given and $\delta = \frac{1}{2} \varepsilon$. Then

$| f(x, y) - f(0, 0) < \varepsilon$, when $\sqrt{x^2 + y^2} < \delta$

EXERCISES

1. $f(x, y) = \dfrac{x^4 - y^4}{x^4 + y^4}$, $(x, y) \neq (0, 0)$ and $f(0, 0) = 0$.

   [**Ans.** Discontinuous]

2. $f(x, y) = \dfrac{x^2 y}{x^3 + y^3}$, $(x, y) \neq (0, 0)$ and $f(0, 0) = 0$.

   [**Ans.** Discontinuous]

   [**Hint.** $\lim_{x \to 0} f(x, mx) = \dfrac{m}{1 + m^3}$, which does not exist.]

3. Discuss the continue of the following function at (0, 0):

   $f(x, y) = \dfrac{x^2 - y^2}{x^2 + y^2}$, $(x, y) \neq (0, 0)$ and $f(0, 0) = 0$.

   [**Ans.** Discontinuous]

   [**Hint.** $\lim_{(x,y)0,0} (x, y)$ does not exist.]

4. $f(x, y) = \dfrac{x^2 y^2}{x^2 + y^2}$, $(x, y) \neq (0, 0)$ and $f(0, 0) = 0$.

   [**Ans.** Continuous]

   Hint. Put $x = r \cos \theta$, $y = r \sin \theta$.

5. If $f(x, y) = \dfrac{x^3 y^3}{x^2 + y^2}$, $(x, y) \neq (0, 0)$ and $f(0, 0) = 0$, then show that f is continuous at (0, 0).

7.8 Partial Derivatives

Let $f : D \to R$, where D is an open subset of R^2, and (a, b) be any point of D.

The **partial derivative of f w.r.t. x** at the point (a, b), denoted as f_x (a, b), is defined as

$$f_x(a, b) = \lim_{h \to 0} \frac{f(a+h, b) - f(a, b)}{h},$$

provided the limit exists.

The **partial derivative of f w.r.t. y** at the point (a, b), denoted as f_y(a, b), is defined as

$$f_y(a, b) = \lim_{k \to 0} \frac{f(a_1 + b + k) - f(a, b)}{k},$$

provided the limit exists.

f_x (a, b), f_y (a, b) are called the **first order partial derivatives** of f at (a, b).

Now, we shall define the **second order partial derivatives** of f at (a, b) viz.

f_{xx}(a, b), f_{yx}(a, b), f_{xy}(a, b), f_{yy}(a, b).

These are defined as

$$f_{xx}(a, b) = \lim_{h \to 0} \frac{f_x(a+h, b) - f_x(a, b)}{h}, \text{ provided the limit exists.}$$

$$f_{yx}(a, b) = \lim_{k \to 0} \frac{f_x(a, b+k) - f_x(a, b)}{k}, \text{ provided the limit exists}$$

$$f_{xy}(a, b) = \lim_{h \to 0} \frac{f_y(a+h, b) - f_y(a, b)}{h}, \text{ provided the limit exists.}$$

$$f_{yy}(a, b) = \lim_{k \to 0} \frac{f_y(a, b+k) - f_y(a, b)}{k}, \text{ provided the limit exists.}$$

[Here we assume that f_x, f_y exist in some *nbd* of (a, b)]

The partial derivatives f_{xy}(a, b) and f_{yx}(a, b) are called the **second order mixed partial derivatives** of f at (a, b). In general

$f_{xy}(a, b) \neq f_{yx}(a, b)$.

Similarly we can define higher order partial derivatives.

7.9 A Mean Value Theorem

If f_x exists throughout a neighbourhood of a point (a, b) and f_x(a, b) exists then for any point (a + h, b + k) of this neighbourhood,

$$f(a + h, b + k) - f(a, b) = hf_x(a + \theta h, b + k) + k[f_y(a, b) + \eta]$$

where $0 < \theta < 1$, and η *is a function of k, tending to zero with k.*

Now,

$$f(a + h, b + k) - f(a, b) = f(a + h, b + k) - f(a, b + k) + f(a, b + k) - f(a, b) \quad ...(1)$$

Since f_x exists in a neighbourhood of (a, b), therefore by Lagrange's mean value theorem,

$$f(a + h, b + k) - f(a, b + k) = hf_x(a + \theta h, b + k), \ 0 < \theta < 1 \quad ...(2)$$

Also $f_y(a, b)$ exist, so that

$$\lim_{k \to 0} \frac{f(a, b+k) - f(a, b)}{k} = f_y(a, b)$$

$$\Rightarrow \quad f(a, b + k) - f(a, b) = k[f_y(a, b) + \eta]. \quad ...(3)$$

where η is a function of k and tends to zero as $k \to 0$.

From (1), (2) and (3) we get the required result.

7.10 A Sufficient Condition for Continuity

A sufficient condition that a function f be continuous at (a, b) is that one of the partial derivatives exists and is bounded in a neighbourhood of (a, b) and that the other exists at (a, b).

Let f_x exist and be bounded in a neighbourhood of (a, b) and let $f_y(a, b)$ exist; then for any point (a + h, b + k) of this neighbourhood, we have

$$f(a + h, b + k) - f(a, b) = hf_x(a + \theta h, b + k) + k[f_y(a, b) + \eta]$$

where $0 < \theta < 1$, and $\eta \to 0$ as $k \to 0$.

Proceeding to limits as $(\eta, k) \to (0, 0)$, since $f_x(a + \theta h, b + k)$ is bounded, we have

$$\lim_{(h,k) \to (0,0)} f(a + h, b + k) = f(a, b)$$

$\Rightarrow$ f is continuous at (a, b).

Note: A sufficient condition that a function be continuous in a *closed region* is that both the partial derivatives exist and are bounded throughout the region.

Example 22:

Let $f(x, y) = \dfrac{x^3y}{x^2 + y^2}$, $x^2 + y^2 \neq 0$, $f(0, 0) = 0$.

Prove or disprove $f_{xy}(0, 0) = f_{yx}(0, 0)$ **(D.U., 1993)**

Solution:

$$f_y(0, 0) = \lim_{k\to 0} [f(0, 0 + k) - f(0, 0)]/k$$

$$= \lim_{k\to 0} \frac{0-0}{k} = 0,$$

$$f_y(h, 0) = \lim_{k\to 0} [f(h, 0 + k) - f(h, 0)]/k$$

$$= \lim_{k\to 0} \frac{h^3}{h^2 + k^2} = h,$$

$$\therefore \quad f_{xy}(0, 0) = \lim_{h\to 0} [f_y(0 + h, 0) - f_y(0, 0)]/h$$

$$= \lim_{h\to 0} \frac{h}{h} = 1.$$

Again $$f_x(0, 0) = \lim_{h\to 0} [f(0 + h, 0) - f(0, 0)]/h$$

$$= \lim_{h\to 0} \frac{0-0}{h} = 0,$$

$$f_x(0, k) = \lim_{h\to 0} [f(0 + h, k) = f(0, k)]/h$$

$$= \lim_{h\to 0} \frac{h^2 k}{h^2 + k^2} = 0,$$

$$\therefore \quad f_{yx}(0, 0) = \lim_{k\to 0} [f_x(0, 0 + k) - f_x(0, 0)]/k$$

$$= \lim_{k\to 0} \frac{0 - 0}{k} = 0.$$

Hence $f_{xy}(0, 0) \neq f_{yx}(0, 0)$.

Example 23:

Show that for the function

$$f(x, y) = \begin{cases} \dfrac{x^2 y^2}{x^2 + y^2}, & (x, y) \neq (0, 0) \\ 0, & (x, y) = (0, 0) \end{cases}$$

$f_{xy}(0, 0) = f_{yx}(0, 0)$. **(D.U., 1989, 82)**

Solution:

$$f_y(0, 0) = \lim_{k\to 0} [f(0, 0 + k) - f(0, 0)]/k$$

$$= \lim_{k\to 0} \frac{0-0}{k} = 0,$$

$$f_y(h, 0) = \lim_{k\to 0} [f(h, 0 + k) - f(h, 0)]/k$$

$$= \lim_{k\to 0} \frac{h^2k}{(h^2+k^2)} = 0,$$

$$\therefore \quad f_{xy}(0, 0) = \lim_{h\to 0} [f_y(0 + h, 0) - f_y(0, 0)]/h$$

$$= \lim_{h\to 0} \frac{0-0}{h} = 0.$$

Again $\quad f_x(0, 0) = \lim_{h\to 0} [f(0 + h, 0) - f(0, 0)]/h$

$$= \lim_{h\to 0} \frac{0-0}{h} = 0.$$

$$f_x(0, k) = \lim_{h\to 0} [f(0 + h, k) - f(0, k)]/h$$

$$= \lim_{h\to 0} \frac{hk^2}{(h^2+k^2)} = 0,$$

$$\therefore \quad f_{yx}(0, 0) = \lim_{k\to 0} [f_x(0, 0 + k) - f_x(0, 0)]/k$$

$$= \lim_{k\to 0} \frac{0-0}{k} = 0.$$

Hence $f_{xy}(0, 0) = f_{yx}(0, 0)$.

7.11 Differentiability at a Point

Definition:

Let $f : D \to R$ be a real-valued function defined on an open set $D \subset R^2$. Then f is said to be *differentiable at a point (a, b)* $\in$ D, if

$$f(a + h, b + k) - f(a, b) = Ah + Bk + \sqrt{h^2+k^2}\, g(h, k),$$

where A and B are real numbers independent of h and k, and g is a real-valued function such that $\lim_{(h,k)\to(0,0)} g(h, k) = 0$.

Equivalently, f is differentiable at (a, b), if

$$f(a + h, b + k) - f(a, b) = Ah + Bk + h\phi(h, k) + k\phi(h, k),$$

where A and B are real numbers independent of h and K and $\phi(h, k)$, $\phi(h, k) \to 0$ as $(h, k) \to (0, 0)$.

Theorem 1:

If a function $f: D \to R$ is differentiable at $(a, b) \in D$, then f has partial derivatives f_x and f_y at (a, b).

Proof:

Since f is differentiable at the point $(a, b) \in D$, so there exist real numbers A and B, and a function g such that

(i) $f(a + h, b + k) - f(a, b) = Ah + Bk + \sqrt{(h^2 + k^2)}g(h, k)$, ...(1)

(ii) A and B are independent of h and k, and

(iii) $\lim\limits_{(h,k)\to(0,0)} g(h, k) = 0$

Putting k = 0 in equation (i) we have

$$f(a + h, b) - f(a, b) = Ah + |h| g(h, 0).$$

Dividing throughout by h, and taking limits as $h \to 0$, then we have

$$\lim_{h\to 0} \frac{f(a+h, b) - f(a, b)}{h} = \lim_{h\to 0}\left[A + \frac{|h|}{h} g(h,0)\right]$$

$$= A + (\pm 1).0 = A,$$

since $g(h, 0) \to 0$ as $h \to 0$.

$\therefore f_x(a, b) = A.$

Putting h = 0 in (1), we have

$$f(a, b + k) - f(a, b) = Bk + |k| g(0, k).$$

Dividing throughout by k, and taking limits as $k \to 0$, we have

$$\lim_{k\to 0} \frac{f(a, b + k) - f(a, b)}{k} = B.$$

$\therefore \quad f_y(a, b) = B.$

Hence f_x and f_y exist at the point (a, b).

Remark 1:

If may be noted that the constants A, B appearing in the definition of differentiability of f are given by:

$$A = f_x(a, b),\ B = f_y(a, b).$$

Remark 2:

If either of f_x, f_y does not exists at (a, b), then f is not differentiable at (a, b).

Remark 3:

The converse of the above example is not true as shown in the following example.

Example 24:

Show that the function $f : R^2 \to R$ defined by setting

$$f(x, y) = \begin{cases} x \sin(1/x) + \sin(1/y), & \text{when } xy \neq 0 \\ x \sin(1/x) & \text{, when } x \neq 0, y = 0 \\ y \sin(1/y) & \text{, when } x = 0, y \neq 0 \\ 0 & \text{, when } x = y = 0 \end{cases}$$

is continuous but not differentiable at (0, 0).

Solution:

f is continuous at (0, 0).

Now $f_x(0, 0) = \lim_{h\to 0} [f(0 + h, 0) - f(0, 0)]/h$

$$= \lim_{h\to 0} \frac{h \sin(1/h) = 0}{h} = \lim_{h\to 0} \sin \frac{1}{h},$$

which does not exist.

Similarly $f_{yx}(0, 0)$ does not exist. Hence f is not differentiable at (0, 0).

Example 25:

Let $f : R^2 \to R$ be defined by setting

$f(x, y) = xy/\sqrt{(x^2 + y^2)}$, when $(x, y) \neq (0, 0)$, $f(0, 0) = 0$.

Show that f_x and f_y exist at (0, 0) but f is not differentiable at (0, 0). Also show that f is continuous at (0, 0). **(D.U., 1994)**

Solution:

$$f_x(0, 0) = \lim_{h\to 0} \frac{f(h, 0) - f(0, 0)}{h} = \lim_{h\to 0} \frac{0-0}{h} = 0.$$

$$f_y(0, 0) = \lim_{k\to 0} \frac{f(0, k) - f(0, 0)}{k} = \lim_{k\to 0} \frac{0-0}{k} = 0.$$

Thus, $f_x(0, 0)$ and $f_y(0, 0)$ both exist and are equal to 0.

Let, if possible, f be differentiable at (0, 0). Then

$$f(h, k) - f(0, 0) = Ah + Bk + \sqrt{(h^2 + k^2)}\, g(h, k), \quad ...(1)$$

where $\lim_{(h,k)\to(0,0)} g(h, k) \to 0$ ...(2)

and $A = f_x(0, 0) = 0, B = f_y(0, 0) = 0.$

From (1), we obtain

$$hk/\sqrt{(h^2 + k^2)} - 0 = 0.h + 0.k + \sqrt{(h^2 + k^2)}\, g(h, k),$$

or $g(h, k) = \dfrac{hk}{h^2 + k^2}.$

Taking k = mh, we see that

$$\lim_{h\to 0} g(h, mh) = \frac{m}{1+m^2}, \text{ which depends on m.}$$

Thus, $\lim_{(h,k)\to(0,0)} g(h, k)$ does not exist, which contradicts (2).

Hence f is not differentiable at (0, 0). f is continuous at (0, 0).

Theorem 2:

If a function $f : D \to R$ is differentiable at a point $(a, b) \in D$, then f is continuous at (a, b).

Proof:

Since f is differentiable at (a, b), so there exist real numbers A and B, and a function g such that

(i) $f(a + h, b + k) - f(a, b) = Ah + Bk + \sqrt{(h^2 + k^2)}\, g(h, k),$ ...(1)

(ii) A and B are independent of h and k, and

(iii) $\lim_{(h,k)\to(0,0)} g(h, k) = 0.$

Since $\lim_{(h,k)\to(0,0)} [Ah + Bk + \sqrt{(h^2 + k^2)}\, g(h, k)] = 0,$

$\therefore \lim_{(h,k)\to(0,0)} [f(a + h, b + k) - f(a, b)] = 0$, by (1)

$\Rightarrow \lim_{(h,k)\to(0,0)} f(a + h, b + k) = f(a, b).$

Hence f is continuous at (a, b).

Remark:

The converse of the above theorem may not be true. A function may be continuous at a point without being differentiable at that point as shown below.

Example 26:

Let $f : R^2 \to R$ be the function defined by setting

$$f(x, y) = |x| + |y|, \text{ for all } (x, y) \in R^2.$$

Show that f is continuous at (0, 0) but is not differentiable at (0, 0).

Solution:

$$|f(x, y) - f(0, 0)| = |x| + |y|.$$

Let $\varepsilon > 0$ be given.

If we take $\delta = \frac{1}{2}\varepsilon$, then

$$|x| < \delta, |y| < \delta \Rightarrow |f(x, y) - f(0, 0)| < \varepsilon.$$

Hence f is continuous at (0, 0).

Now we show that f is not differentiable at (0, 0). We have

$$\lim_{h\to 0} \frac{f(h, 0) - f(0, 0)}{h} = \lim_{h\to 0} \frac{|h|}{h},$$

which does not exist. Therefore $f_x(0, 0)$ does not exist. Similarly $f_y(0, 0)$ does not exist.

Hence f is not differentiable at (0, 0).

Theorem 3:

If a function $f: D \to R$ has continuous partial derivatives f_x and f_y at $(a, b) \in D$, then f is differentiable at (a, b). **(D.U., 1995)**

Proof:

Since f_x and f_y are continuous at (a, b), so f_x and f_y both exist in a nbd. $N \equiv N((a, b), \delta)$ of (a, b). Let (a + h, b + k) be any point of N. We have

$$f(a + h, b + k) - f(a, b) = [f(a + h, b + k) - f(a + h, b)] + [f(a + h, b) - f(a, b)]. \quad ...(1)$$

We define two functions G and H as given below

$$G(y) = f(a + h, y) \quad \forall \quad y \in [b, b + k],$$

and $\quad H(x) = f(x, b) \quad \forall \quad x \in [a, a + h].$

Since f_y exists in N, so the function G satisfies both the conditions of

Lagrange's Mean Value Theorem in [b, b + k]. Thus there exists a θ_1 ($0 < \theta_1 < 1$) such that

$$G(b + k) - G(b) = kG'(b + \theta_1 k)$$

or $f(a + h, b + k) - f(a + h, b) = kf_y(a + h, b + \theta_1 k)$. ...(2)

Similarly, since H satisfies both the conditions of Lagrange's mean value theorem in [a, a + h], there exists θ_2 ($0 < \theta_2 < 1$) such that

$$f(a + h, b) - f(a, b) = h\, f_x(a + \theta_2 h, b). \qquad ...(3)$$

Using (2) and (3) in (1), we have

$$f(a + h, b + k) - f(a, b) = hf_x(a + \theta_2 h, b) + kf_y(a + h, b + \theta_1 k)$$

$$= hf_x(a, b) + kf_y(a, b) + h\,[f_x(a + \theta_2 h, b) - f_x(a, b)]$$
$$+ k\,[f_y(a + h, b + \theta_1 k) - f_y(a, b)]$$

$$= hf_x(a, b) + kf_y(a, b) + \sqrt{(h^2 + k^2)}.\ F(h, k), \qquad ...(4)$$

where $f(h, k) = h[f_x(a + \theta_2 h, b) - f_x(a, b)]/\sqrt{(h^2 + k^2)}$

$$+ k[f_y(a + h, b + \theta_1 k) - f_y(a, b)]/\sqrt{(h^2 + k^2)}. \qquad ...(5)$$

Since f_x and f_y are both continuous at (a, b), so

$$\left.\begin{aligned} &\lim_{(h,k)\to(0,0)} f_x(a+\theta_2 h, b) = f_x(a,b), \\ \text{and} \quad &\lim_{(h,k)\to(0,0)} f_y(a+h, b+\theta_1 k) = f_y(a,b), \end{aligned}\right\} \qquad ...(6)$$

From (5) and (6), we have

$$\lim_{(h,k)\to(0,0)} F(h, k) = 0, \qquad ...(7)$$

since $\left|\dfrac{h}{\sqrt{(h^2+k^2}}\right| \le 1, \left|\dfrac{k}{\sqrt{(h^2+k^2}}\right| \le 1,$

for all values of (h, k), such that $h^2 + k^2 \neq 0$.

From (4) and (7), we get

$$f(a + h, b + k) - f(a, b) = h\, f_x(a, b) + k\, f_y(a, b) + \sqrt{(h^2 + k^2)}\, F(h, k),$$

where $F(h, k) \to 0$ as $(h, k) \to (0, 0)$.

Hence f is differentiable at (a, b).

Remark 1:

The conditions of the above theorem are sufficient, but are not necessary. If a function f is differentiable at a point (a, b), then f_x, f_y may not be continuous at (a, b) as shown below:

Remark 2:

If a function f is not differentiable at (a, b), then f_x and f_y cannot be continuous at (a, b).

Example 27:

Find the value of k for which the function f where

$$f(x, y) = \{ | x + y | + (x + y)\}^2$$

is everywhere differentiable.

Solution:

For $k \le 0$, the given function is not defined if

$x + y \le 0$ $[\because$ then $| x + y | = -(x + y)]$

For $k > 0$, $f(x, y) = \{2 (x + y)\}^2$, if $x + y > 0$,

$= 0$, if $x + y \le 0$.

Let $0 < k < 1$. Now $f_x(0, 0) = \lim_{h\to 0} [f(h, 0) - f(0, 0)]/h$

$$\text{or } f_x(0, 0) \lim_{h\to 0} \frac{2^k . h^k}{h} = \lim_{h\to 0} \frac{1}{h^{1-2}},$$

which does not exist.

Similarly, $f_y(0, 0)$ does not exist.

Hence, f cannot be differentiable if $0 < k < 1$.

Suppose $k \ge 1$. In this case

$$f_x(x, y) = f_y(x, y) = \begin{cases} k, 2^k (x+y)^{k-1}, & \text{if } x + y > 0, \\ 0, & \text{if } x + y \le 0. \end{cases}$$

Since f_x and f_y are both continuous at each point $(x, y) \in r^2$, so f is differentiable everywhere is $k \ge 1$.

Example 28:

Investigate the differentiability of the function f where $f(x, y) = (xy)^{1\ 3}$ at all points $(x, y) \in R^2$.

Solution:

Let $x \ne 0$, $y \ne 0$. Then

$$f_x(x, y) = \frac{1}{2} (y^{1/3}/x^{3/2}),\ f_y(x, y) = \frac{1}{3} (x^{1/3}/y^{2/3}).$$

Since f_x and f_y are both continuous at each point $(x, y) \ne (0, 0)$, so f is differentiable at each point $(x, y) \ne (0, 0)$.

Now, $f_x(0, 0) = \lim_{h\to 0} [f(h, 0) - (f(0, 0)]/h = \lim_{h\to 0} \frac{0-0}{h} = 0,$

$f_y(0, 0) = \lim_{k\to 0} [f(0, k) - f(0, 0)]/k = 0.$

Suppose f is differentiable at (0, 0). Then

$$f(h, k) - f(0, 0) = 0h + 0k + \sqrt{h^2+k^2}\ \phi(h, k),$$

where $\phi(h, k) \to 0$ as $(h, k) \to (0, 0)$. ...(1)

For $h = k \neq 0$, we obtain

$$f(h, h) = \sqrt{2}\ h\ \phi(h, h) \quad \text{or} \quad \phi(h, h) = (hh)^{1/3}/\sqrt{2}h$$

$$\therefore \lim_{h\to 0} \phi(h, h) - = \lim_{h\to 0} \left(\frac{1}{\sqrt{2}h^{1/3}}\right),$$ which does not exits.

This contradicts (1). Hence f is differentiable at all points (x, y) except (0, 0).

Example 29:

Prove that the function $f(x, y) = \sqrt{|xy|}$ is not differentiable at the point (0, 0), but that f_x and f_y both exist at the origin and have the value 0. Hence deduce that these two partial derivatives are continuous except at the origin.

Solution:

$$f_x(0, 0) = \lim_{h\to 0} \frac{f(h,0) - f(0, 0)}{h} = \lim_{h\to 0} \frac{0}{h} = 0,$$

$$f_y(0, 0) = \lim_{k\to 0} \frac{f(0,k) - f(0, 0)}{k} = \lim_{k\to 0} \frac{0}{k} = 0.$$

Thus, f_x, f_y both exist at (0, 0).

If the function f is differentiable at (0, 0), then by definition

$$f(h, k) - f(0, 0) = 0h + 0k + \sqrt{h^2+k^3}\ g(h, k)$$

where $g(h, k) \to 0$ as $(h, k) \to (0, 0)$.

or $|hk|^{1/2} = \sqrt{h^2+k^2}\ g(h, k).$

Putting $h = r \sin\theta$, $k = r \sin\theta$ and dividing by r, we get

$|\cos\theta \cos\theta|^{1/2} = g(h, k)$

Now for arbitrary θ, $r \to 0$ implies that $(h, k) \to (0, 0)$.

Taking the limit as $r \to 0$, we get

$$|\cos\theta \sin\theta|^{1/2} = 0,$$ which is impossible for all values of θ.

Hence, the function f is not differentiable at (0, 0), and consequently the partial derivatives f_x, f_y cannot be continuous at (0, 0), for otherwise the function would be differentiable thereat.

Ex. 1:

Show that the function $f(x, y) = \sqrt{|xy|}$ *is continuous at the origin but is not differentiable thereat.*

[**Hint.** $|f(x, y) - f(0, 0)| = |\sqrt{|xy|} - 0| \leq |xy|$]

$= |x||y| < \varepsilon$, if $|x| < \delta$, $|y| < \delta$ $(\delta = \sqrt{\varepsilon} > 0)$]

Hence, f is continuous at the origin.

Ex. 2:

Show that $f(x, y) = \sqrt{|xy|}$ *is not derivable at (0, 0).* **(D.U., 1990)**

Example 30:

Show that the function

$$f(x, y) = \begin{cases} x^2y/(x^2+y^2), & \text{when } x^2+y^2 \neq 0 \\ 0 & \text{, when } x^2+y^2 = 0 \end{cases}$$

is continuous but not differentiable at (0, 0).

Solution:

Putting $x = r\cos\theta$, $y = r\sin\theta$; we get

$$|f(x, y) = f(0, 0)| = \left|\frac{r^2\cos^2\theta \,.\, r\sin\theta}{r^2} - 0\right|$$

$$= r\,|\cos\theta|\;|\cos\theta|\;|\sin\theta|$$

$$< r = \sqrt{x^2 + y^2}$$

Let $\varepsilon > 0$ be given. Choose $\delta = \varepsilon$. Then

$$|f(x, y) - f(0, 0)| < \varepsilon \text{ if } \sqrt{x^2 + y^2} < \delta.$$

Hence, f is continuous at the origin.

Now $f_x(0, 0) = \lim_{h \to 0} [f(h, 0) - f(0, 0)]/h = \lim_{h \to 0} \dfrac{0 - 0}{h} = 0.$

Similarly $f_y(0, 0) = 0$.

Let, if possible, f be differentiable at (0, 0). Then

$$f(h, k) - f(0, 0) = Ah + Bk + \sqrt{h^2 + k^2}\, g(h, k),$$

where $A = f_x(0, 0)$, $B = f_y(0, 0)$ and $g(h, k) \to 0$ as $(h, k) \to (0, 0)$. ...(1)

$$\therefore \frac{h^2 k}{h^2 + k^2} = \sqrt{h^2 + k^2}\, g(h, k) \Rightarrow g(h, k) = \frac{h^2 k}{(h^2 + k^2)^{3/2}}.$$

Now $\lim_{h \to 0} g(h, mh) = \dfrac{m}{(1 + m^2)^{3/2}}$ (k = mh)

$$\therefore \lim_{(h,k) \to (0,0)} g(h, k) = \frac{m}{(1 + m^2)^{3/2}}.$$

which depends on m and so the limit does not exist. This contradicts (1). Hence f is not differentiable at (0, 0).

Ex.:

Discuss the continuity and differentiability of the following function at (0, 0):

$$f(x, y) = \begin{cases} xy^2/(x^2 + y^2), & (x, y) \neq (0, 0) \\ 0 & , (x, y) = (0, 0) \end{cases}$$ **(D.U., 1985)**

[**Hint.** Show that the given function is continuous but not differentiable at (0, 0)].

Example 31:

Let $f : R^2 \to R$ be defined by setting

$f(x, y) = x^2 \sin(1/x) + y^2 \sin(1/y)$, when $xy \neq 0$,

$f(x, 0) = x^2 \sin(1/x)$, when $x \neq 0$,

$f(0, y) = y^2 \sin(1/y)$, when $y \neq 0$, $f(0, 0) = 0$.

Show that (i) f_x and f_y both exist at (0, 0) but are not continuous at (0, 0) ; (ii) f is differentiable at (0, 0).

Solution:

We have

$$f_x(x, y) = 2x \sin(1/x) - \cos(1/x),\ x \neq 0, \quad ...(1)$$

$f_x(0, y) = 0.$

Again $f_y(x, y) = 2y \sin(1/y) - \sin(1/y), y \neq 0,$

$f_y(x, 0) = 0.$

Since $\lim_{x \to 0} \cos(1/x)$ does not exist, so by (1), f_x is not continuous at (0, 0). Similarly f_y is not continuous at (0, 0).

$$\text{Now } f_x(0, 0) = \lim_{x \to 0} \frac{f(h, 0) - f(0, 0)}{h} = \lim_{h \to 0} \frac{h^2 \sin(1/h) - 0}{h}$$

$$= \lim_{h \to 0} h \sin(1/h) = 0.$$

Similarly $f_y(0, 0) = 0$. Thus, f_x, f_y both exist at (0, 0) but are not continuous at (0, 0).

Now we show that f is differentiable at (0, 0).

We shall verify that

$$f(h, k) - f(0, 0) = Ah + Bk + \sqrt{h^2 + k^2}\ g(h, k) \qquad ...(2)$$

where $\lim_{(h, k) \to (0, 0)} g(h,k) = 0$

we know $A = f_x(0, 0) = 0$, $B = f_y(0, 0) = 0$. Also $f(0, 0) = 0$ ...(3)

From (2) and (3), we obtain

$$h^2 \sin(1/h) + k^2 \sin(1/k) = \sqrt{h^2 + k^2}\ g(h, g)$$

$$\text{or } g(h, k) = \frac{h}{\sqrt{h^2 + k^2}} . h \sin(1/h) + \frac{k}{\sqrt{h^2 + k^2}} . k \sin(1/k)$$

$$\leq h \sin(1/h) + k \sin(1/k) \qquad [\because h, k \leq \sqrt{h^2 + k^2}\,]$$

$$\therefore \lim_{(h,k) \to (0,0)} g(h,k) = 0.$$

$$[\because \lim_{h \to 0} h \sin(1/h) = 0 = \lim_{k \to 0} k \sin(1/k)]$$

Hence f is differentiable at (0, 0).

Theorem 4:

(Young's Theorem): *Let f be a real-valued function defined on an open set $D \subset R^2$ and $(a, b) \in D$. If f_x and f_y are both differentiable at (a, b), then $f_{xy}(a, b) = f_{yx}(a, b)$.*

Proof:

Since f_x and f_y are differentiable at (a, b), it follows that f_x and f_y both exist in a neighbourhood N ≡ N ((a, b), d) of (a, b), and that f_{xx}, f_{yx}, f_{xy}, f_{yy} all exist at (a, b). Let (a + h, b + h) ∈ N and that

$$G(y) = f(a + h, y) - f(a, y) \quad \forall \; y \in (b, b + h). \qquad ...(1)$$

If $F = G(b + h) - G(b)$, then

$$F = [f(a + h, b + h) - f(a, b + h)] - [f(a + h, b) - f(a, b)].$$

Since f_y exists in N, so G is differentiable in [b, b + h]. By Lagrange's Mean Value Theorem, there exists a $\theta \in (0, 1)$ such that

$$F = hG'(b + \delta h) = h(a + h, b + \delta h) - f_y(a, b + \delta h)], \text{ by (1)} \qquad ...(2)$$

Since f_y is differentiable at (a, b), therefore

$$f_y(a + h, b + \delta h) - f_y(a, b) = Ah + B\theta h + h\delta_1(h, h) + \delta h\, \phi_2 (h, h), \qquad ...(3)$$

where $\phi_1(h, h) \to 0$, $\phi_2 (h, h) \to 0$ as $h \to 0$ and

$$A = f_{xy} (a, b),\ B = f_{yy} (a, b). \qquad ...(4)$$

Again, since f_y is differentiable at (a, b), therefore

$$f_y(a, b + \delta h) - f_y(a, b) = A.0 + B.\delta h + 0.\phi_3(h, h) + \delta h \phi_4 (h, h) \qquad ...(5)$$

where $\phi_2 (h, h) \to 0$, $\phi_4 (h, h) \to 0$ as $h \to 0$.

From equations (2) – (5), we obtain

$$F/h^2 = f_{xy} (a, b) + \phi_1 (h, h) + \delta\phi_2 (h, h) - \delta\phi_4(h, h), \qquad ...(6)$$

where $\phi_1, \phi_2, \phi_4 \to 0$ as $h \to 0$.

Similarly if we take $F = H(a + h) - H(a)$, where

$$H(x) = f(x, b + h) - f(x, b), \quad \forall \; x \in [a, a + h],$$

we can show that there exists a $\delta' \in (0, 1)$ such that

$$F/h^2 = f_{yx}(a, b) + \psi_1 (h, h) + \delta'\psi_2 (h, h) - \delta'\psi_4 (h, h), \qquad ...(7)$$

where $\psi_1, \psi_2, \psi_4 \to 0$ as $h \to 0$.

From (6) and (7), we obtain

$$\lim_{h\to 0} F/h^2 = f_{xy}(a, b) = f_{yx} (a, b).$$

Hence $f_{xy} (a, b) = f_{yx} (a, b)$.

Remark:

The conditions of Schwarz's and Young's Theorems are sufficient but they are not necessary conditions. The following example shows that the conditions are not necessary.

7.12 Schwarz's and Young's Theorems

In this section we shall discuss sufficient conditions so that

f_{xy} (a, b) = f_{yx}(a, b).

These sufficient conditions are due to Schwarz and Young.

Theorem 5:

(Schwarz's Theorem): *Let f be a real-valued function defined on an open set $D \subset R^2$ and (a, b) $\in$ D. If*

(i) *f_x exists in a neighbourhood of (a, b),*

(ii) *f_{xy} is continuous at (a, b),*

then f_{yx} (a, b) exists and is equal to f_{xy} (a, b). **(D.U., 1995)**

Proof:

The hypotheses of the theorem imply that f_x, f_y and f_{xy} all exist in a neighbourhood N $\equiv$ N ((a, b), δ) of (a, b).

Let (a + h, b + k) $\in$ N and that

$$G(y) = f(a + h, y) - f(a, y) \quad \forall \ y \in [b, b + k]. \qquad ...(1)$$

If F = G(b + k) – G(b), then by (1),

$$F = [f(a + h, b + k) - f(a, b + k)] - [f(a + h, b) - f(a, b)]. \qquad ...(2)$$

Since f_y exists in N, so by (1), G is differentiable in [b, b + k]. By Lagrange's Mean Value Theorem, there exists a $\delta \in (0, 1)$ such that

$$F = kG'(b + \delta k) = k\,[f_y(a + h, b + \delta k) - f_y(a, b + \delta k)], \text{ by (1)}$$

Again, since f_{xy} exists in N, the function $f_y(x, b + \delta k)$ is differentiable in [a, a + h]. Using Lagrange's Mean Value Theorem, there exists a $\delta' \in$ (0, 1) such that

$$F = hk\, f_{xy}(a + \delta' h, b + \delta k). \qquad ...(3)$$

From (2) and (3), we obtain

$$\frac{1}{k}\left[\frac{f(a + h, b + k) - f(a, b + k)}{h} - \frac{f(a + h, b) - f(a, b)}{h}\right] = f_{xy}(a + \delta' h, b + \delta k).$$

Since f_x exists in N, so on taking the limit as h $\to$ 0, we obtain

$$\frac{1}{k}\,[f_x(a, b + k) - f_x(a, b) = \lim_{h\to 0} f_{xy}(a + \delta' h, b + \delta k).$$

$$\Rightarrow \lim_{k\to 0} [f_x(a, b + k) - f_x(a, b)]/k = \lim_{k\to 0}\ \lim_{h\to 0} f_{xy}(a + \delta' h, b + \delta k)$$

$\Rightarrow$ $f_{yx}(a, b) = f_{xy}(a, b)$, since f_{xy} is continuous at (a, b).

Hence, f_{yx} (a, b) exists and is equal to f_{xy} (a, b).

Corollary:

If f_{xy} and f_{yx} are both continuous at (a, b), then

$f_{xy}(a, b) = f_{yx}(a, b)$.

Example 32:

If $f(x, y) = xy\,(x^2 - y^2)/(x^2 + y^2)$, when $(x, y) \neq (0, 0)$ and $f(0, 0) = 0$, show that $f_{xy}(0, 0) \neq f_{yx}(0, 0)$. **(D.U., 1995, 1987, 83)**

Solution:

We have

$$f_{xy}(0, 0) = \lim_{h\to 0} \frac{f_y(0+h,0) - f_y(0,0)}{h}$$

Now, $$f_y(0, 0) = \lim_{k\to 0} \frac{f(0, 0 + k) - f(0, 0)}{k} \lim_{k\to 0} \frac{0}{k} = 0,$$

and $$f_y(h, 0) = \lim_{k\to 0} \frac{f(h, 0 + k) - f(h, 0)}{k} \lim_{k\to 0} \frac{hk(h^2 - k^2)}{k(h^2 + k^2)} = h,$$

$$\therefore \quad f_{xy}(0, 0) = \lim_{h\to 0} \frac{h - 0}{h} = 1.$$

Again $$f_{yx}(0, 0) = \lim_{k\to 0} \frac{f_x(0, 0 + k) - f_x(0, 0)}{k}$$

Now, $$f_x(0, 0) = \lim_{h\to 0} \frac{f(0+h,0) - f(0,0)}{h} = 0,$$

and $$f_x(0, k) = \lim_{h\to 0} \frac{f(0+h,k) - f(0,k)}{h}$$

$$= \lim_{h\to 0} \frac{hk(h^2 - k^2)}{h(h^2 + k^2)} = -k.$$

$$\therefore \quad f_{yx}(0, 0) = \lim_{k\to 0} \frac{-k-0}{k} = -1.$$

Thus, $f_{xy}(0, 0) \neq f_{yx}(0, 0)$.

Example 33:

Let $f(x, y) = 2xy/\sqrt{(x^2 + y^2)}$, if $(x, y) \neq (0, 0)$ and $f(0, 0)$. Show that $f_{yx}(0, 0) \neq f_{xy}(0, 0)$.

Solution:

$$f_y(0, 0) = \lim_{k\to 0} [f(0, 0 + k) - f(0, 0)]/k = \lim_{k\to 0} \frac{0-0}{k} = 0,$$

$$f_y(h, 0) = \lim_{k\to 0} [f(h, 0 + k) - f(h, 0)]/k = \lim_{k\to 0} \frac{2h-0}{\sqrt{h^2 + k^2}}$$

$$= \frac{2h}{|h|} = \pm 2.$$

$$\therefore f_{xy}(0, 0) = \lim_{h\to 0} [f_y(0 + h, 0) - \phi(0, 0)]/h = \lim_{h\to 0} \left(\pm \frac{2}{4}\right),$$

which does not exist.

$$f_x(0, 0) = \lim_{h\to 0} [f(0 + h, 0) - f(0, 0]/h = \lim_{h\to 0} \frac{0 - 0}{h} = 0,$$

$$f_x(0, k) = \lim_{h\to 0} [f(0 + h, 0) - f(0, k)]/h = \lim_{h\to 0} \frac{k}{\sqrt{h^2 + k^2}} = \pm 2,$$

$$\therefore f_{yx}(0, 0) = \lim_{k\to 0} [f_x(0, 0 + k) - f_x(0, 0)]/h = \lim_{k\to 0} \left(\pm \frac{2}{k}\right),$$

which does not exist.

Thus $f_{xy}(0, 0) \neq f_{yx}(0, 0)$.

Example 34:

Examine the equality of $f_{xy}(0, 0)$ and $f_{yx}(0, 0)$ if

$$f(x, y) = \sqrt{(x^2 + y^2)} \sin 2\phi,$$

where $f(0, 0) = 0$ *and* $\phi = \tan^{-1}(y/x)$. **(D.U., 1990)**

Solution:

Clearly $x = 0 \Rightarrow \phi = \pm \frac{1}{2}\pi$ and $y = 0 \Rightarrow \phi = 0$.

$\therefore f(x, y) = 0$, if $xy = 0$

For $xy \neq 0$, $\sin 2\phi = \dfrac{2\tan\phi}{1 + \tan^2\phi} = \dfrac{2y/x}{1+(y^2/x^2)} = \dfrac{2xy}{x^2+y^2}$.

$$\therefore \qquad f(x, y) = \begin{cases} \dfrac{2xy}{\sqrt{x^2+y^2}}, & \text{if } xy \neq 0 \\ 0 \quad , & \text{if } xy = 0 \end{cases}$$

By Example 33, it follows that

$f_{xy}(0, 0) \neq f_{yx}(0, 0)$.

Example 35:

Prove or disprove : $f_{xy}(0, 0) = f_{yx}(0, 0)$, where

$$f(x, y) = \frac{(x^2y + xy^2)\sin(x - y)}{x^2 + y^2}, \quad (x, y) \neq (0, 0)$$

$f(0, 0) = 0.$ **(D.U., 1994)**

Solution:

$$f_y(0, 0) = \lim_{k\to 0} [f(0, 0 + k) - f(0, 0)]/k = \lim_{k\to 0} \frac{0 - 0}{k} = 0,$$

$$f_y(h, 0) = \lim_{k\to 0} [f(h, 0 + k) - f(h, 0)]/k$$

$$= \lim_{k\to 0} \frac{[(h^2k + hk^2)\sin(h-k) - 0]}{k(h^2+k^2)}$$

$$= \lim_{k\to 0} \frac{(h^2 + hk)\sin(h-k)}{(h^2+k^2)} = \sin h.$$

$$\therefore f_{xy}(0, 0) = \lim_{h\to 0} [f_y(0 + h, 0) - f_y(0, 0)]/h = \lim_{h\to 0} \frac{\sin h}{h} = 1.$$

$$\text{Now } f_x(0, 0) = \lim_{h\to 0} [f(0 + h, 0) - f(0, 0)]/h$$

$$= \lim_{h\to 0} \frac{0 - 0}{h} = 0,$$

$$f_x(0, k) = \lim_{h\to 0} [f(0 + h, 0) - k(0, k)]/h$$

$$= \lim_{h\to 0} \frac{(h^2k + hk^2)\sin(h-k)}{h(h^2 + k^2)} = -\sin k.$$

$$\therefore \; f_{xy}(0, 0) = \lim_{k\to 0} [f_x(0, 0 + k) - f_x(0, 0)]/k$$

$$= \lim_{k\to 0}\left(\frac{-\sin k}{k}\right) = -1$$

Hence $f_{xy}(0, 0) \neq f_{yx}(0, 0)$.

Example 36:

If $f(x, y) = x^2 + y^2)\ \tan^{-1}(y/x)$, *when* $x \neq 0$

and $f(0, k) = \pi y^2/2$,

Show that $f_{xy}(0, 0) \neq f_{yx}(0, 0)$. **(D.U., 1993, 88, 86)**

Solution:

$$f_x(0, 0) = \lim_{h\to 0}\frac{f(h, 0) - f(0, 0)}{h} = \lim_{h\to 0}\frac{0-0}{h} = 0.$$

$$f_y(0, 0) = \lim_{k\to 0}\frac{f(0, k) - f(0, 0)}{k} = \lim_{k\to 0}\left(\frac{\pi k}{2}\right) = 0.$$

For $k \neq 0$,

$$f_x(0, k) = \lim_{h\to 0}\ [f(h, k) - f(0, k)]/h$$

$$= \lim_{h\to 0}\frac{(h^2 + k^2)\tan^{-1}(k/h) - \pi k^2/2}{h}\left(\frac{0}{0}\right)$$

$$= \lim_{h\to 0}\ 2h\tan^{-1}(k/h) + (h^2 + k^2)\ .\ \frac{1}{1 + (k^2/h^2)}\left(\frac{-k}{h^2}\right)$$

(by L' Hospital's Rule)

$$= 0\left(\pm\frac{\pi}{2}\right) - k = -k.$$

$$\therefore\ f_{yx}(0, 0) = \lim_{k\to 0}\ [f_x(0, k) - f_x(0, 0)]/k = \lim_{k\to 0}\ (-k)/k = -1.$$

For $h \neq 0$, $f_y(h, 0) = \lim_{k\to 0}\ [f(h, k) - f(h, 0)]/k$

$$= \lim_{k\to 0}\ [(h^2 + k^2)\tan^{-1}(k/h) - 0]/k$$

$$= \lim_{k\to 0}\frac{h^2 + k^2}{h}\ \frac{\tan^{-1}(k/h)}{(k/h)} = h\ .\ 1 = h$$

$\therefore f_{xy}(0, 0) = \lim_{h\to 0} [f_y(h, 0) - f_x(0, 0)]/h = \lim_{h\to 0} h/h = 1.$

Hence $f_{xy}(0, 0) \neq f_{yx}(0, 0)$.

Ex.:

If $f(x, y) = (x^2 + y^2) \tan^{-1}(x/y)$, when $y \neq 0$ and $f(x, 0) = \pi x^2/2$, show that $f_{yx}(0, 0) \neq f_{xy}(0, 0)$.

Solution:

Do your self.

Example 37:

Let

$$f(x, y) = \begin{cases} x^2y^2/(x^2+y^2), & \text{if } (x, y) \neq (0, 0), \\ 0, & \text{if } (x, y) = (0, 0) \end{cases}$$

Show that $f_{xy}(0, 0) = f_{yx}(0, 0)$, but the conditions of Schwarz's and Young's Theorems are not satisfied. **(D.U., Physics (H) 1995)**

Solution:

$f_{xy}(0, 0) = f_{yx}(0, 0)$, we have

For $(x, y) \neq (0, 0)$, we have

$f_x(x, y) = [(x^2 + y^2)(2xy^2) - x^2y^2(2x)]/(x^2 + y^2)^2$

$\Rightarrow f_x(x, y) = 2xy^4/(x^2 + y^2)^2$ ...(1)

$\therefore f_{yx}(x, y) = 8xy^3(x^2 + y^2)^{-2} - (4y)(2xy^4)(x^2 + y^2)^{-3}$

$\Rightarrow f_{yx}(x, y) = 8x^3y^3/(x^2 + y^2)^3$.

Similarly, for $(x, y) \neq (0, 0)$,

$f_y(x, y) = 2x^4y/(x^2 + y^2)^2$,

and $f_{xy}(x, y) = 8x^3y^3/(x^2 + y^2)^3$

Since $\lim_{(x,y)\to(0,0)} \dfrac{x^3y^3}{(x^2+y^2)^3} = \dfrac{m^3}{(1+m^2)^3}$, (along the path $y = mx$),

so $\lim f_{xy}(x, y)$ and $\lim f_{yx}(y, x)$ do not exist.

Hence f_{xy} and f_{yx} cannot be continuous at (0, 0) and consequently the conditions of Schwarz's Theorem are not satisfied.

Now we show that f_x and f_y are not differentiable at (0, 0).

We can verify that $f_{xx}(0, 0) = 0$. $f_{yy}(0, 0) = 0$.

Let, if possible, f_x be differentiable at (0, 0). Then

$$f_x(h, k) - f_x(0, 0) = h\, f_{xx}(0, 0) + k\, f_{yx}(0, 0) + h\phi(h, k) + k\psi(h, k) \quad ..(2)$$

or $f_x(h, k) = h\phi(h, k) + k\psi(h, k)$,

where $\phi(h, k), \psi(h, k) \to 0$ as $(h, k) \to (0, 0)$

Taking $h = k \neq 0$ in (1) and (2), we get

$$2h.\, h^4/(h^2 + h^2)^2 = h\, \phi(h, h) + h\, \psi(h, h).$$

or $$\frac{1}{2} = \phi(h, h) + \psi(h, h) \Rightarrow \frac{1}{2} = 0 \text{ as } h \to 0.$$

This is impossible. Hence f_x cannot be differentiable at (0, 0). Similarly, we can show that f_y cannot be differentiable at (0, 0). Hence, the conditions of Young's theorem are not satisfied.

Example 38:

Let f(x, y) be defined by

$$f(x, y) = \begin{cases} (x^2 + y^2) \log (x^2 + y^2), & (x, y) \neq (0, 0), \\ 0 & , (x, y) = (0, 0). \end{cases}$$

Prove that f_{xy} and f_{yx} are continuous at (0, 0) but $f_{xy}(0, 0) = f_{yx}(0, 0)$. **(D.U., 1984)**

Solution:

For $x \neq 0$, $y \neq 0$, we have

$$f_x = 2x \log (x^2 + y^2) + \frac{x^2 + y^2}{(x^2 + y^2)} 2x$$

$\therefore$ $$f_x = 2x\, [1 + \log (x^3 + y^2)].$$

Similarly, $$f_y = 2y\, [1 + \log (x^2 + y^2)].$$

Also $$f_{xy} = \frac{4xy}{(x^2 + y^2)} \quad \forall\ (x, y) \neq (0, 0)$$

Now , $$f_x(0, 0) = \lim_{h \to 0} [f(0 + h, 0) - f(0, 0)/h = \lim_{h \to 0} \frac{h^2 \log h^2}{h}$$

$$= \lim_{h \to 0} 2h \log h = 2 \lim_{h \to 0} \frac{\log h}{1/h} \left(\frac{\infty}{\infty}\right)$$

$$= 2 \lim_{h \to 0} \frac{-h^2}{h} = 0. \text{ Similarly } f_y(0, 0) = 0.$$

Now, $f_x(0, k) = \lim_{h\to 0} [f(0 + h, k) - f(0, k)]/h$

$$= \lim_{h\to 0} \frac{(h^2 + k^2)\log(h^2 + k^2) - k^2 \log k^2}{h} \left(\frac{0}{0}\right)$$

$$= \lim_{h\to 0} 2h \log(h^2 + k^2) + 2h = 0 \text{ (L' Hospital's Rule)}$$

$\therefore$ $f_{yx}(0, 0) = \lim_{k\to 0} [f_x(0, 0 + k) - f_x(0, 0)]/k$

$$= \lim_{k\to 0} \frac{0 - 0}{k} = 0$$

Similarly, we can show that $f_{xy}(0, 0) = 0$.

Thus, $f_{xy}(0, 0) = f_{yx}(0, 0)$

Now, $= \lim_{(x,y)\to(0,0)} f_{xy} = \lim_{(x,y)\to(0,0)} \frac{4xy}{x^2 + y^2}$

$$= \lim_{x\to 0} \frac{4mx^2}{x^2(1+m^2)}, \text{ along the path } y = mx$$

$\therefore \lim_{(x,y)\to(0,0)} f_{xy} = \frac{4m}{1 + m^2}$, and so $\lim_{(x,y)\to(0,0)} f_{xy}$ does not exist.

Similarly, $\lim_{(x,y)\to(0,0)} f_{yx}$ does not exist.

Hence, f_{yx} and f_z are not continuous at (0, 0).

Example 39:

If $f(x, y) = x^2 \tan^{-1}(y/x) - y^2 \tan^{-1}(x/y)$, when $x \neq 0$, $y \neq 0$ and $f(x, y) = 0$, otherwise; show that $f_{xy}(0, 0) \neq f_{yx}(0, 0)$

Solution:

We have

$f_x(0, 0) = \lim_{h\to 0} [f(h, 0) - f(0, 0)]/h \lim_{h\to 0} (0 - 0)/h = 0$

and $f_y(0, 0) = \lim_{k\to 0} [f(0, k) - f(0, 0)]/k \lim_{k\to 0} (0 - 0)/k = 0$

For $k \neq 0$, we have

$$f_x(0, k) = \lim_{h\to 0} [f(h, k) - f(0, k)]/h = \lim_{h\to 0} f(h, k)/h$$

$$= \lim_{h\to 0} [h^2 \tan^{-1} (k/h) - k^2 \tan^{-1} (h/k)]/h$$

$$= \lim_{h\to 0} [h \tan^{-1}\left(\frac{k}{h}\right) - \frac{k^2}{h} \tan^{-1}\left(\frac{h}{k}\right)$$

$$= 0\left(\pm\frac{\pi}{2}\right) - k^2 \lim_{h\to 0} \frac{\tan^{-1}(h/k)}{h} \quad (\text{as } k > 0,\ k < 0)$$

$$= -k^2 \lim_{h\to 0} \frac{k^2}{h^2 + k^2}\left(\frac{1}{k}\right) 1$$

(By L'Hospital Rule)

$$= (-k^2)(1/k) = -k.$$

$\therefore$ $\quad f_{yx}(0, 0) = \lim_{k\to 0} [f_x(0, k) - f_x(0, 0)]/k = \lim_{k\to 0} (-k)/k = -1$

For $\quad h \neq 0,\ f_y(h, 0) = \lim_{k\to 0} [f_a(h, k) - f(h, 0)]/k$

$$= \lim_{k\to 0} [h^2 \tan^{-1}(k/h) - k^2 \tan^{-1}(h/k)/k$$

$$= \lim_{k\to 0}\left[h.\frac{\tan^{-1}(k/h)}{k/h} - k \tan^{-1}(h/k)\right]$$

$$= h \,.\, 1 - 0\,(\pm\pi/2) = h.$$

$\therefore$ $\quad f_{xy}(0, 0) = \lim_{h\to 0} [f_y(h, 0) - f_y(0, 0)]/h = \lim_{h\to 0} h/h = 1$

Hence $f_{xy}(0, 0) \neq f_{yx}(0, 0)$

Example 40:

Let $f(x, y) = (xy^5 - x^2y^3)/(x^2 + xy^2 + y^4)$ when $(x, y) \neq (0, 0)$ and $f(0, 0) = 0$. Show that $f_{xy}(0, 0) \neq f_{xy}(0, 0)$.

Solution:

$$f_y(0, 0) = \lim_{h\to 0} [f(h, 0) - f(0, 0)]/h = 0$$

and $\quad f_y(0, 0) = \lim_{k\to 0} [f(0, k) - f(0, 0)]/k = 0$

For $k \neq 0$, $f_x(0, k) = \lim_{h\to 0} [f(h, k) - f(0, k)]/h = \lim_{h\to 0} f(h, k)/h$

$$= \lim_{h\to 0} [hk^5 - h^2k^3)/h(h^2 + hk^2 + k^4) = k$$

$$= \lim_{h\to 0} \frac{k^3(k^2 - h)}{h^2 + hk^2 + k^4}$$

$$= \lim_{h\to 0} \frac{k^3(k^6 - h^3)}{(h^2 + hk^2 + k^2)} = k$$

$$[\because (k^6 - h^3) = (k^2 - h)(k^4 + k^2h + h^2)]$$

$$\therefore \quad f_{yx}(0, 0) \lim_{k\to 0} [f_x(0, k) - f_x(0, 0)]/k = \lim_{k\to 0} (k - 0)/k = 1$$

For $h \neq 0$, $f_x(h, 0) = \lim_{k\to 0} [f(h, k) - f(h, 0)]/k = \lim_{k\to 0} f(h, k)/k$

$$= \lim_{k\to 0} (hk^3 - h^2k^2)/k(h^2 + hk^2 + k^4)$$

$$= \lim_{k\to 0} \frac{hk^2(k^6 - h^3)}{(h^2 + hk^2 + k^4)^2} = 0$$

$$\therefore \quad f_{xy}(0, 0) = \lim_{h\to 0} [f_y(h, 0) - f_y(0, 0)]/h\ (h, k)/h = \lim_{h\to 0} (0, 0)/h = 0$$

Hence $f_{yx}(0, 0) \neq f_{xy}(0, 0)$

Example 41:

Let ϕ be a function of two variables defined as

$\phi(x, y) = (x^3 + y^3)/(x - y)$, *when* $x \neq y$

$\phi(x, y) = 0)$, *when* $x = y$

Show that ϕ is discontinuous at the origin, but the first order partial derivatives exist at that point.

Solution:

ϕ is discontinuous at (0, 0) Now

$$\phi_x(0, 0) = \lim_{h\to 0} [\phi(0 + h, 0) - \phi(0, 0)]/h = \lim_{h\to 0} \frac{h^2 - 0}{h} = 0$$

$$\phi_y(0, 0) = \lim_{k\to 0} [\phi(0, 0 + k) - \phi(0, 0)]/k = \lim_{k\to 0} \frac{-k^2 - 0}{k} = 0$$

Hence the result.

7.13 Directional Derivatives

Let f be a real-valued function defined on an open set $D \subset R^2$ and let $p = (a, b)$ be a point of D. Let $u = (h, k)$ be such that $h^2 + k^2 = 1$. Since D is an open set containing p, so there exists; an open disc $N(p, \delta)$ such that $N(p, \delta) \subset D$ for some $\delta > 0$. Thus for sufficiently small values of t, $(a + ht, b + kt) \in D$.

The directional derivative of f in the direction u at the point p denoted by $d_u f(p)$ or $f_u(p)$, is defined as

$$f_u(p) = \lim_{t \to 0} \frac{f(p + tu) - f(p)}{t},$$

provided the limit exists.

Equivalently,

$$f_u(p) = \lim_{t \to 0} \frac{f(a + ht, b + kt) - f(a, b)}{t}.$$

Remark 1:

If $u = (1, 0)$, then

$$f_u(p) = \lim_{t \to 0} \frac{f(a + t, b) - f(a, b)}{t} = f_x(a, b).$$

Again if $u = (0, 1)$, then

$$f_u(p) = \lim_{t \to 0} \frac{(a, b + t) - f(a, b)}{t} = f_y(a, b).$$

If follows that the partial derivatives are simply the directional derivatives along the axes.

Remark 2:

A function may possess a directional derivative in every direction at a point, but may fail to be continuous at that point as shown in the following example:

Example 42:

Let f be the function defined by setting

$$f(x,y) = \begin{cases} \dfrac{x^2 + y^2}{x - y}, & \text{if } x \neq y \\ 0, & \text{if } x = y \end{cases}$$

Show that f possesses a directional derivative in every direction at (0, 0) but is not continuous at (0, 0).

Solution:

Let u = (cos θ, sin θ). Then

$$\frac{f(0+tu)-f(0)}{t} = t^2\left(\frac{\cos^2\theta+\sin^2\theta}{\cos\theta-\sin\theta}\right)$$

At $t \to 0$, we obtain

$f_u(0) = 0$, if $\cos\theta \neq \sin\theta$.

Thus, the given function has a directional derivative in every direction at (0, 0). f is not continuous (0, 0).

Example 43:

Let f be the function defined by setting

$$f(x, y) = \begin{cases} \dfrac{xy}{x^2+y^2}, & \text{if } (x, y) \neq (0, 0) \\ 0 \quad, & \text{if } (x, y) = (0, 0). \end{cases}$$

Show that f possesses a directional derivative in every direction at (0, 0) but is not continuous at (0, 0).

Also show that f is not differentiable at (0, 0).

Solution:

Let u = (cos θ, sin θ). Now

$$\frac{f(0+tu)-f(0)}{t} = \frac{f(t\cos\theta, t\sin\theta)-f(0,0)}{t}$$

$$= \cos\theta\sin\theta,\ t \neq 0$$

At $t \to 0$, we see that $fu(0) = \cos\theta\sin\theta$.

Thus, the given function has a directional derivative in every direction at (0, 0).

Now $\lim\limits_{x\to 0} f(x, mx) = \dfrac{m}{1+m^2}$.

which is different for different values of m.

Thus $\lim\limits_{(x,y)\to(0,0)} \dfrac{xy}{x^2+y^2}$ does not exist and so the given function is not continuous at (0, 0).

Let, if possible, f be differentiable at (0, 0). Then f is continuous at (0, 0), which is a contradiction. Hence f is not differentiable at (0, 0).

Theorem 6:

Taylor's Theorem for a function of two variables **(D.U., 1993, 89)**

If f(x, y) is a function of two independent variables x and y which possesses continuous partial derivatives of the nth order in some neighbourhood of a point (a, b), and if (a + h, b + k) is any point of this neighbourhood, then there exists some θ, 0 < θ < 1, such that

$$f(a + h, b + k) = f(a, b) + df(a, b) + \frac{1}{2!} d^2 f(a, b) + \ldots$$

$$+ \frac{1}{(n-1)!} d^{n-1} f(a,. b) + \frac{1}{n!} d^n f(a + \theta h, b + \theta k),$$

where $d = \left(h\frac{\partial}{\partial x} + k\frac{\partial}{\partial y}\right), \ldots, d^n = \left(\frac{\partial}{\partial x} + k\frac{\partial}{\partial y}\right)^n.$

Proof:

Let x = a + ht, y = b + kt, where $0 \le t \le 1$.

$\therefore \frac{dx}{dt} = h$ and $\frac{dy}{dt} = k$.

Now, f(x, y) = f(a + ht, b + kt) = g(t) (say) ...(1)

Thus, g(t) is a function of a single variable t.

Since f possesses continuous partial dervatives of order n, so the function $g^n(t)$ is continuous in [0, 1]. Now

$$g'(t) = \frac{dg}{dt} = \frac{\partial g}{\partial x}\frac{dx}{dt} + \frac{\partial g}{\partial y}\frac{dy}{dt} = h\frac{\partial g}{\partial x} + k\frac{\partial g}{\partial y}$$

$$\therefore \quad g'(t) = \left(h\frac{\partial}{\partial x} + k\frac{\partial}{\partial y}\right)g.$$

Again $g''(t) = \frac{d^2 g}{dt^2} = \frac{d}{dt}\left(h\frac{\partial}{\partial x} + k\frac{\partial}{\partial y}\right)g$

$$= \left(h\frac{\partial}{\partial x} + k\frac{\partial}{\partial y}\right)\frac{dg}{dt}.$$

$$\therefore \quad g''(t) = \left(h\frac{\partial}{\partial x} + k\frac{\partial}{\partial y}\right)^2 g \text{ and so on.}$$

Thus $g^k(t) = \left(h\dfrac{\partial}{\partial x} + k\dfrac{\partial}{\partial y}\right)^k g = d^k g \ (k = 1, 2, \ldots, n),$ ...(2)

where $d = h\dfrac{\partial}{\partial x} + k\dfrac{\partial}{\partial y}.$

since $g^n(t)$ is continuous in [0, 1], so by Maclaurin's theorem, there exists a number θ, $0 < \theta < 1$, sucht aht

$$g(t) = g(0) + tg'(0) + \frac{t^2}{2!} g''(0) + \ldots + \frac{t^{n-1}}{(n-1)} g^{n-1}(0) + \frac{1}{n!} g^n(\theta t)$$

Putting t = 1, we get

$$g(1) = g(0) + g'(0) + \frac{1}{2!} g''(0) + \ldots + \frac{1}{(n-1)!} g^{n-1}(0) + \frac{1}{n!} g^n(\theta)$$

$$\left.\begin{aligned} &\text{From (1), } g(1) = f(a + h, b + k),\ g(0) = f(a, b). \\ &\text{From (2), } g^k(0) = \left(h\frac{\partial}{\partial x} + k\frac{\partial}{\partial y}\right)^k g(0) = d^k f(a, b). \end{aligned}\right\} \quad \ldots(4)$$

From (3) and (4), we obtain

$$f(a + h, b + k) = f(a, b) + df(a, b) + \frac{1}{2!} d^2 f(a, b) + \ldots$$

$$+ \frac{1}{(n-1)!} d^{n-1} f(a, b) + \frac{1}{n!} d^n f(a + \theta h, b + \theta k).$$

Remark:

For t = 1, we have x = a + h, y = b + k.

Thus Taylor's Theorem can be restated as :

$$f(x, y) = f(a, b) + \left[(x - a)\frac{\partial}{\partial x} + (y - b)\frac{\partial}{\partial y}\right] f(a, b)$$

$$+ \frac{1}{2}\left[(x - a)\frac{\partial}{\partial x} + (y - b)\frac{\partial}{\partial y}\right]^2 f(a, b) + \ldots$$

$$+ \frac{1}{(n-1)!}\left[(x - a)\frac{\partial}{\partial x} + (y - b)\frac{\partial}{\partial y}\right]^{n-1} f(a, b) + R_n,$$

Where $R_n = \frac{1}{n!}\left[(x-a)\frac{\partial}{\partial x}+(y-b)\frac{\partial}{\partial y}\right]^n f(a + (x - a)\theta, b + (y - b)\theta)$.

This is called the **Taylor's expansion of f(x, y)** *about the point (a, b)' in powers of x – a and y – b.*

Example 44:

Prove that the first four terms of the Maclaurin Expansion of e^{ax} cos by are

$$1 + ax + \frac{1}{2!}(a^2x^2 - b^2y^2) + \frac{1}{3!}(a^3x^3 - 3ab^2 xy^2).$$

Solution:

Let $f(x, y) = e^{ax} \cos by$; $a = b = 0$; $f(0, 0) = 1$

$\therefore\ f_x(x, y) = ae^{ax} \cos by \Rightarrow f_x(0, 0) = a$

$f_y(x, y) = -be^{ax} \sin by \Rightarrow f_y(0, 0) = 0$

$f_{xx}(x, y) = a^2e^{ax} \cos by \Rightarrow f_{xx}(0, 0) = a^2$

$f_{yy}(x, y) = -b^2e^{ax} \cos by \Rightarrow f_{yy}(0, 0) = -b^2$

$f_{xy}(x, y) = -abe^{ax} \sin by \Rightarrow f_{xy}(0, 0) = 0$

$f_{xxx}(x, y) = a^2e^{ax} \cos by \Rightarrow f_{xxx}(0, 0) = a^3$

$f_{yyy}(x, y) = -b^3e^{ax} \sin by \Rightarrow f_{yyy}(0, 0) = 0$

$f_{xxy}(x, y) = a^2be^{ax} \sin by \Rightarrow f_{xxy}(0, 0) = 0$

$f_{xyy}(x, y) = -ab^2e^{ax} \cos by \Rightarrow f_{xyy}(0, 0) = -ab^2$.

By Taylor's expansion,

$$f(x, y) = f(0, 0) + \left(x\frac{\partial}{\partial x} + y\frac{\partial}{\partial y}\right) f(0, 0)$$

$$+ \frac{1}{2!}\left(x\frac{\partial}{\partial x} + y\frac{\partial}{\partial y}\right)^2 f(0, 0) + \frac{1}{3!}\left(x\frac{\partial}{\partial x} + y\frac{\partial}{\partial y}\right)^3 f(0, 0).$$

Hence $e^{ax} \cos by = 1 + ax + \frac{1}{2!}(a^2x^2 - b^2y^2) + \frac{1}{3!}(a^3x^3 - 3ab^2xy^2)$.

Example 45:

Expand $x^2y + 3y - 2$ in powers of $x - 1$ and $y + 2$.

(D.U., Physics (II) 1994)

Solution:

We take a = 1, b = –2.

Let $f(x, y) = x^2y + 3y - 2 \Rightarrow f(1, -2) = -10,$

$f_x(x, y) = 2xy \Rightarrow f_x(1, -2) = -4,$

$f_y(x, y) = x^2 + 3 \Rightarrow f_y(1, -2) = 4,$

$f_{xx}(x, y) = 2y \Rightarrow f_{xx}(1, -2) = -4,$

$f_{xy}(x, y) = 2x \Rightarrow f_{xy}(1, -2) = 2,$

$f_{yy}(x, y) = 0 \Rightarrow f_{yy}(1, -2) = 0,$

$f_{xxx}(x, y) = 0 = f_{yyy}(x, y),\ f_{yxx}(1, -2) = 2 = f_{xxy}(1, -2).$

All higher derivatives are zero.

By Taylor's Theorem,

$$f(x, y) = f(a, b) + [(x - a) f_x(a, b) + (y - b) f_y(a, b)] + \frac{1}{2!}[(x - a)^2 f_{xx}(a, b) + 2(x - a)(y - b) f_{xy}(a, b) + (y - b)^2 f_{yy}(a, b)] + \ldots$$

$$\therefore x^2y + 3y - 2 = -10 - 4(x - 1) + 4(y + 2) + \frac{1}{2}[-4(x - 1)^2 + 4(x - 1)(y + 2) + 0] + \frac{1}{3!} 3(x - 1)^2 (y + 2)(2) + 0$$

Hence $x^2y + 3y - 2 = -10 - 4(x - 1) + 4(y + 2) - 2(x - 1)^2 + 2(x - 1)(y + 2) + (x - 1)^2 (y + 2).$

Example 46:

Expand $x^4 + x^2y^2 - y^4$ about the point (1, 1) upto the terms of the second degree. **(D.U., Physics (H) 1995, 1993)**

Solution:

Let $f(x, y) = x^4 + x^2y^2 - y^4$; a = b = 1; f(1, 1) = 1

$f_x(x, y) = 4x^2 + 2xy^2 \Rightarrow f_x(1, 1) = 6,$

$f_y(x, y) = 2x^2y - 4y^2 \Rightarrow f_y(1, 1) = -2$

$f_{xx}(x, y) = 12x^2 - 2y^2 \Rightarrow f_{xx}(1, 1) = 14,$

$f_{yy}(x, y) = 2x^2 - 12y^2 \Rightarrow f_{yy}(1, 1) = -10,$

$f_{xy}(x, y) = 2xy \Rightarrow f_{xy}(1, 1) = 4.$

By Taylor's Thoerem

$$f(x, y) = f(1, 1) + \left[\left\{(x-1)\frac{\partial}{\partial x} + (y-1)\frac{\partial}{\partial y}\right\} f(x, y)\right]_{(1, 1)}$$

$$+ \frac{1}{2!}\left[\left\{(x-1)\frac{\partial}{\partial x} + (y-1)\frac{\partial}{\partial y}\right\}^2 f(x, y)\right]_{(1, 1)} + \ldots..$$

or $x^6 + x^2y^2 - y^4 = 1 + 6(x-1) - 2(y-1)$

$+ \frac{1}{2}[14(x-1)^2 - 10(y-1)^2 + (8(x-1)(y-1)] + \ldots.$

Example 47:

Show that for $0 < \theta < 1$,

$\sin x \sin y = xy - \frac{1}{6}[(x^3 + 3xy^2)\cos\theta x \sin\theta y + (y^3 + 3x^2y)\sin\theta x \cos\theta y]$.

Solution:

Let $f(x, y) = \sin x \sin y$. Then $f(0, 0) = 0$.

$f_x(x, y) = \cos x \sin y \Rightarrow f_x(0, 0) = 0,$

$f_y(x, y) = \sin x \sin y \Rightarrow f_y(0, 0) = 0,$

$f_{xx}(x, y) = \sin x \sin y \Rightarrow f_{xx}(0, 0) = 0,$

$f_{yy}(x, y) = -\sin x \sin y \Rightarrow f_{yy}(0, 0) = 0,$

$f_{xy}(x, y) = \cos x \cos y \Rightarrow f_{xy}(0, 0) = 1,$

$f_{xxx}(x, y) = -\cos x \sin y,\ f_{yyy}(x, y) = -\sin x \cos y,$

$f_{xxy} = -\sin x \cos y,\ f_{xyy} = -\cos x \sin y.$

Now $\left[\left(x\frac{\partial}{\partial x} + y\frac{\partial}{\partial y}\right) f(x, y)\right]_{(0, 0)} = x.0 + y.0 = 0,$

$$\left[\left(x\frac{\partial}{\partial x} + y\frac{\partial}{\partial y}\right)^2 f(x, y)\right]_{(0, 0)}$$

$$= \left[\left(x^2\frac{\partial^2}{\partial x^2} + 2xy\frac{\partial^2}{\partial x\partial y} + y^2\frac{\partial^2}{\partial y^2}\right) f(x, y)\right]_{(0, 0)} = 2xy,$$

Also $\left[\left(x\frac{\partial}{\partial x} + y\frac{\partial}{\partial y}\right)^2 f(x, y)\right]_{(\theta x, \theta y)}$

$$= \left[\left(x^3 \frac{\partial^3}{\partial x^3} + 3x^2y \frac{\partial^3}{\partial x^3 \partial y} + 3xy^2 \frac{\partial^3}{\partial x \partial y^2} + y^3 \frac{\partial^3}{\partial y^3}\right) f(x, y)\right]_{(\theta x, \theta y)}$$

$= [(-x^3 - 3xy^2) \cos x \sin y + (-y^3 - 3x^2y) \sin x \cos] y(\theta x, \theta y)$

$= -[(x^3 + 3xy^2) \cos (\theta x) \sin (\theta y) + (y^3 + 3x^2y) \sin (\theta x) \cos (\theta y)]$.

Putting these values in

$$f(x, y) = f(0, 0) + \left(x \frac{\partial}{\partial y} + y \frac{\partial}{\partial y}\right) f(0, 0) + \frac{1}{2!}\left(x \frac{\partial}{\partial x} + y \frac{\partial}{\partial y}\right)^2 f(0, 0)$$

$$+ \frac{1}{3!}\left(x \frac{\partial}{\partial x} + y \frac{\partial}{\partial y}\right)^3 f(\theta x, \theta y),$$ we get the desired result.

Example 48:

Find the expansion of the polynomial $x^2 + 2x^2y + y^2$ at (1, 1) upto 2nd degree only. **(D.U., 1995)**

Solution:

Let $f(x, y) = x^3 + 2x^2y + y^3$; $a = b = 1$; $f(a, b) = 4$

$f_x(x, y) = 3x^2 + 4xy \Rightarrow f_x(1, 1) = 7,$

$f_y(x, y) = 2x^2 + 3y^2 \Rightarrow f_y(1, 1) = 5,$

$f_{xx}(x, y) = 6x + 4y \Rightarrow f_{xx}(1, 1) = 10,$

$f_{xy}(x, y) = 4x \Rightarrow f_{xy}(1, 1) = 4,$

$f_{yy}(x, y) = 6y \Rightarrow f_{xy}(1, 1) = 6,$

By Taylor's Theorem,

$$f(x, y) = f(1, 1) + \left[\left\{(x - 1) \frac{\partial}{\partial x} + (y - 1) \frac{\partial}{\partial y}\right\} f(x, y)\right]_{(1, 1)}$$

$$+ \frac{1}{2!}\left[\left\{(x - 1) \frac{\partial}{\partial y} + (y - 1) \frac{\partial}{\partial y}\right\}^2 f(x, y)\right]_{(1, 1)} + \ldots\ldots$$

$$x^3 + 2x^2y + y^3 = 4 + [7(x - 1) + 5(y - 1)]$$

$$+ \frac{1}{2} [10(x - 1)^2 + 6(y - 1)^2 + 8(x - 1)(y - 1)] + \ldots..$$

Example 49:

Prove that

$$\log(x-y) = x-(y+1) - \frac{1}{2}x^2 + x(y+1) - \frac{1}{2}(y+1)^2 + \frac{[x-(y+1)]^2}{3[\theta x + 1 - \theta(y+1)^2]}, \ 0 < \theta < 1.$$

Solution:

Let $f(x, y) = \log(x-y)$; $a = 0$, $b = -1$, $f(a, b) = 0$.

$$\therefore \ f_x(x, y) = \frac{1}{x-y} \quad \Rightarrow f_x(0, -1) = 1,$$

$$f_y(x, y) = -\frac{1}{x-y} \quad \Rightarrow f_y(0, -1) = -1,$$

$$f_{xx}(x, y) = -\frac{1}{(x-y)^2} \quad \Rightarrow f_{xx}(0, -1) = -1,$$

$$f_{yy}(x, y) = -\frac{1}{(x-y)^2} \quad \Rightarrow f_{yy}(0, -1) = -1,$$

$$f_{xy}(x, y) = \frac{1}{(x-y)^2} \quad \Rightarrow f_{xy}(0, -1) = 1,$$

$$f_{xxx} = \frac{2}{(x-y)^3}, \ f_{xxy} = -\frac{2}{(x-y)^3}, \ f_{xyy} = \frac{2}{(x-y)^3},$$

$$f_{yyy} = -\frac{2}{(x-y)^3}.$$

We have

$$\left[\left\{x\frac{\partial}{\partial x} + (y+1)\frac{\partial}{\partial y}\right\} f(x, y)\right]_{(0,-1)} = x-(y+1);$$

$$\left[\left\{x\frac{\partial}{\partial x} + (y+1)\frac{\partial}{\partial y}\right\}^2 f(x, y)\right]_{(0,-1)} \quad (xy+1)^2(-1) + 2x(y+1).1$$

$$\left\{x\frac{\partial}{\partial x} + (y+1)\frac{\partial}{\partial y}\right\}^2 (0+(x-0)\theta, -1+(y+1)\theta,$$

$$= \frac{2x^3 + 3x^2(y+1)(-2) + 3x(y+1)^2(2) + (y+1)^2(-2)}{(x-y)^3 \ [x = \theta x, y = -1 + (y+1)\theta}$$

$$= \frac{2\{x-(y+1)\}^3}{[\theta x + 1 - \theta(y+1)]^3}.$$

Putting these values in the Taylor's expansion

$$f(x, y) = f(0, -1) + \left[\left\{x\frac{\partial}{\partial x} + (y+1)\frac{\partial}{\partial y}\right\} f(x, y)\right]_{(0,-1)}$$

$$+ \frac{1}{2!}\left[\left\{x\frac{\partial}{\partial x} + (y+1)\frac{\partial}{\partial y}\right\}^2 f(x, y)\right]_{(0,-1)}$$

$$+ \frac{1}{3!}\left[\left\{x\frac{\partial}{\partial x} + (y+1)\frac{\partial}{\partial y}\right\}^2 f(\theta x, -1 + (y+1)\theta)\right],$$

we get the required result.

Example 50:

Prove that

$$y^x = 1 + 2(y-1) + (x-2)(y-1) + (y-1)^2 + \ldots\ldots$$

Solution:

Let $f(x, y) = y^x$ and $a = 2$, $b = 1$. Then $f(a, b) = 1$.

$f_x(x, y) = y^x \log y \quad \Rightarrow f_x(2, 1) = 0,$

$f_y(x, y) = xy^{x-1} \quad \Rightarrow f_y(2, 1) = 2,$

$f_{xx}(x, y) = y^x(\log y)^2 \quad \Rightarrow f_{xx}(2, 1) = 0,$

$f_{yy}(x, y) = x(x-1)\, y^{x-2} \quad \Rightarrow f_{yy}(2, 1) = 2,$

$f_{xy}(x, y) = y^{x-1} + xy^{x-1}\log y \Rightarrow f_{xy}(2, 1) = 1.$

By Taylor's expansion,

$$f(x, y) = f(a, b) + \left[(x-a)\frac{\partial}{\partial x} + (y-b)\frac{\partial}{\partial y}\right] f(a, b)$$

$$+ \frac{1}{2!}\left[(x-a)\frac{\partial}{\partial x} + (y-b)\frac{\partial}{\partial y}\right]^2 f(a, b) + \ldots$$

or $\quad y^x = 1 + [(x-2).0 + (y-1).2] + \dfrac{1}{2!}\,[(x-2)^2 f_{xx}(2, 1)$

$$+ 2(x-2)(y-1) f_{xy}(2, 1) + (y-1)^2 f_{yy},(2, 1)] + \ldots\ldots$$

$$= 1 + 2\,(y - 1) + \frac{1}{2}\,[0 + 2(x - 2)\,(y - 1).1 + (y - 1)^3.2] + ...$$

Hence, $y^x = 1 + 2(y - 1) + (x - 2)\,(y - 1) + (y - 1)^2 + ...$

7.14 Implicit Function Theorem

If f be a real-valued function defined on an open subset D of R^2 and that $(a, b) \in D$. If

(i) $f(a, b) = 0$,

(ii) f_x *and* f_y *are both continuous in some neighbourhood of (a, b),*

(iii) $f_y(a, b) \neq 0$,

then there exist real numbers h and k such that for all $x \in [a - h, a + h]$, the equation $f(x, y) = 0$ determines a unique solution $y = g(x)$ in the interval $[b - k, b + k]$ with the following properties;

(i) $b = g(a)$,

(ii) $f(x, g(x)) = 0 \quad \forall\, x \in [a - h, a + h]$,

(iii) *g' eists and is continuous is $[a - h, a + h]$, where*

$g'(x) = -\,[f_x(x, y)/f_y(x, y)]$.

The proof is omitted.

Note: The above theorem tells us that for an *implicitly defined* function of the form $f(x, y) = 0$, when is it possible to obtain a unique solution $y = g(x)$ such that $f(x, g(x)) = 0$.

Example 51:

Examine the equation $y^2 + yx^2 + x^3 = 0$ for the existence of a unique solution of the form $y = g(x)$ near the origin, and verify by direct calculation.

Solution:

Let $f(x, y) = y^2 + yx^3 + x^2$. Then $f_y(x, y) = 2y + x^2$.

Since $f_y(0, 0) = 0$, the function f does not satisfy the conditions of the Implicit function theorem and so f may or may not possess a unique solution in any nbd. of (0, 0).

Solving $y^2 + yx^3 + x^2 = 0$ for y, we obtain

$$y = \frac{1}{2}\,\{-x^3 \pm \sqrt{(x^4 - 4x^2)}\}$$

or $$y = \frac{1}{2}\{-x^3 \pm |x| \sqrt{(x^4 - 4)}\}.$$

Now, $|x| < \sqrt{2} \Rightarrow x^4 - 4 < 0$. Thus

$x^6 - 4x^2 < 0$ for $0 < |x| < \sqrt{2}$.

So $f(x, y) = 0$ has no solution of $0 < |x| < \sqrt{2}$.

Hence the equation $f(x, y) = 0$ does not possess a solution by any nbd. of $(0, 0)$.

Example 52:

Examine the equation $x^2 + y^2 - 1 = 0$ for existence of unique solutions at the points $(\pm 1, 0)$ and $(0, \pm 1)$. Also find the solutions whenever they exist.

Solution:

Let $f(x, y) = x^2 + y^2 - 1$ so that $f_y(x, y) = 2y$.

At $(0, 1)$:

(i) $f(0, 1) = 0$,

(ii) $f_y(0, 1) = 2 \neq 0$,

(iii) obviously f_x, f_y are continuous in a nbd. of $(0, 1)$.

Thus $f(x, y) = 0$ has a unique solution near $(0, 1)$

Solving $x^2 + y^2 = 1$, $y = \pm \sqrt{1 - x^2}$. But $y = g_1(x) = \sqrt{1 - x^2}$ in the unique solution, since $g(0) = 1$.

At $(0, -1)$: it can similarly be shown that $y = g_2(x) = -\sqrt{1 - x^2}$ is the unique solution, since $g(0) = 1$.

At $(1, 0)$, we notice that $f_y(1, 0) = 0$. Thus the function f does not satisfy the conditions of Implicit function theorem and so f may or may not possess a unique solution in any nbd. of $(0, 0)$.

Solving $x^2 + y^2 - 1 = 0$, $y = \pm \sqrt{(1 - x^2)}$,

Since neither of the functions $y = \sqrt{(1 - x^2)}$ and $y = -\sqrt{(1 - x^2)}$ is defined for $|x| < 1$, so $f(x, y) = 0$ does not have any solution in any *nbd.* of $(1, 0)$.

Similarly, it can be shown that $f(x, y) = 0$ does not have any solution in any nbd. of $(-1, 0)$.

Example 53:

Show that the equation $y^2 - yx^2 - 2x^5 = 0$ has a unique solution $y = g(x)$ near the point (1, –1), and find g'(x) at (1, 1). Verify by direct calculation. **(D.U., 1993, 88)**

Solution:

Let $f(x, y) = y^2 - yx^2 - 2x^5$

$\therefore f_x(x, y) = -2xy - 10x^4$, $f_y(x, y) = 2y - x^2$. Now

(i) $f(1, -1) = 0$,

(ii) $f_y(1, -1) = -3 \neq 0$,

(iii) f_x and f_y are continuous for all $(x, y) \in R^2$ and so in a nbd. of $(1, -1)$.

Thus $f(x, y) = 0$ has a unique solution $y = g(x)$ in some nbd. of $(1, -1)$.

By solving the equation $y^2 - yx^2 - 2x^5 = 0$ for y in terms of x,

$$y = \frac{1}{2}\left(x^2 \pm \sqrt{x^4 + 8x^5}\right) = \frac{1}{2}x^2\left(1 \pm \sqrt{1 + 8x}\right), \ x > -\frac{1}{3}.$$

Out of these two possible solutions,

$$y = \frac{1}{2}x^2\left(1 - \sqrt{1 + 8x}\right) = g(x)$$

is the unique solution of $f(x, y) = 0$ for $x > -\frac{1}{8}$, since $g(1) = -1$.

Now $g'(x) = -\dfrac{f_x(x, y)}{f_y(x, y)} = \dfrac{2xy + 10x^4}{2y - x^2}$.

At $(1, -1)$, $g'(x) = -8/3$.

Example 54:

Show that the equation $2xy - \log xy = 2$ determines a unique solution near (1, 1). Find the first derivative of the solution at (1, 1).

Solution:

Let $f(x, y) = 2xy - \log xy - 2$.

$\therefore f_x(x, y) = 2y - (1/x)$, $f_y(x, y) = 2x - (1/y)$. Now

(i) $f(1, 1) = 0$,

(ii) $f_y(1, 1) = 1 \neq 0$

(iii) Clearly f_x and f_y are continuous in a *nbd.* of (1, 1).

Hence by the Implicit function theorem, f(x, y) = 0 has a unique solution near (1, 1). We have

$$g'(x) = -\frac{f_x}{f_y} = \frac{2y - 1/x}{2x - 1/y} = \frac{y}{x}.$$

At (1, 1); g'(x) = 1.

Example 55:

Show that the equation $x^3 - y^3 - 3xy - y = 0$ has a unique solution y = g(x) near the point (2, 1), and find g'(x).

Solution:

Let $f(x, y) = x^3 - y^2 - 3xy - y$.

$\therefore$ $f_x(x, y) = 3x^3 - 3y$, $f_y(x, y) = -3y^2 - 3x - 1$. Now

(i) $f_x(2, 1) = 0$,

(ii) $f_y(2, 1) = -10 \neq 0$,

(iii) f_x and f_y are continuous for all $(x, y) \in R^2$ and so in a nbd of (2, 1). Hence by the Implicit function theorem, f(x, y) = 0 has a unique solution y = g(x) in some nbd. of (2, 1). We have

$g'(x) = -f_x/f_y = 3(x^2 - y)/(3y^2 + 3x + 1)$.

Example 56:

Show that the equation

$$ax + by + Ax^2 + Hxy + By^2 = 0 \qquad (b \neq 0)$$

determines a unique solution near (0, 0). Find the first derivative of the solution at (0, 0). **(D.U., 1986)**

Solution:

Let $f(x, y) = ax + by + Ax^2 + Hxy + By^2$.

$\therefore$ $f_x(x, y) = a + 2Ax + Hy$, $f_y(x, y) = b + Hx + 2By$. Now

(i) f(0, 0) = 0,

(ii) $f_y(0, 0) = b \neq 0$,

(iii) f_x and f_y are obviously continuous in a nbd. of (0, 0). So by the Implicit function theorem, f(x, y) = 0 has a unique solution y = g(x) near (0, 0). Now

$$g'(x) = -\frac{f_x}{f_y} = \frac{a + 2Ax + Hy}{b + Hx + 2By}.$$

At (0, 0), g'(x) = –a/b.

Example 57:

Show that the equation $x^3 + xy + y^2 - 1 = 0$ determines a unique implicit function in a neighbourhood of (1, 0). **(D.U., 1990)**

Solution:

Let $f(x, y) = x^3 + xy + y^2 - 1$.

$\therefore \quad f_x = 3x^2 + y,\ f_y = x + 2y$. Now

(i) $f(1, 0) = 0$, (ii) $f_y(1, 0) = 1 \neq 0$,

(iii) f_x and f_y are continuous for all $(x, y) \in R^2$ and so in a nbd. of (1, 0)

Hence by the Implicit function theorem, f(x, y) = 0 has a unique solution y = g(x) in some nbd. of (1, 0). Solving $x^3 + xy + y^2 - 1 = 0$ for y in terms of x, we obtain

$$y = \tfrac{1}{2}\left(-x \pm \sqrt{[x^2 - 4(x^5 - 1)]}\right)$$

Out of these two possible solutions,

$$y = \tfrac{1}{2}\left(-x + \sqrt{[x^2 - 4(x^3 - 1)]}\right) = g(x)$$

is the unique solution of f(x, y) = 0, since g(1) = 0.

Example 58:

Show that the equation $y^3 + x^3 - 3xy + y = 0$ determines a unique implicit function near the origin satisfying the properties laid down in the implicit function theorem. **(D.U., 1989, 87, 82)**

Solution:

Let $f(x, y) = y^3 + x^3 - 3xy + y$.

$\therefore \quad f_x = 3x^2 - 3y,\ f_y = 3y^2 - 3x + 1$. Now

(i) $f(0, 0) = 0$, (ii) $f_y(0, 0) = 1 \neq 0$,

(iii) f_x and f_y are continuous for all $(x, y) \in R^2$ and so in a nbd of (0, 0).

Hence by the Implicit Function Theorem, f(x, y) = 0 has a unique solution y = s (x) in some nbd. of (0, 0).

7.15 Maxima and Minima of Functions of Two Variables

Definition 1:

A function f(x, y) is said to have a *maxima* at a point (a, b) if there exists a nbd. N of (a, b) such that

$$f(a + h, b + k) < f(a, b) \quad \forall\ (a + h, b + k) \in N \sim \{(a, b)\}.$$

The set N ~ {(a, b)} is called a *deleted nbd.* of (a, b).

Definition 2:

A function f(x, y) is said to have a *minimum* a point (a, b) if there exists a nbd. N of (a, b) such that

$$f(a + h, b + k) < f(a, b) \ \forall \ (a + h, b + k) \in N \sim \{(a, b)\}.$$

Definition 3:

A function f(x, y) is said to have an *extremum* at a point (a, b) if it has either a maximum or minimum at (a, b).

Theorem 7:

If a function f(x, y) has an extremum at any point (a, b) of the domain of f and if f admits of partial derivatives f_x and f_y at (a, b), then

$$f_x(a, b) = 0 \text{ and } f_y(a, b) = 0.$$

Proof:

Suppose f(x, y) has a maximum at (a, b). Then the function f(x, b) of a single variable x must have a maximum at x = a and so

$$f_x(x, b) = 0 \text{ when } x = a \text{ i.e., } f_x(a, b) = 0.$$

Again the function f(a, y) of a single variable y must have a maximum at y = b and so

$$f_y(a, y) = 0 \text{ when } y = b \text{ i.e., } f_x(a, b) = 0.$$

Exactly similar is the case when f has a minimum at (a, b).

Remark:

The conditions of the above theorem are only necessary but are not sufficient.

Let $f(x, y) = x^2 - y^2$. Then

$$f_x(0, 0) = 0, \ f_y(0, 0) = 0. \text{ Also } f(0, 0) = 0.$$

Let N be any nbd. of (0, 0). Let $\varepsilon > 0$.

We take two points of N as A $(\frac{1}{2}\varepsilon, 0)$ and B $(0, \frac{1}{2}\varepsilon)$.

At the point A, $f(\frac{1}{2}\varepsilon, 0) = \frac{1}{4}\varepsilon^2 - 0 = \frac{1}{4}\varepsilon^2$.

At the point B, $f(0, \frac{1}{2}\varepsilon) = 0 - \frac{1}{4}\varepsilon^2 = -\frac{1}{4}\varepsilon^2$.

Thus, $f(\frac{1}{2}\varepsilon, 0) - f(0, 0) = \frac{1}{4}\varepsilon^2 - 0 = \frac{1}{4}\varepsilon^2 > 0,$

$$f(0, \tfrac{1}{2}\varepsilon) - f(0, 0) = -\tfrac{1}{4}\varepsilon^2 - 0 = -\tfrac{1}{4}\varepsilon^2 < 0$$

or $f(\frac{1}{2}\varepsilon, 0) > f(0, 0)$ and $f(0, \frac{1}{2}\varepsilon) < f(0, 0)$,

where $(\frac{1}{2}\varepsilon, 0), (0, \frac{1}{2}\varepsilon) \in N$.

Hence the given function is neither maximum nor minimum at (0, 0).

Definition 4:

A point (a, b) is said to be a *critical point* for a function f(x, y) if f_x and f_y both exist at (a, b) and

$$f_x(a, b) = 0, \quad f_y(a, b) = 0.$$

If (a, b) is a critical point for a function f(x, y), then f(a, b) is called a *stationary value* of f.

Remark:

If f has an extremum at (a, b), then (a, b) is a cretical point but not conversely.

Theorem 8:

Let $f_x(a, b) = 0$, $f_y(a, b) = 0$;

and $f_{xx}(a, b) = A$, $f_{xy}(a, b) = B$, $f_{yy}(a, b) = C$. *Then*

(i) f has minimum at (a, b), if $AC - B^2 > 0$, $A > 0$,

(ii) f has maximum at (a, b), if $AC - B^2 > 0$, $A < 0$,

(iii) f has neither maximum value nor minimum value at (a, b), if $AC - B^2 < 0$.

(In this case, the point (a, b) is called a saddle point.)

(iv) f may or may not have an extremum at (a, b) if $AC - B^2 = 0$.

(The case (iv) is often called the doubtful case, which needs further investigation.)

Example 59:

Show that the function $f(x, y) = 2x^4 - 3x^2y + y^2$ *has neither a maximum nor a minimum at (0, 0).*

Solution:

$$f_x(x, y) - 8x^3 - 6xy, \quad f_y(x, y) = -3x^2 + 2y.$$

$\therefore$ $f_x(0, 0) = 0 = f_y(0, 0)$

Now $f_{xx}(x, y) = 24x^2 - 6y = 0$, at (0, 0)

$f_{xy}(x, y) = -6x = 0$, at (0, 0)

$f_{yy}(x, y) = 2$, at (0, 0)

At (0, 0), $f_{xx}, f_{yy} - (f_{xy})^2 = 0$.

Thus it is a doubtful case, and so requires further examination.

We have $f(x, y) = (x^2 - y)(2x^2 - y), f(0, 0) = 0$.

$$\therefore \quad f(x, y) - f(0, 0) = (x^2 - y)(2x^2 - y)$$
$$> 0, \text{ for } y < 0 \text{ or } x^2 > y > 0$$
$$< 0, \text{ for } y > x^2 > \tfrac{1}{2}y > 0.$$

We observe that $f(x, y) - f(0, 0)$ does not keep the same sign near the origin. Hence the given function has neither a maximum nor a minimum value at (0, 0).

Example 60:

Investigate the maxima and minima of

$$f(x, y) = x^2 + 3xy + y^2 + x^3 + y^2.$$

Solution:

$$f_x = 2x + 3y + 3x^2, \; f_y = 3x + 2y + 3y^3,$$
$$f_{xx} = 2 + 6x, \; f_{yy} = 2 + 6y, \; f_{xy} = 3.$$

For critical points, we have $f_x = 0$ and $f_y = 0$

$$\Rightarrow \quad f_x - f_y = 0 \quad \Rightarrow (y - x) - 3(y^2 - x^2) = 0$$
$$\Rightarrow \quad (y - x)(1 - 3y - 3x) = 0 \quad \Rightarrow x = y \text{ or } 3x + 3y = 1.$$

Taking $x = y$ in $f_x = 0, f_y = 0$, we obtain

$$5x + 3x^2 = 0 \quad \Rightarrow x = 0, \; -5/3.$$

So two critical points are (0, 0), (–5/3, –5/3).

It can be verified that on solving

$$\{3x + 3y = 1, f_x = 0\}, (3x + 3y = 1, f_y = 0\},$$

we get imaginary points (and so are discarded).

At (–5/3, –5/3), $A = f_{xx} = -8 < 0$ and

$$AC - B^2 = (-8)(-8) - 9 = 55 > 0.$$

Hence the given function has a maximum at (–5/3, –5/3).

At (0, 0), $AC - B^2 = 4 - 9 = -5 < 0$.

Hence the given function has neither a maximum nor a minimum at (0, 0).

Example 61:

Show that the function $f(x, y) = x^2 - 3xy^2 + 2y^4$ has neither a maximum nor a minimum value at the origin. **(D.U., 1990)**

Solution:

$$f_x = 2x - 3y^2, \qquad f_y = -6xy + 8y^2, \; f_{xx} = 2,$$

$$f_{yy} = -6x + 24y^2, \qquad f_{xy} = -6y.$$

At (0, 0), $AC - B^2 = f_{xx}, \; f_{yy} - (f_{xy})^2 = 0.$

Thus, it is a *doubtful* case and so needs further investigation. We have $f(0, 0) = 0$.

Clearly $f(x, y) = (x - y^2)(x - 2y^2)$

$\therefore \quad f(x, y) - f(0, 0) = (x - y^2)(x - 2y^2)$

$$> 0, \begin{cases} \text{for } y^2 > x \text{ (and so } 2y^2 > x) \\ \text{or} \\ \text{for } y^2 < \frac{1}{2}x \text{ (and so } y^2 < x) \end{cases}$$

$$< 0, \text{ for } \tfrac{1}{2}x < y^2 < x.$$

Thus, we observe that $f(x, y) - f(0, 0)$ does not keep the same sign near the origin and so the given function has neither a maximum nor a minimum value at the origin.

Example 62:

Prove that the function

$$f(x, y) = x^2 - 2xy + y^2 + x^4 + y^4$$

has a minima at the origin.

Solution:

$$f_x = 2x - 2y + 4x^3, \; f_y = -2x + 2y + 4y^3$$

$$f_{xx} = 2 + 12x^3, \; f_{xy} = -2, \; f_{yy} = 2 + 12y^2.$$

At the origin, $AC - B^2 = f_{xx} f_{yy} - (f_{xy})^2 = 4 - 4 = 0.$

Thus, it is a *doubtful* case and so requires further investigation. We have

$f(x, y) = (x - y)^2 + x^4 + y^4.$ Also $f(0, 0) = 0.$

Clearly $f(x, y) > f(0, 0)$ for all values of (x, y) in a deleted nbd of (0, 0). Hence the given function has a minimum at the origin.

Example 63:

Find the maxima and minima of the function

$f(x, y) = x^2 + y^2 - 3x - 12y + 20.$ **(D.U., Physics (H) 1994)**

Solution:

We have

$$f_x = 3x^2 - 3,\ f_y = 3y^2 - 12,\ f_{xx} = 6x,\ f_{yy} = 6y,\ f_{xy} = 0.$$

To determine the critical points, we solve

$$f_x = 0,\ f_y = 0 \ \Rightarrow 3x^2 - 3 = 0,\ 3y^2 - 12 = 0$$

$\Rightarrow$ $x = \pm 1,\ y = \pm 2$. Hence the critical points are

$(1, 2), (1 - 2), (-1, 2), (-1, -2)$.

At $(1, 2)$, $A = f_{xx} = 6 > 0$.

$$AC - B^2 = f_{xx} \cdot f_{yy} - (f_{xy})^2 = 6 \times 12 - 0 = 72 > 0.$$

Thus f has minimum at (1, 2).

At $(1, -2)$, $AC - B^2 = 6 \times - 12 - 0 < 0$

Thus, f has neither maximum nor minimum at (1, – 2).

At $(-1, 2)$, $AC - B^2 = -6 \times - 12 - 0 < 0$.

Thus, f has neither maximum nor minimum at (–1, 2)

At $(-1, -2)$, $A = - 6 < 0$,

$$AC - B^2 = (-6)(-12) - 0 > 0.$$

Thus, f has maximum at (–1, –2).

Ex.:

Discuss the nature of the critical values of

$f(x, y) = x^3 + y^3 - 3x - 12y + 20 = 0.$ **(D.U., 1994)**

Example 64:

Show that the function

$$f(x, y) = (x - y)^2 + x^3 - y^3 + x^5$$

has neither a maximum nor a minimum at the origin.

(D.U., Physics (H) 1995)

Solution:

We have

$$f_x = 2(x - y) + 3x^2 + 5x^4,\ f_y = - 2(x - y) - 3y^2,$$

$$f_{xx} = 2 + 6x + 20x^3,\ f_{xy} = - 2,\ f_{yy} = 2 - 6y.$$

At (0, 0), $f_x = 0,\ f_y = 0,\ f_{xx} = 2,\ f_{xy} = -2,\ f_{yy} = 2$, so that

$f_{xx}\, f_{yy} - (f_{xy})^2 = 0.$

Thus, it is a doubtful case. We have $f(0, 0) = 0$.

The given function can be rewritten as

$$f(x, y) = (x - y)^2 + (x - y)(x^2 + xy + y^2) + x^5.$$

For points along the line $y = x$, we have

$$f(x, y) = x^5.$$

Since the power of x is odd here, so in every neighbourhood of (0, 0), there are points at which f(x, y) is positive and there are points at which f(x, y) is negative. [e.g., $(\varepsilon, \varepsilon)$, $(-\varepsilon, -\varepsilon)$ where $\varepsilon > 0$].

Hence, f(0, 0) is neither a maximum nor a minimum value.

Example 65:

Discuss the nature of the critical values of

$f(x, y) = e^x \sin y$. **(D.U., 1993)**

Solution:

$$f_x = e^x \sin y,\ f_y = e^x \cos y,\ f_{xx} = e^x \sin y,$$
$$f_{xy} = e^x \cos y,\ f_{yy} = -e^x \sin y.$$

The critical values are given by solving

$$f_x = 0,\ f_y = 0 \Rightarrow f_x - f_y = 0$$

$$\Rightarrow e^x(\sin y - \cos y) = \ \Rightarrow \sin y = \cos y \ (\because e^x \neq 0 \ \forall\ x)$$

Thus, $(x, \frac{1}{4}\pi)$ is a critical point, where $x \in R$ is arbitrary.

We know that $e^x > 0 \ \forall\ x \in R$.

$$\text{At } y = \tfrac{1}{4}\pi,\ A = f_{xx} = \frac{1}{\sqrt{2}} e^x > 0 \quad \forall \quad x \in R.$$

$$AC - B^2 = f_{xx} f_{yy} - (f_{xy})^2 = e^{zn} \left[\frac{1}{\sqrt{2}}\left(-\frac{1}{\sqrt{2}}\right) - \left(\frac{1}{\sqrt{2}}\right)^2\right]$$

$$= -e^{zn} < 0.$$

Hence, the given function is neither maximum nor minimum at the critical points $\left(x, \frac{1}{4}\pi\right)$, $x \in R$.

Example 66:

Show that $f(x, y) = y^2 + x^2y + x^4$, has a minimum at (0, 0).

Solution:

We have $f_x = 2xy + 4x^3$, $f_y = 2y + x^2$, $f_{xx} = 2y + 12x^2$, $f_{xy} = 2x$, $f_{yy} = 2$.

At (0, 0); $f_x = 0$, $f_y = 0$, $f_{xx} = 0$, $f_{xy} = 0$, $f_{yy} = 2$.

Thus, at the origin $f_{xx} f_{xy} - (f_{xy})^2 = 0$, so that it is a *doubtful* case and requires further investigation. The given function can be rewritten as

$$f(x, y) = (y + \tfrac{1}{2}x^2)^2 + \tfrac{3}{4}x^4 \text{ . Also } f(0, 0) = 0.$$

$$\therefore f(x, y) - f(0, 0) = (y + \tfrac{1}{2}x^2)^2 + \tfrac{3}{4}x^4 > 0,$$

for all (x, y) in a deleted nbd of (0, 0).

Hence, f has a minimum at (0, 0).

Example 67:

Show that the function $(y - x)^4 + (x - 2)^4$ has a minimum at (2, 2).

(D.U., Physics (H) 1993)

Solution:

Let $f(x, y) = (y - x)^4 + (x - 2)^4$. Then

$$f_x = -4(y - x)^3 + 4(x - 2)^2,\ f_y = 4(y - x)^3,\ f_{yy} = 12(y - x)^2,$$

$$f_{xx} = 12(y - x)^2 + 12(x - 2)^2,\ f_{xy} = -12(y - x)^2.$$

At (2, 2); $f_x = 0$, $f_y = 0$, $f_{xx} = 0$, $f_{yy} = 0$, $f_{xy} = 0$.

$\therefore \quad f_{xx} f_{yy} - (f_{xy})^2 = 0.$

Thus, it is a doubtful case and requires further investigation. We have $f(2, 2) = 0$.

Now, $f(x, y) - f(2, 2) = (y - x)^4 + (x - 2)^4 > 0$ i.e., $f(x, y) < f(2, 2)$ for all values of (x, y) in a deleted nbd. of (2, 2). Hence the given function has a minimum at (2, 2).

Example 68:

Find all the maxima and minima of the function given by

$f(x, y) = x^3 + y^3 - 63(x + y) + 12xy$. **(D.U., Physics (H) 1996)**

Solution:

We have

$$f_x(x, y) = 3x^2 - 63 + 12y,\quad f_y(x, y) = 3y^2 - 63 + 12x;$$

$$f_{xx}(x, y) = 6x,\ f_{xy}(x, y) = 12,\ f_{yy}(x, y) = 6y.$$

The critical points of f are given by $f_x = 0$, $f_y = 0$.

Thus, $\quad f_x(x, y) = 3x^3 - 63 + 12y = 0$

$$f_x(x, y) - f_y(x, y) = 3(x^2 - y^2) + 12(y - x) = 0$$

or $\quad x^2 - 21 + 4y = 0,\ (y - x)(4 - y - x) = 0$

$$\therefore \begin{cases} x^2 - 21 + 4x = 0 \text{ (if } y = x) \text{ and} \\ (x + 7)(x - 3) = 0 \end{cases} \quad \begin{cases} x^2 - 21 + 4y = 0, \\ 4 - y - x = 0. \end{cases}$$

Solving these, the four critical points are

$(-7, -7), (3, 3), (5, -1), (-1, 5).$

At $(-7, -7)$,

$A = f_{xx} = -42 < 0$ and $AC - B^2 = f_{xx} f_{yy} - (f_{xy})^2 = 1620 > 0$

so that the function is maximum at $(-7, -7)$.

At $(3, 3)$,

$A = f_{xx} = 18 > 0$ and $AC - B^2 = f_{xx} f_{yy} - (f_{xy})^2 = 180 > 0$,

so that the function is minimum at $(3, 3)$.

At each of the other points $(5, -1)$ and $(-1, 5)$,

$AC - B^2 = f_{xx} f_{yy} - (f_{xy})^2 = -324 < 0$

so that the function is neither a maximum nor a minimum.

Example 69:

Find the maxima and minima of

$f(x, y) = x^4 + y^4 = 2(x - y)^2.$

Solution:

$f_x = 4x^3 - 4(x - y),\ f_y = 4y^2 + 4(x - y),$

$f_{xx} = 12x^2 - 4,\ f_{yy} = 12y^2 - 4,\ f_{xy} = 4,$

For critical points, $f_x = 0$ and $f_y = 0$

$\Rightarrow \quad f_x - f_y = 0 \Rightarrow 4(x^3 - y^3) - 8(x - y) = 0$

$\Rightarrow \quad 4(x - y)(x^2 + y^2 + xy - 2) = 0$

Similarly, $f_x + f_y = 0 \Rightarrow 4(x^3 + y^2) = 0$

$\Rightarrow \quad (x + y)(x^2 - xy + y^2 = 0).$

Taking $x = y$ in $f_x = 0$ and $f_y = 0$, we see that $(0, 0)$ is a critical point for $f(x, y)$.

Taking $y = -x$ in $f_x = 0$ and $f_y = 0$, we see that

$4x^3 - 8x = 0$ and $4y^3 - 8y = 0$

$\Rightarrow \quad x(x^2 - 2) = 0$ and $y(y^2 - 2) = 0$

or $\quad x = \pm\sqrt{2},\ y = \pm\sqrt{2}.$

Thus $(\sqrt{2}, -\sqrt{2}), (-\sqrt{2}, \sqrt{2})$ are the other critical points for $f(x, y)$.

At $(\sqrt{2}, -\sqrt{2})$, $A = f_{xx} = 20 > 0$

And $AC - BC_2 = 20 \times 20 - 4 > 0$

Thus, f has a minimum at $(\sqrt{2}, -\sqrt{2})$.

At $(\sqrt{2}, -\sqrt{2})$, A $f_{xx} = 20 > 0$ and $AC - B^2 > 0$ and so f also has a minimum at $(\sqrt{2}, \sqrt{2})$.

At (0, 0), $AC - B^2 = 0$. So it is a doubtful case which needs further investigation. We see that f(0, 0) = 0.

Since f(x, y) – f(0, 0) does not keep the same sign near the origin, so the given function has neither a maximum nor a minimum at the origin.

Example 70:

Show that of all triangles inscribed in a circle, the ones with maximum area are equilateral.

Solution:

Let ABC be a triangle inscribed in a circle of radius r. Let $\angle A = x$, $\angle B = y$, $\angle C = z$ so that $x + y + z = \pi$. If S denotes the area of ΔABC, then $S = \frac{1}{2} ab \sin z$, where a, b denote the lengths of the sides containing the angle z. We have

$$S = + \tfrac{1}{2}(2r \sin x)(2r \sin y) \sin z$$

or $$\frac{S}{2r^2} = \sin x \sin y \sin (x + y), \text{ since } z = p - (x + y).$$

To maximize S, we shall maximize

$$f(x, y) = \sin x \sin y \sin (x + y)$$

f(x, y) has a maximum at $x = y = \frac{1}{3}\pi$.

Thus, $z = \pi - (x + y) \Rightarrow z = \frac{1}{3}\pi$. Hence ΔABC is equilateral.

Example 71:

Show that the function

$$f(x, y) = \sin x \sin y \sin (x + y)$$

has a maximum at $(\frac{1}{3}\pi, \frac{1}{3}\pi)$ *in the interval* $(0, \pi)$.

Solution:

$$f(x, y) = \sin x \sin y \sin (x + y).$$

$\therefore$ $f_x = \{\cos x \sin (x + y) + \sin x \cos (x + y)\} \sin y$

or $f_x = \sin (2x + y) \sin y.$ ...(1)

Now, $f_y = \{\cos yu \sin(x + y) + \sin y \cos(x + y)\} \sin x$

or $f_y = \sin(x + 2y) \sin x.$ ...(2)

Again, $f_{xx} = 2 \cos(2x + y) \sin y$, $f_{yy} = 2 \cos(x + 2y) \sin x$,

$f_{xy} = \sin(2x + y) \cos y + \sin y \cos(2x + y)$

or $f_{xy} = \sin(2x + 2y).$

Now, $f_x = f_y = 0 \Rightarrow \sin(2x + y) = 0 = \sin(x + 2y).$

$(\because \sin x \neq 0, \sin y \neq 0 \text{ for } 0 < x, y < \pi)$

Thus, $2x + y = \pi$, $x + 2y = \pi$ ...(3)

$(\because 2x + y = 0 = x + 2y \Rightarrow x + y = 0 \Rightarrow \sin(x + y) = 0$

$\Rightarrow f(x, y) = 0.)$

From (3), we get $x = y = \frac{1}{3}\pi$.

Now, $A = f_{xx}\left(\frac{1}{3}\pi, \frac{1}{3}\pi\right) = 2 \cos \pi \sin \frac{1}{3}\pi = -\sqrt{3},$

$B = f_{xy}\left(\frac{1}{3}\pi, \frac{1}{3}\pi\right) = \sin(4\pi/3) = -\frac{1}{2}\sqrt{3},$

$C = f_{yy}\left(\frac{1}{3}\pi, \frac{1}{3}\pi\right) = 2 \cos \pi \sin \frac{1}{3}\pi = -\sqrt{3}.$

$\therefore$ $AC - B^2 = 9/4 > 0$ and $A < 0$.

Hence, $f(x, y)$ has a maximum at $\left(\frac{1}{3}\pi, \frac{1}{3}\pi\right)$.

7.16 Maxima and Minima Subject to Constraints

First we discuss the method of finding the maxima and minima of a function of three variables.

7.17 Extreme Points of a Function of Three Variables

The extreme point of a function $f(x, y, z)$ of three independent variables x, y, z can be determined by the following:

Rule:

If $df(a, b, c) = f_x\, dx + f_y\, dy + f_z\, dz,$

$d^2f(a, b, c) = f_{xx}(dx)^2 + f_{yy}(dy)^2 + f_{zz}(dz)^2$

$+ 2 f_{xy}\, dx\, dy + f_{yz}\, dy\, dz + 2f_{xz}\, dz\, dx$, *then*

(i) $f(x, y, z)$ *has a* ***maximum*** *at a point* (a, b, c) *if*

$df = 0$ *(i.e.,* $f_x = f_y = f_z = 0$*) and* $d^2 f > 0.$

(ii) $f(x, y, z)$ *has a* ***minimum*** *at a point* (a, b, c) *if*

$df = 0$ *(i.e.,* $f_x = f_y = f_z = 0$*) and* $d^2f > 0.$

7.18 Lagrange's Method of Undetermined Multipliers

Suppose we have to find the stationary points of a function f(x, y, z) = 0 subject to the conditions (constraints).

$$\phi_1(x, y, z) = 0 \quad \text{and } \phi_2(x, y, z) = 0. \qquad ...(1)$$

The stationary points of the given function f are given by determining the stationary points of the function F, where

$$F(x, y, z) = f + \lambda_1 \phi_1 + \lambda_2 \phi_2.$$

and considering the variables x, y, z as independent variables. The constants λ_1 and λ_2 are known as **Lagrange's Undetermined Multipliers.**

The stationary points of the function F are given by

$$dF = F_x\, dx + F_y\, dy + F_z\, dz = 0 \text{ (i.e., } F_x = F_y = F_z = 0). \qquad ...(2)$$

The equations (1) and (2) help us to determine the undertermined multipliers λ_1 and λ_2 as well as the extreme points of the given function f(x, y, z).

We shall illustrate the Lagrange's method of undetermined multipliers with the help of examples.

Example 72:

Find the shortest distance from the origin to the hyperbola

$$x^2 + 8xy + 7y^2 = 225,\ z = 0.$$

Solution:

We have to find the minimum value of $x^2 + y^2$ (the square of the distance from the origin to any point in the xy-plane) subject to the constraint $x^2 + 8xy + 7y^2 = 225$.

Consider the function

$$F = x^2 + y^2\, \lambda(x^2 + 8xy + 7y^2 - 225). \qquad ...(1)$$

Now, $F_x = (2x + 2x\lambda + 8y\lambda)$, $F_y = (2y + 8x\lambda + 14y\lambda)$

$$\left.\begin{array}{l} F_x = 0 \Rightarrow (1 + \lambda)\, x + 4\lambda y = 0 \\ F_y = 0 \Rightarrow 4\lambda x + (1 + 7\lambda)y = 0 \end{array}\right\} \quad \because \lambda = 1; -\frac{1}{9}.$$

Putting, $\lambda = 1$, $x = -2y$ in $x^2 + 8xy + 7y^2 = 225$, we get

$y^2 = -45$. This is impossible

Putting, $\lambda = -\frac{1}{9}$, $y = 2x$ in $x^2 + 8xy + 7y^2 = 225$, we get

$$x^2 = 5,\ y^2 = 20, \text{ and so } x^2 + y^2 = 25.$$

Now, $A = F_{xx} = 2(1 + \lambda)$, $B = F_{xy} = 8\lambda$, $C = F_{yy} = 2(1 + 7\lambda)$.

At $\lambda = -1/9$, $A = 16/9 > 0$, $C = 4/9$, $B = -8/9$.

$\therefore$ $AC - B^2 = 0$. Putting $\lambda = -1/9$ in (1), we get

$F = \frac{2}{9}(2x - y)^2 + 25 > 0$ for all values of (x, y).

Hence the function $x^2 + y^2$ has a minimum value 25.

Example 73:

Show that the maximum and minimum values of r^2, where

$$r^2 = a^2x^2 + b^2y^2 + c^2z^2, \quad \text{...(1)}$$

$$x^2 + y^2 + z^2 = 1 \quad \text{and} \quad lx + my + nz = 0 \quad \text{...(2)}$$

are given by the equation

$$\frac{l^2}{a^2 - r^2} + \frac{m^2}{b^2 - r^2} + \frac{n^2}{c^2 - r^2} = 0.$$

Solution:

Let us consider the function

$$F = (a^2x^2 + b^2y^2 + c^2z^2) + \lambda_1 (x^2 + y^2 + z^2 - 1) + \lambda_2 (lx + my + nz)$$

$\therefore$ $dF = F_a\,dx + f_y\,d_y + F_z\,dz$,

where $F_x = 2a^2x + 2\lambda_1 x + \lambda_z l$,

$F_y = 2b^2y + 2\lambda_1 y + \lambda_2 m$,

$F_z = 2c^2z + 2\lambda_1 z + \lambda_2 n$.

At stationary points.

$F_x = F_z = F_2 = 0$.

$\therefore$ $xF_x + yF_z + zF_2 = 0$ given us

$2(a^2x^2 + b^2y^2 + c^2z^2) + 2\lambda_1 (x^2 + y^2 + z^2) + \lambda_2 (lx + my + nz) = 0$

or $2r^2 + 2\lambda_1 = 0$ or $\lambda_1 = -r^2$, using (1) and (2)

Putting $\lambda_1 = -r^2$ in $F_x = 0$, we obtain

$2(a^2 - r^2)x + \lambda_2 l = 0$ or $x = -\lambda_2 l/2(a^2 = r^2)$ etc.

$$\therefore \quad x = \frac{-\lambda_2 l}{2(a^2 - r^2)}, \; y = \frac{-\lambda_2 m}{(b^2 - r^2)}, \; z = \frac{-\lambda_2 n}{(c^2 - r^2)}$$

Substituting these values in $lx + my + nz = 0$, we obtain

$$\frac{l^2}{a^2 - r^2} + \frac{m^2}{b^2 - r^2} + \frac{n^2}{c^2 - r^2} = 0. \quad (\because \lambda_2 \neq 0)$$

Example 74:

Prove that the volume of the greatest rectangular parallelopiped, that can be inscribed in the ellipsoid

$$\frac{x^2}{a^2}+\frac{y^2}{b^2}+\frac{z^2}{c^2}=1, \text{ is } \frac{8abc}{3\sqrt{3}}$$ **(D.U., 1993, 87)**

Solution:

The volume of a rectangular parallelopiped of sides 2x, 2y, 2z is 2x. 2y. 2z = 8xyz.

We have to find the greates value of 8xyz subject to the condition

$$x^2/a^2 + y^2/b^2 + z^2/c^2 = 1, (x > 0, y > 0, z > 0). \quad ...(1)$$

We consider a function F of three independent variables x, y, z;

$$F = 8xyz + \lambda \left(\frac{x^2}{a^2}+\frac{y^2}{b^2}+\frac{z^2}{c^2}=1\right).$$

We know $dF = F_x\, dx + F_y\, dy + F_z\, dz$

or $$dF = \left(8yz+\frac{2x\lambda}{a^2}\right) dx + \left(8zx+\frac{2y\lambda}{b^2}\right) dy + \left(8xy+\frac{2z\lambda}{c^2}\right) dz$$

For stationary points, $F_y = F_a = F_z = 0$

$$\Rightarrow \quad 8yz+\frac{2x\lambda}{a^2} = 0,\ 8zx+\frac{2y\lambda}{b^2} = 0,\ 8xy+\frac{2zy}{c^2} = 0, \quad ...(2)$$

Multiplying these by x, y, z respectively and adding, we get

$$24xyz + 2\lambda \left(\frac{x^2}{a^2}+\frac{y^2}{b^2}+\frac{z^2}{c^2}\right) = 0 \quad \text{or} \quad \lambda = -12xyz., \quad \text{using (1)}$$

Putting this value of λ in (2), we get $8a^2yz - 24xyz = 0$ etc.

$$\therefore\ x = a/\sqrt{3},\ y = b/\sqrt{3},\ z = c/\sqrt{3}, \text{ and so } \lambda = -4abc/\sqrt{3} \quad ...(3)$$

Now $d^2F = Fxx\, dx^2 + F_{yy}\, dy^2 + F_{zz}\, dz^2 + 2F_{xy}\, dx\, dy + 2F_{yz}\, dy\, dz + 2F_{zx}\, dz\, dx$

$$= 2\lambda \left(\frac{dx^2}{a^2}+\frac{dy^2}{b^2}+\frac{dz^2}{c^2}\right) + 16z\, dx\, dy + 16x\, dy\, dz + 16y\, dz\, dx$$

Using (3), we get

$$d^2F = -\frac{8abc}{\sqrt{3}} \sum \frac{1}{a^2} dx^2 + \frac{16}{\sqrt{3}} \sum c\, dx\, dy \quad ...(4)$$

From (1) and (3), we have

$$2x \frac{dx}{a^2} + 2y \frac{dy}{b^2} + 2z \frac{dz}{c^2} = 0 \quad \text{or} \quad \frac{2}{\sqrt{3}}\left(\frac{dx}{a} + \frac{dy}{b} + \frac{dz}{c}\right) = 0$$

$$\therefore \left(\frac{dx}{a} + \frac{dy}{b} + \frac{dz}{c}\right)^2 = 0 \quad \Rightarrow \quad \sum \frac{ax^2}{a^2} + 2\sum \frac{dx\,dy}{ab} = 0$$

Multiplying it throughout by abc, we get

$$abc \sum \frac{dx^2}{a^2} = -2\sum c\,dx\,dy \qquad ...(5)$$

From (4) and (5), we obtain

$$\therefore \quad d^2F = -\frac{16}{\sqrt{3}} abc \sum \frac{dx^2}{a^2} < 0.$$

Hence $\left(\frac{a}{\sqrt{3}}, \frac{b}{\sqrt{3}}, \frac{c}{\sqrt{3}}\right)$ is a point of maxima and the maximum value of 8xyz is $\frac{8abc}{3\sqrt{3}}$.

Ex.: *Find the maximum value of xyz when*

$x^2/a^2 + y^2/b^2 + z^2/c^2 = 1.$ **(D.U., 1981) [Ans.** $abc/3\sqrt{3}$]

Example 75:

Find the maximum and minimum radii vectors of the section of the surface $(x^2 + y^2 + z^2)^2 = a^2x^2 + b^2y^2 + c^2z^2$ ...(1)

by the plane $lx + my + nz = 0.$ ...(2)

Solution:

We have to find the stationary values of $r^2 = x^2 + y^2 + z^2$ subject to the conditions (1) and (2). Let us consider the function

$$F = (x^2 + y^2 + z^2) + \lambda_2 \{(x^2 + y^2 + z^2)^2 - (a^2x^2 + b^2y^2 + c^2z^2)\} + \lambda_2 (lx + mx + nz)$$

$\therefore dF = F_x dx = F_y dy + F_z dz$, where

$F_x = [2x + \lambda_1 \{4x (x^2 + y^2 + z^2) - 2a^2x\} + \lambda_2 l]$,

$F_y = [2y + \lambda_1 \{4y (x^2 + y^2 + z^2) - 2b^2y\} + \lambda_2 m]$,

$F_z = [2z + \lambda_1 \{4z (x^2 + y^2 + z^2) - 2c^2z\} + \lambda_2 n]$.

At stationary points, $F_x = F_y = F_z = 0$

or $\quad xF_x = yF_y + zF_z = 0$

or $\quad 2(x^2 + y^2 + z^2) + 2\lambda_1 [2(x^2 + y^2 + z^2)^2 - (a^2x^2 + b^2y^2 + c^2z^2)] + \lambda_2 (lx + mx + nz) = 0$

or $\quad (x^2 + y^2 + z^2) + \lambda_1 (x^2 + y^2 + z^2)^2 = 0$; using (1) and (2)

$$\therefore \quad r^2 + \lambda_1 r^4 = 0 \Rightarrow \lambda_1 = -\frac{1}{r^2}$$

Now, $\quad F_x = 0 \Rightarrow 2x - \frac{1}{r^2}(4r^2x - 2a^3x) = -\lambda_2 l.$

$$\therefore \quad x = -\frac{\lambda_2 l r^2}{2(a^2 - r^2)}. \text{ Similarly } y = -\frac{\lambda_2 m r^2}{2(b^2 - r^2)},\ z = -\frac{\lambda_2 n r^2}{2(c^2 - r^2)}$$

Substituting these values in lx + my + zx = 0, we get

$$\frac{l^2 r^2}{a^2 - r^2} + \frac{m^2 r^2}{b^2 - r^2} + \frac{n^2 r^2}{c^2 - r^2} = 0$$

Ex. 1: *Find the maxima and minima of $x^2 + y^2 + z^2 = r^2$ subject to the conditions $ax^2 + by^2 + cz^2 = l$ and $lx + my + nz = 0$.*

$$\left[\textbf{Ans. } \frac{l^2}{ar^2 - 1} + \frac{m^2}{br^2 - 1} + \frac{n^2}{cr^2 - 1} = 0\right]$$

Ex. 2: *Find the stationary value of* $x^2/a^4 + y^2/b^4 + z^2/c^4 = r^2$, when $lx + my + nz = 0$ and $x^2/a^2 + y^2/b^2 + z^2/c^2 = 1$.

$$\left[\textbf{Ans. } \frac{l^2 a^4}{1 - a^2r^2} + \frac{m^2 b^4}{1 - b^2r^2} + \frac{n^2 c^4}{1 - a^2r^2} = 0\right]$$

Example 76:

Show that the length of the axes of the section of the ellipsoid $\frac{x^2}{a^2} + \frac{y^2}{b^2} + \frac{z^2}{c^2} = 1$ *by the plane $lx + my + nz = 0$ are the roots of the quadratic in r^2,*

$$\frac{l^2 a^2}{r^2 - a^2} + \frac{m^2 b^2}{r^2 - b^2} + \frac{n^2 c^2}{r^3 - c^2} = 0 \qquad \textbf{(D.U., 1988, 84, 82)}$$

Solution:

We have to find the stationary values of the function r^2, where $r^2 = x^2 + y^2 + z^2$, subject to the two conditions :

$$\frac{x^2}{a^2} + \frac{y^2}{b^2} + \frac{z^2}{c^2} = 1, \quad lx + my + nz = 0 \qquad ...(1)$$

Consider a function F of independent variables x, y, z :

$$F = x^2 + y^2 + z^2 + \lambda_1 \left(\frac{x^2}{a^2} + \frac{y^2}{b^2} + \frac{z^2}{c^2} = 1\right) + 2\lambda_2 (lx + my + nz)$$

Then $dF = F_x\, dx + F_y\, d_y + F_z\, dz$

$$\text{or } dF = 2\left(x + \frac{x\lambda_1}{a^2} + \lambda_2 l\right) dx + 2\left(y + \frac{x\lambda_1}{b^2} + \lambda_2 m\right) dy$$

$$+\left(z + \frac{z\lambda_1}{c^2} + \lambda_2 n\right) dz$$

For stationary points, $F_x = F_y = F_z = 0$

$$x + \frac{x}{a^3}\lambda_1 + l\lambda_2 = 0,\ y + \frac{y}{b^2}\lambda_1 + m\lambda_2 = 0,\ z + \frac{z}{c^2}\lambda_1 + n\lambda_2 = 0, \quad ...(2)$$

Multiplying these by x, y, z respectively and adding, we get

$$(x^2 + y^2 + z^2)\ \lambda_1 \left(\frac{x^2}{a^2} + \frac{y^2}{b^2} + \frac{z^2}{c^2}\right) + \lambda^2 (lx + my + nz) = 0$$

or $r^2 + \lambda_1 = 0$, [using (1)]. $\therefore\ \lambda_1 = r^2$.

Putting $\lambda_1 = r^2$ in (2), $\quad x\left(1 - \frac{r^2}{a^2}\right) + l\lambda_2 = 0$

$$\Rightarrow \quad x = \frac{a^2 l\lambda_2}{r^2 - a^2} \text{ etc.}$$

$$\text{Thus} \quad x = \frac{a^2 l\lambda_2}{r^2 - a^2},\ y = \frac{b^2 m\lambda_2}{r^2 - b^2},\ z = \frac{c^2 n\lambda_2}{r^2 - c^2}$$

$$\therefore \quad 0 = l_x + m_y + n_z = \lambda_2 \left\{\frac{a^2 l^2}{r^2 - a^2} + \frac{b^2 m^2}{r^2 - b^2} + \frac{c^2 n^2}{r^2 - c^2}\right\}$$

$$\text{Hence,} \quad \frac{a^2 l^2}{r^2 - a^2} + \frac{b^2 m^2}{r^2 - b^2} + \frac{c^2 n^2}{r^2 - c^2} = 0 \qquad (\because\ \lambda_2 \neq 0)$$

Example 77:

Show that the maximum and minimum values of the function $f(x, y) = x^2 + xy + y^2$, when $ax^2 + by^2 = ab$ will be the values of θ given by the equation $4(\theta - a)(\theta - b) = ab$. **(D.U., 1983)**

Solution:

Let $F = x^2 + xy + y^2 + \lambda\,(ax^2 + by^2 - ab)$,

$\therefore\quad F_x = 2x + y + 2\lambda ax,\ F_y = x + 2y + 2\lambda by.$

At a stationary value, $F_x = F_y = 0$ ...(1)

$\therefore\quad xF_x + yF_y = 0$ gives that

$2(x^2 + xy + y^2) + 2\lambda\,(ax^2 + by^2) = 0$

or $\quad 2f + 2\lambda ab = 0,\quad$ where $f = x^2 + xy + y^2$.

or $\quad \lambda = -\,f/ab$. Putting this value of λ in (1).

$2x + y - 2fx/b = 0,\ x + 2y - 2fy/a = 0$

or $\quad 2bx + by = 2fx,\ ax + 2ay = 2fy$

or $\quad 2(b - f)x - by,\ ax = 2(f - a)y$

$$\therefore\quad \frac{2(b-f)}{a} = \frac{-b}{2(f-a)}$$

or $\quad 4(f - b)\,(f - a) = ab$

Replacing f by θ, we get

$4\,(\theta - a)\,(\theta - b) = ab$

where θ gives the maximum and minimum values of f.

Example 78:

If the variables x, y z satisfy the equation

$$\phi(x)\phi(y)\phi(z) = k^2$$

and $\phi(a) = \kappa \neq 0$, $\phi'(a) \neq 0$, *show that the function*

$$f(x) + f(y) + f(z)$$

has a maximum, when $x = y = z = a$, *provided that*

$$f'(a)\left\{\frac{\phi'(a)}{\phi'(a)} - \frac{\phi'(a)}{\phi'(a)}\right\} > f'(a).$$ **(D.U., 1993)**

Solution:

Let $F = f(x) + f(y) + f(z) + \lambda\{\phi(x)\phi(y)\phi(z) - k^2\}$

$\therefore\quad dF = F_x dx + F_y dy + F_z dz$

or $\quad dF = \Sigma\ \{f(x) + \lambda\phi'(x)\phi(y)\phi(z)\}dx.$

At stationary points.

$$\left.\begin{aligned} f'(x)+\lambda\phi'(x)\phi(y)\phi(z) &= 0, \\ f'(y)+\lambda\phi'(y)\phi(z)\phi(x) &= 0, \end{aligned}\right\}$$

$$f'(z) + \lambda\phi'(z)\phi(x)\phi(y) = 0$$

Since the function has a maximum at (a, a, a), therefore

$$f'(a) + \lambda\phi'(a)\phi(a)\phi(a) = 0$$

or $$\lambda = -\frac{f'(a)}{\phi'(a)\{\phi(a)\}^2} = \frac{f'(a)}{k^2\phi'(a)} \quad [\because \phi'(a) \neq 0, \phi(a) = k \neq 0] \quad ...(1)$$

Now, $d^2 F = \Sigma F_{xx} dx^2 + 2\, \Sigma F_{xy} dxdy$

or $d^2 F = \Sigma\{f''(a) + \lambda\phi''(x)\phi(y)\phi(z)\}\, dx^2 + 2\lambda\Sigma\phi'(x)\phi'(y)\phi(z)\, dx\, dy.$

At the stationary point (a, a, a),

$$d^2 F = \{f''(a) + \lambda k^2\phi''(a)\}\, \Sigma dx^2 + zk\, [\phi'(a)]^2\, \Sigma\, dx\, dy \quad (\because \phi(a) = k) \quad ...(2)$$

Now, $\phi(x)\phi(y)\phi(z) = k^2 \Rightarrow \Sigma\phi'(x)\phi(y)\phi(z)\, dx = 0$

$\Rightarrow$ $k^2\phi'(a)(dx + dy + dz) = 0$, at (0, a)

$\Rightarrow$ $dx + dy + dz = 0 \qquad (\because k \neq 0, \phi'(a) \neq 0)$

$\Rightarrow$ $(dx + dy + dz)^2 = 0$

$\Rightarrow$ $2\Sigma d\;\; y = -\, \Sigma dx^2.$ ...(3)

From (2) and (3), we obtain

$$d^2 F = \{f''(a) + \lambda k^2\phi''(a)\}\, \Sigma dx^2 - \lambda k\, [\phi'(a)]^2\, \Sigma dx^2$$

$$= \left[f''(a) - f'(a)\frac{\phi''(a)}{\phi'(a)} + f'(a)\frac{\phi'(a)}{\phi(a)}\right]\sum dx^2, \text{ using (1)}$$

For a maximum value at (a, a, a), we have $d^2F<0$

$$\Rightarrow \quad f''(a) - f'(a)\left\{\frac{\phi''(a)}{\phi'(a)} - \frac{\phi'(a)}{\phi(a)}\right\} < 0$$

$$\Rightarrow \quad f''(a) < f'(a)\left\{\frac{\phi''(a)}{\phi'(a)} - \frac{\phi'(a)}{\phi(a)}\right\}.$$

Hence, $$f'(a)\left\{\frac{\phi''(a)}{\phi'(a)} - \frac{\phi'(a)}{\phi(a)}\right\} > f''(a).$$

Example 79:

If lx + my + mz = 1; l, m, n are positive constants, show that the stationary value of xy + yz + zx is

$$[2\,lm + 2mn + 2nl - l^2 - m^2 - n^2]^{-1}$$

and this value is maximum when it is positive. **(D.U., 1990)**

Solution:

Let $F = xy + yz + zx + \lambda\,(lx + my + nz - 1)$. Then

$$F_x = y + z + \lambda l,\ F_y = z + x + \lambda m,\ F_z = x + y + \lambda n.$$

At a stationary value, $F_x = F_y = F_z = 0$. ...(1)

Now $xF_x = yF_y = zF_z = 0$ gives that

$$2\,(xy + yz + zx) + \lambda\,(lx + my + nz) = 0.$$

$\therefore \lambda = -2f$, where $f = (xy + yz + zx)$. Putting in (1), we get

$$y + z = 2lf,\ z + x = 2mf,\ x + y = 2nf. \quad ...(2)$$

Adding these equations, we obtain

$$x + y + z = (l + m + n)\,f. \quad ...(3)$$

from (2) and (3), we get

$$x = (m + n - l)f\ (n + l - m)f,\ z = (l + m - n)f.$$

Putting these values of x, y, z in $f = xy + yz + zx$, we get

$$f = [(m + n - l)\,(n + l - m)(l + m - n) + (1 + m - n)(\,m + n - l)f^2$$

$$= (2mn + 2nl + 2lm - l^2 - m^2 - n^2)^2.$$

Hence, $f = [2lm + 2mn + 2nl - l^2 - m^2 - n^2]^{-1}$. $(\because f \neq 0)$

We have

$$F_{xx} = F_{yy} = F_{zz} = 0,\ F_{xy} = F_{yz} = F_{zx} = 1.$$

If $\phi = lx + my + nz$, then

$$\phi_x = l,\ \phi_y = m, = \phi_z,\ a$$

For maximum value of xy + yz + zx, we have $d^2F < 0$,

Equivalently, $$\begin{vmatrix} F_{xx} & F_{xy} & F_{xz} & \phi_x \\ F_{yx} & F_{yy} & F_{yz} & \phi_y \\ F_{zx} & F_{zy} & F_{zz} & \phi_z \\ \phi_x & \phi_y & \phi_z & 0 \end{vmatrix} < 0$$

$$\Rightarrow \begin{vmatrix} 0 & 1 & 1 & 1 \\ 1 & 0 & 1 & m \\ 1 & 1 & 0 & n \\ l & m & n & 0 \end{vmatrix} < 0 \Rightarrow [l^2 + m^2 + n^2 - 2mn - 2nl - 2lm] < 0$$
$$\Rightarrow [2mn + 2nl + 2lm - l^2 - m^2 - n^2] > 0$$

Hence $[2mn + 2nl + 2lm - l^2 - m^2 - n^2]^{-1} > 0$

Example 80:

If $1/x + 1/y + 1/z = 1$, show that the stationary value of $a^3x^2 + b^3y^2 + c^3z^2$ is given by

$x = (a + b + c)/a$, $y = (a + b + c)/b$, $z = (a + b + c)/c$.

Solution:

Let $F = (a^3x^2 + b^3y^2 + c^3z^2) + \lambda(1/x + 1/y + 1/z)$.

$\therefore Fx = 2a^3x - \lambda(1/x^2)$, $F_y = 2b^3y - \lambda(1/y^2)$, $Fz = 2c^3z - (1/z^2)$ ...(1)

As a stationary value, $F_x = F_y = F_z = 0$. ...(2)

$\therefore$ $xF_x = yF_y = zF_z = 0$ gives that

$2(a^3x^2 + b^3y^2 = c^3z^2) - \lambda(1/x + 1/y + 1/z) = 0$

or $2f - \lambda = 0 \Rightarrow \lambda = 2f$, where $f = a^3x^2 + b^3y^2 = c^3z^2$.

From (1) and (2),

$2a^3x^3 + 2b^3y^3 = c^3z^2 (= \lambda)$

$\therefore$ $ax = by = cz = (k, \text{say})$ ...(3)

$\Rightarrow$ $1/x = a/k$, $1/y = b/k$, $1/z = c/k$

$\Rightarrow$ $1 = 1/x + 1/y, 1/z + 1/k\,(a + b + c)$

$\Rightarrow$ $k = (a + b + c)$.

Putting this value of k in (3), we get

$x = (a + b + c)/a$, $y = (a + b + c)/b$, $z = (a + b + c)/c$.